Computer Algorithms for Solving Linear Algebraic Equations

The State of the Art

NATO ASI Series

Advanced Science Institutes Series

A series presenting the results of activities sponsored by the NATO Science Committee, which aims at the dissemination of advanced scientific and technological knowledge, with a view to strengthening links between scientific communities.

The Series is published by an international board of publishers in conjunction with the NATO Scientific Affairs Division

A Life Sciences B Physics	Plenum Publishing Corporation London and New York
C Mathematical and Physical Sciences D Behavioural and Social Sciences E Applied Sciences	Kluwer Academic Publishers Dordrecht, Boston and London
F Computer and Systems Sciences G Ecological Sciences H Cell Biology I Global Environmental Change	Springer-Verlag Berlin Heidelberg New York London Paris Tokyo Hong Kong Barcelona

NATO-PCO DATABASE

The electronic index to the NATO ASI Series provides full bibliographical references (with keywords and/or abstracts) to more than 30000 contributions from international scientists published in all sections of the NATO ASI Series. Access to the NATO-PCO DATABASE is possible in two ways:

- via online FILE 128 (NATO-PCO DATABASE) hosted by ESRIN, Via Galileo Galilei, I-00044 Frascati, Italy.
- via CD-ROM "NATO-PCO DATABASE" with user-friendly retrieval software in English, French and German (© WTV GmbH and DATAWARE Technologies Inc. 1989).

The CD-ROM can be ordered through any member of the Board of Publishers or through NATO-PCO, Overijse, Belgium.

Series F: Computer and Systems Sciences Vol. 77

Computer Algorithms for Solving Linear Algebraic Equations

The State of the Art

Edited by

Emilio Spedicato

Department of Mathematics, University of Bergamo
Piazza Rosate 2, 24100 Bergamo, Italy

Springer-Verlag
Berlin Heidelberg New York London Paris Tokyo
Hong Kong Barcelona Budapest
Published in cooperation with NATO Scientific Affairs Division

Proceedings of the NATO Advanced Study Institute on Computer Algorithms for Solving Linear Algebraic Equations: The State of the Art, held at Il Ciocco, Barga, Italy, September 9–21, 1990.

Library of Congress Cataloging-in-Publication Data
NATO Advanced Study Institute on Computer Algorithms for Solving Linear Equations: the State of the Art (1990: Il Ciocco, Italy) Computer algorithms for solving linear algebraic equations: the state of the art/edited by Emilio Spedicato. p. cm.—(NATO ASI series. Series F, Computer and systems sciences: vol. 77) "Proceedings of the NATO Advanced Study Institute on Computer Algorithms for Solving Linear Equations: the State of the Art, held at Il Ciocco, Barga, Italy, September, 9–12, 1990"—T.p. verso. Includes indexes.

ISBN-13: 978-3-642-76719-7 e-ISBN-13: 978-3-642-76717-3
DOI: 10.1007/978-3-642-76717-3

1. Equations—Nummerical solutions—Data processing—Congresses. 2. Algebras, Linear—Congresses. I. Spedicato, E. II. Title. III. Series. QA218.N18 1990 512.9'4—dc20 91-20698

Softcover reprint of the hardcover 1st edition 1991

Typesetting: camera ready by authors

45/3140-543210 – Printed on acid-free-paper

Preface

The NATO Advanced Study Institute on "Computer algorithms for solving linear algebraic equations: the state of the art" was held September 9-21, 1990, at Il Ciocco, Barga, Italy. It was attended by 68 students (among them many well known specialists in related fields!) from the following countries: Belgium, Brazil, Canada, Czechoslovakia, Denmark, France, Germany, Greece, Holland, Hungary, Italy, Portugal, Spain, Turkey, UK, USA, USSR, Yugoslavia.

Solving linear equations is a fundamental task in most of computational mathematics. Linear systems which are now encountered in practice may be of very large dimension and their solution can still be a challenge in terms of the requirements of accuracy or reasonable computational time. With the advent of supercomputers with vector and parallel features, algorithms which were previously formulated in a framework of sequential operations often need a completely new formulation, and algorithms that were not recommended in a sequential framework may become the best choice.

The aim of the ASI was to present the state of the art in this field. While not all important aspects could be covered (for instance there is no presentation of methods using interval arithmetic or symbolic computation), we believe that most important topics were considered, many of them by leading specialists who have contributed substantially to the developments in these fields.

Of the 14 contributions presented at the ASI (40 hours of lectures) and available in this volume, some relate to algorithms in their general sequential formulation, some consider both sequential and multiprocessor aspects, some (namely those of M. Cosnard, I. Duff and Y. Robert) are especially devoted to the algorithm implementation on particular multiprocessor architectures.

The first paper, by C. Broyden, considers some classic results on linear equations, proved through a novel approach. V. Pan gives a comprehensive review of computational complexity results. Algorithms for linear least squares are reviewed by Å. Björck, while the recently proposed class of ABS methods is presented by E. Spedicato and J. Abaffy. The multigrid method is considered by W. Hackbusch. The hierarchical basis method is studied by H. Yserentant. Analysis of preconditioning methods for block H-matrices is given by O. Axelsson. Domain decomposition methods for large symmetric sparse linear systems are considered by G. Meurant. A special splitting method for a problem arising in the context of image reconstruction is described by I. Galligani. Parallel factorization algorithms on a shared memory computer are considered by M. Cosnard. Analysis of Gaussian elimination on distributed memory architectures is given by Y. Robert. Several parallel techniques for non-structured sparse systems are described by I. Duff. Finally, L. Dixon discusses a number of intriguing issues in linear systems related to nonlinear optimization.

The ASI was organized with the help of the codirectors Professors M. Cosnard, L. Dixon, I. Galligani and J. Stoer, to whom the workshop director is indebted for useful advice. The organizing committee included Dr. M.T. Vespucci and Prof. M. Bertocchi, of the University of Bergamo, and Prof. Z. Xia, visiting Bergamo from the Technological University of Dalian, whose help in practical questions was instrumental for the smooth running of the meeting. Thanks are also due to the management at Il Ciocco, particularly to Dr. B. Giannasi, for providing the appreciated quality of the facilities and a friendly atmosphere.

The ASI was organized mainly with the financial support of NATO, Scientific Affairs Division. Thanks are due, for additional important financial help, to the University of Bergamo, particularly to its Department of Mathematics, to CNR (GNIM and Progetto Finalizzato Sistemi Informatici e Calcolo Parallelo), to IBM, ISMES, Unione Industriali di Bergamo and Banca Popolare di Bergamo.

This volume is dedicated to Professor Ben Noble, whose beautifully written and stimulating book "Applied Linear Algebra" was instrumental in guiding me in the field of linear algebra and optimisation.

Bergamo, April 1991

Emilio Spedicato

Course Director

Table of Contents

1. Classical Methods for Linear Equations....................1
Charles G. Broyden

2. Complexity of Algorithms for Linear
Systems of Equations......................................27
Victor Pan

3. Algorithms for Linear Least Squares Problems.............57
Åke Björck

4. ABS Algorithms for General Linear Systems................93
Emilio Spedicato

5. ABS Algorithms for Sparse Linear Systems................111
Jozsef Abaffy

6. Multi-Grid Algorithms. Introduction,
Survey and Convergence Analysis.........................133
Wolfgang Hackbusch

7. Two Multi-Level Methods for Nonuniformely
Refined Grids...161
Harry Yserentant

8. Preconditioning Methods for Block H-Matrices............169
Owe Axelsson

9. Domain Decomposition Methods for Solving Large
Sparse Linear Systems...................................185
Gérard Meurant

10. A Splitting Method for Large Sparse Linear
Systems on a Multivector Computer.......................207
Ilio Galligani

11. Scheduling Parallel Factorization Algorithms
on a Shared Memory Computer.............................217
Michel Cosnard

12. Gaussian Elimination on Distributed Memory
Architectures...253
Yves Robert

13. Parallel Algorithms for General Sparse Systems..........277
Iain S. Duff

14. The Solution of Partially Separable Linear
Equations on Parallel Processing Systems................299
Lawrence C.W. Dixon

Author Index...339

Subject Index..343

List of Participants.....................................349

Classical Methods for Linear Equations

C. G. Broyden
University of Bologna
Department of Mathematics
Piazza Porta S. Donato 5
40127 Bologna, Italy

Part 1. Direct Methods

1.1 Introduction

We consider in this section the solution by direct methods of the equation

$$\mathbf{Ax} = \mathbf{b} \tag{1.1}$$

where $\mathbf{A} = [a_{ij}]$ is an nth-order non-singular real matrix and $\mathbf{b} = [b_i]$ an arbitrary nth-order real vector. By "direct methods" we mean methods that would, in the absence of rounding error, give an exact solution of equation (1.1) in a finite number of arithmetic operations. Since in linear algebra computations an addition or subtraction is normally preceded by a multiplication or division, the number of arithmetic operations is almost exactly proportional to the number of multiplications and divisions so that this number may be taken as a measure of the work involved in implementing a particular algorithm. For all methods considered in Part 1 the number of multiplications plus divisions is a polynomial in n. If n is large the amount of work required to solve equation (1.1) will depend essentially on the leading term of this polynomial and we shall denote that term by μ. Thus, for a particular method, μ is a rough measure of the time taken by a computer to solve equation (1.1) for a given size of matrix $\mathbf{A}$. Note that, in contrast with purely iterative methods, the amount of work required is essentially independent of the values of the elements comprising $\mathbf{A}$.

NATO ASI Series, Vol. F 77
Computer Algorithms for Solving Linear Algebraic Equations
Edited by E. Spedicato

1.2 Triangular matrices

Before considering the general problem we consider the case where **A** = **U** and **U** = $[u_{ij}]$ is upper triangular, that is $u_{ij} = 0$ for i > j. This case is important since in many direct methods equation (1.1) is transformed into

$$\mathbf{Ux} = \mathbf{c}, \tag{1.2}$$

where $\mathbf{A}^{-1}\mathbf{b} = \mathbf{U}^{-1}\mathbf{c}$. The solutions of equations (1.1) and (1.2) are then identical so that the required result is merely the solution of equation (1.2). If $n = 3$ this equation may be represented schematically by

$$\begin{bmatrix} \times & \times & \times \\ 0 & \times & \times \\ 0 & 0 & \times \end{bmatrix} \begin{bmatrix} ? \\ ? \\ ? \end{bmatrix} = \begin{bmatrix} \times \\ \times \\ \times \end{bmatrix}$$

where × denotes an arbitrary but known scalar and ? denotes a scalar whose value has not yet been determined. If **U** is non-singular none of its diagonal elements is equal to zero and it is easily seen in this case that equation (1.2) can be solved simply by computing x_3 followed by x_2 and then x_1. For arbitrary n the same principles apply. This method, which is known as *back-substitution,* is reasonably numerically stable (see [Wi],[Br]) and "triangular systems are usually solved to high accuracy" (see [Hi]). However Dixon [Di] claims that many sets of triangular equations arising from optimisation problems cannot be solved accurately unless some form of pivoting (see section 3 below) is used in the decomposition phase of the algorithm, such pivoting being performed precisely with the view of stabilising the final, substitution, phase of the algorithm. For triangular systems the factor μ is equal to $n^2/2$. Equations involving lower triangular matrices may be solved in a similar manner.

1.3 General matrices - Gaussian Elimination and LU Decomposition

We consider now one of the earliest, simplest and best of all the direct methods for transforming equation (1.1) into equation (1.2) where **A** is a general matrix with no special properties. This is the method of *Gaussian elimination*. To see how it works we take an example where $n = 5$ and look at an intermediate stage of the process, one where two stages of elimination have already been completed. Schematically the resulting equation looks like

$$\begin{bmatrix} \times & \times & \times & \times & \times \\ 0 & \times & \times & \times & \times \\ 0 & 0 & \Rightarrow\times & \times & \times \\ 0 & 0 & \times & \times & \times \\ 0 & 0 & \times & \times & \times \end{bmatrix} \begin{bmatrix} ? \\ ? \\ ? \\ ? \\ ? \end{bmatrix} = \begin{bmatrix} \times \\ \times \\ \times \\ \times \\ \times \end{bmatrix} \qquad (1.3)$$

The element indicated by the arrow, i.e. the intersection of the diagonal with the first full column of the matrix, is called the *pivot* and provided that it is nonzero it is possible, by subtracting suitable multiples of row three from rows four and five, to obtain a matrix equation that looks like

$$\begin{bmatrix} \times & \times & \times & \times & \times \\ 0 & \times & \times & \times & \times \\ 0 & 0 & \times & \times & \times \\ 0 & 0 & 0 & \times & \times \\ 0 & 0 & 0 & \times & \times \end{bmatrix} \begin{bmatrix} ? \\ ? \\ ? \\ ? \\ ? \end{bmatrix} = \begin{bmatrix} \times \\ \times \\ \times \\ \times \\ \times \end{bmatrix} \qquad (1.4)$$

and is thus one stage nearer equation (1.2). If the same multiples of the third element of the right-hand side are subtracted from elements four and five of the right-hand side the solution of the equations is unchanged. Clearly n-1 applications of this technique will yield a set of equations where the matrix of coefficients is upper triangular and whose solution is identical to that of (1.1). If a pivot should happen to be zero it can be made non-zero by *row interchanges*. In equation (1.3) this means interchanging rows three and four or rows three and five (and the corresponding elements of the right-hand side) so that the element occupying the arrowed position is non-zero (if **A** is nonsingular then at least one of the possible pivots is non-zero - see [Wi])

Assume now that the transformation of the equations from the form of equation (1.3) to that of equation (1.4) is brought about by subtracting m_{43} and m_{53} times row three from rows four and five respectively. It is easily seen that the transformation is equivalent to multiplying equation (1.3) by $\mathbf{M}_3$ where

$$\mathbf{M}_3 = \begin{bmatrix} 1 & 0 & 0 & 0 & 0 \\ 0 & 1 & 0 & 0 & 0 \\ 0 & 0 & 1 & 0 & 0 \\ 0 & 0 & -m_{43} & 1 & 0 \\ 0 & 0 & -m_{53} & 0 & 1 \end{bmatrix} \tag{1.5}$$

and that the complete transformation from equation (1.1) to equation (1.2) may be expressed as

$$\mathbf{M}_4\mathbf{M}_3\mathbf{M}_2\mathbf{M}_1\mathbf{Ax} = \mathbf{Ux} = \mathbf{M}_4\mathbf{M}_3\mathbf{M}_2\mathbf{M}_1\mathbf{b} = \mathbf{c}. \tag{1.6}$$

where each of the matrices $\mathbf{M}_r$, r = 1,2,3,4, is merely the unit matrix but with the appropriate negative multipliers forming the sub-diagonal elements of the *r*th column. If we now define $\mathbf{L}$ by $\mathbf{L} = (\mathbf{M}_4\mathbf{M}_3\mathbf{M}_2\mathbf{M}_1)^{-1}$ we find that it is given by

$$\mathbf{L} = \begin{bmatrix} 1 & 0 & 0 & 0 & 0 \\ m_{21} & 1 & 0 & 0 & 0 \\ m_{31} & m_{32} & 1 & 0 & 0 \\ m_{41} & m_{42} & m_{43} & 1 & 0 \\ m_{51} & m_{52} & m_{53} & m_{54} & 1 \end{bmatrix}$$

and it is not difficult to show that if $\mathbf{L}$ is an $n \times n$ matrix defined by

$$\mathbf{L} = (\mathbf{M}_{n-1}\mathbf{M}_{n-2}\ldots\mathbf{M}_1)^{-1} \tag{1.7}$$

where $\mathbf{M}_r$, $1 \le r \le n-1$, is as defined above, then the same pattern holds. Since $\mathbf{M}_{n-1}\mathbf{M}_{n-2}\ldots\mathbf{M}_1\mathbf{A} = \mathbf{U}$ it follows from equation (1.7) that

$$\mathbf{A} = \mathbf{LU} \tag{1.8}$$

so that Gaussian elimination is equivalent to factorising the matrix $\mathbf{A}$ into the product of a unit lower triangular matrix $\mathbf{L}$ and an upper triangular matrix $\mathbf{U}$. Regarded in this way the process is known as *LU, or triangular, decomposition.*

In order for Gaussian elimination to be possible it is

necessary, as we have seen, for all the pivots to be non-zero but this does not necessarily guarantee numerical stability. To make this more likely we need to choose the pivots so that the absolute values of all the multipliers m_{ij} are less than or equal to unity, and as before this is achieved by interchanging the last (n-r) rows of the matrix, where r is the number of elimination stages already completed, so that the pivot has the largest absolute value of all the possible pivotal candidates (see [Wi],[GvL]).

We now consider an intermediate stage of Gaussian elimination regarded as LU decomposition. Define $\mathbf{V}_r$ to be the matrix of coefficients after r stages of elimination, so that

$$\mathbf{V}_r = \mathbf{M}_r\mathbf{M}_{r-1}\ldots\mathbf{M}_1\mathbf{A} \tag{1.9}$$

(the matrices appearing in equations (1.3) and (1.4) being thus $\mathbf{V}_2$ and $\mathbf{V}_3$ respectively). Define

$$\mathbf{K}_r = (\mathbf{M}_r\mathbf{M}_{r-1}\ldots\mathbf{M}_1)^{-1} \tag{1.10}$$

so that

$$\mathbf{K}_{r+1} = \mathbf{K}_r\mathbf{M}_{r+1}^{-1} \tag{1.11}$$

$$\mathbf{K}_r\mathbf{V}_r = \mathbf{A} \tag{1.12}$$

and

$$\mathbf{V}_{r+1} = \mathbf{M}_{r+1}\mathbf{V}_r \tag{1.13}$$

and return to our 5 x 5 example. Assume that $\mathbf{K}_2$ and $\mathbf{V}_2$ are partially known so that equation (1.12) with r = 2 may be represented schematically by

$$\underbrace{\left[\begin{array}{cc|ccc} 1 & 0 & 0 & 0 & 0 \\ \times & 1 & 0 & 0 & 0 \\ \hline \times & \times & 1 & 0 & 0 \\ \times & \times & 0 & 1 & 0 \\ \times & \times & 0 & 0 & 1 \end{array}\right]}_{\mathbf{K}_2} \underbrace{\left[\begin{array}{cc|ccc} \times & \times & \times & \times & \times \\ 0 & \times & \times & \times & \times \\ \hline 0 & 0 & ? & ? & ? \\ 0 & 0 & ? & ? & ? \\ 0 & 0 & ? & ? & ? \end{array}\right]}_{\mathbf{V}_2} = \underbrace{\left[\begin{array}{cc|ccc} \times & \times & \times & \times & \times \\ \times & \times & \times & \times & \times \\ \hline \times & \times & \times & \times & \times \\ \times & \times & \times & \times & \times \\ \times & \times & \times & \times & \times \end{array}\right]}_{\mathbf{A}}$$

This equation may be written

$$\begin{bmatrix} \mathbf{K}_{11} & \mathbf{0} \\ \mathbf{K}_{21} & \mathbf{I} \end{bmatrix} \begin{bmatrix} \mathbf{V}_{11} & \mathbf{V}_{12} \\ \mathbf{0} & \mathbf{V}_{22} \end{bmatrix} = \begin{bmatrix} \mathbf{A}_{11} & \mathbf{A}_{12} \\ \mathbf{A}_{21} & \mathbf{A}_{22} \end{bmatrix} \tag{1.14}$$

where the submatrices correspond to the indicated partitions. Thus $\mathbf{K}_{11}$, $\mathbf{K}_{21}$, $\mathbf{V}_{11}$ and $\mathbf{V}_{12}$ are known but, in contrast with Gaussian elimination, $\mathbf{V}_{22}$ is as yet undetermined. In order to carry out the next step of the decomposition it is necessary to determine the next pivot together with its associated set of multipliers, and to do this we need the first column of $\mathbf{V}_{22}$. From (1.14) it follows that

$$\mathbf{V}_{22} = \mathbf{A}_{22} - \mathbf{K}_{21}\mathbf{V}_{12}$$

so that if we denote the first column of $\mathbf{V}_{22}$ by $\mathbf{v}$ we have

$$\mathbf{v} = \text{firstcol}(\mathbf{A}_{22}) - \mathbf{K}_{21}.\text{firstcol}(\mathbf{V}_{12})$$

and since all matrices on the right-hand side of this equation are known, $\mathbf{v}$ may be readily computed. If the first element of $\mathbf{v}$ is not the element of largest absolute value it should be exchanged with the one that is, and the corresponding interchanges made to the rows of $\mathbf{K}_{21}$, $\mathbf{A}_{21}$ and $\mathbf{A}_{22}$ (the unit submatrix remaining unchanged). Then, since the pivot is now known, the multipliers may be determined and these yield $\mathbf{M}_3$. Now, from (1.11) and (1.13), $\mathbf{K}_3 = \mathbf{K}_2\mathbf{M}_3^{-1}$ and $\mathbf{V}_3 = \mathbf{M}_3\mathbf{V}_2$, and using these equations to evaluate as many elements of $\mathbf{K}_3$ and $\mathbf{V}_3$ as possible enables us to write equation $\mathbf{K}_3\mathbf{V}_3 = \mathbf{A}$ schematically as

$$\underset{\mathbf{K}_3}{\left[\begin{array}{ccc|cc} 1 & 0 & 0 & 0 & 0 \\ \times & 1 & 0 & 0 & 0 \\ \times & \times & 1 & 0 & 0 \\ \hline \times & \times & \times & 1 & 0 \\ \times & \times & \times & 0 & 1 \end{array}\right]} \underset{\mathbf{V}_3}{\left[\begin{array}{ccc|cc} \times & \times & \times & \times & \times \\ 0 & \times & \times & \times & \times \\ 0 & 0 & \times & ? & ? \\ \hline 0 & 0 & 0 & ? & ? \\ 0 & 0 & 0 & ? & ? \end{array}\right]} = \underset{\mathbf{A}}{\left[\begin{array}{ccc|cc} \times & \times & \times & \times & \times \\ \times & \times & \times & \times & \times \\ \times & \times & \times & \times & \times \\ \hline \times & \times & \times & \times & \times \\ \times & \times & \times & \times & \times \end{array}\right]}$$

Let now the submatrices of equation (1.14) refer to the partitions indicated in the above equation. Then $\mathbf{V}_{12}$ is known except for its last row but since $\mathbf{K}_{11}$ is unit lower triangular this may be computed simply from

$$\text{lastrow}(\mathbf{K}_{11}).\mathbf{V}_{12} = \text{lastrow}(\mathbf{A}_{12})$$

Thus, as at the beginning of the elimination stage, $\mathbf{K}_{11}$, $\mathbf{K}_{12}$, $\mathbf{V}_{11}$ and $\mathbf{V}_{12}$ are all known but the submatrices now refer to $\mathbf{K}_3$

and $\mathbf{V}_3$ rather than to $\mathbf{K}_2$ and $\mathbf{V}_2$, and repetition of this procedure n-1 times yields the factorisation **LU** = **PA**, where **P** is a permutation matrix (i.e. the unit matrix with re-ordered rows) introduced to take into account the interchanges due to pivoting. The solution of equation (1.1) is then obtained by solving **Ly** = **Pb** for **y** and **Ux** = **y** for **x**.

In the absence of rounding error both Gaussian elimination and LU decomposition yield identical factors since the individual arithmetic operations are the same in each case, but they are performed in different orders in the two calculations. To see this, consider the calculation of **v** using equation (1.14). Each element of **v** is computed by

$$v_r = a_{rs} - \sum_{j=1}^{j=t} l_{rj} v_{js}$$

where r, s and t are integers whose precise value need not concern us here. In LU decomposition each element of **v** is computed before the calculation of the next element is begun so if a double length accumulator is used, the inner product may be formed effectively double-length, rounding to single- length only when the element has been fully computed and potentially damaging cancellations have been completed. This *double length accumulation of inner products* is far more accurate than single length accumulation (see [Wi]) and is to be preferred whenever possible. In Gaussian elimination **v** is effectively computed by adding one term of the sum of the right-hand side into the accumulator, storing, adding one term into the next sum and so on. To derive the benefits of double-length accumulation in this case would require substantial quantities of double-length storage rather than a single accumulator.

Since Gaussian elimination and LU decomposition are arithmetically identical the amount of work needed in each case is the same, and $\mu = n^3/3$. This represents effectively the amount of work involved in the decomposition since the back-substitution phase requires $O(n^2)$ multiplications. Both methods are numerically satisfactory for most matrices **A** provided that the pivoting strategy described above is carried out (see [Wi],[St],[GvL]).

1.4 Symmetric positive definite matrices - Choleski's Method

In constructing algorithms it is always desirable, if at all possible, to exploit any structure inherent in the problem and the symmetry of a matrix is perhaps the most obvious structural feature that beckons the algorist. However simple row pivoting destroys symmetry and simple row/column pivoting, although it preserves symmetry, can sometimes lead to a seriously unstable computation or even total breakdown of the algorithm. If, though, the symmetric matrix is also positive definite a simple and elegant factorisation is open to us. This is the method of *Choleski*, who showed that such a matrix could be expressed as the product of a real lower triangular matrix $\mathbf{L}$ and its transpose, i.e.

$$\mathbf{A} = \mathbf{L}\mathbf{L}^T \tag{1.15}$$

We can derive the Choleski decomposition by writing equation (1.15) as

$$\begin{bmatrix} \mathbf{L}_{11} & \mathbf{0} \\ \mathbf{L}_{21} & \mathbf{L}_{22} \end{bmatrix} \begin{bmatrix} \mathbf{L}_{11}^T & \mathbf{L}_{21}^T \\ \mathbf{0} & \mathbf{L}_{22}^T \end{bmatrix} = \begin{bmatrix} \mathbf{A}_{11} & \mathbf{A}_{12} \\ \mathbf{A}_{21} & \mathbf{A}_{22} \end{bmatrix} \tag{1.16}$$

Assume that we know only $\mathbf{L}_{11}$. Then the first row of $\mathbf{L}_{21}$ is obtained immediately by solving

$$\mathbf{L}_{11}.\text{firstcol}(\mathbf{L}_{21}^T) = \text{firstcol}(\mathbf{A}_{12})$$

and it remains, in order to compute the next larger $\mathbf{L}_{11}$, to compute the first diagonal element of $\mathbf{L}_{22}$. Now from (1.16) it follows that

$$\begin{aligned} \mathbf{L}_{22}\mathbf{L}_{22}^T &= \mathbf{A}_{22} - \mathbf{L}_{21}\mathbf{L}_{21}^T \\ &= \mathbf{A}_{22} - \mathbf{A}_{21}\mathbf{A}_{11}^{-1}\mathbf{A}_{12} \end{aligned}$$

Denote by λ the first diagonal element of $\mathbf{L}_{22}$. Since $\mathbf{L}_{22}$ is lower triangular it follows that λ^2 is equal to the first diagonal element of $\mathbf{A}_{22} - \mathbf{L}_{21}\mathbf{L}_{21}^T$, an easily computable quantity since the first column of $\mathbf{L}_{21}^T$ has already been computed. But λ^2 is also the first diagonal element of $\mathbf{A}_{22} - \mathbf{A}_{21}\mathbf{A}_{11}^{-1}\mathbf{A}_{12}$, a Schur complement of $\mathbf{A}$, and since all Schur complements of a positive definite matrix are themselves positive definite (see appendix) it follows that $\lambda^2 > 0$. Hence λ is real and the factorisation (1.15) is established by induction.

For the Choleski decomposition $\mu = n^3/6$, i.e. only half the work needed for LU decomposition. No pivoting is needed, double length accumulation is possible and the method is superbly stable ([Wi],[MPW],[GvL]). This must be one of the most satisfying and satisfactory algorithms in the whole of numerical analysis, its only disadvantage being the need to compute n square roots. Once $\mathbf{L}$ is known the solution of equation (1.1) is obtained by solving $\mathbf{Ly} = \mathbf{b}$ for $\mathbf{y}$ and $\mathbf{L}^T\mathbf{x} = \mathbf{y}$ for $\mathbf{x}$. Another version of this method performs the factorisation $\mathbf{A} = \mathbf{LDL}^T$, where $\mathbf{L}$ is now unit lower triangular and $\mathbf{D}$ is a diagonal matrix whose diagonal elements are strictly positive. Although in principle this factorisation could be performed using real matrices for indefinite $\mathbf{A}$ provided that all its leading principal submatrices were nonsingular the danger of numerical instability is too great to make this a viable method. It is necessary to replace the diagonal matrix $\mathbf{D}$ by a tridiagonal matrix $\mathbf{T}$ and to carry out some form of pivoting to achieve stability if $\mathbf{A}$ is indefinite and this is precisely what is done in the following method.

1.5 Symmetric indefinite matrices - Aasen's method

We consider now the case where $\mathbf{A}$ is symmetric but not positive definite and describe Aasen's method [Aa]. This is based on the decomposition

$$\mathbf{LTL}^T = \mathbf{A} \tag{1.17}$$

where $\mathbf{L}$ is unit lower triangular and $\mathbf{T} = [t_{ij}]$ is *tridiagonal*, i.e. $t_{ij} = 0$ if $|i-j| > 1$, but much use is made of the form

$$\mathbf{TL}^T = \mathbf{L}^{-1}\mathbf{A}$$

where $\mathbf{TL}^T = \mathbf{H} = [h_{ij}]$ is *upper Hessenberg,* i.e. $h_{ij} = 0$ if $i > j+1$. In this respect the method closely resembles LU decomposition and a further resemblence stems from the form of **L**, which is defined to be (compare equation (1.7))

$$\mathbf{L} = (\mathbf{M}_{n-1}\mathbf{M}_{n-2}\ldots\mathbf{M}_2)^{-1} \tag{1.18}$$

where the matrices $\mathbf{M}_r$, $2 \le r \le n-1$, are as defined in section 3 above. Note, however, that the factor $\mathbf{M}_1$ is missing from equation (1.18) since $\mathbf{TL}^T$ is Hessenberg rather than triangular.

Let then $\mathbf{K}_r$ and $\mathbf{S}_r$ denote the intermediate forms of **L** and **T** that are found after r-1 stages of the factorisation. Thus

$$\mathbf{K}_r = (\mathbf{M}_r\mathbf{M}_{r-1}\ldots\mathbf{M}_2)^{-1} \tag{1.19}$$

so that, if

$$\mathbf{A} = \mathbf{K}_r\mathbf{S}_r\mathbf{K}_r^T \tag{1.20}$$

and $\mathbf{V}_r$ is defined by

$$\mathbf{S}_r\mathbf{K}_r^T = \mathbf{V}_r \tag{1.21}$$

it follows from (1.20) that

$$\mathbf{K}_r\mathbf{V}_r = \mathbf{A} \tag{1.22}$$

and that

$$\mathbf{K}_{r+1} = \mathbf{K}_r\mathbf{M}_{r+1}^{-1} \tag{1.23}$$

$$\mathbf{V}_{r+1} = \mathbf{M}_{r+1}\mathbf{V}_r \tag{1.24}$$

and

$$\mathbf{S}_{r+1} = \mathbf{M}_{r+1}\mathbf{S}_r\mathbf{M}_{r+1}^T \tag{1.25}$$

To see how the method works, consider again the 5 × 5 example after two stages of reduction. Assume that $\mathbf{S}_3$ and $\mathbf{K}_3$ are partially known as indicated below so that equation (1.21) with r=3 may be written schematically as

$$\left[\begin{array}{cc|ccc} \times & \times & 0 & 0 & 0 \\ \times & \times & \times & 0 & 0 \\ \hline 0 & \times & ? & ? & ? \\ 0 & 0 & ? & ? & ? \\ 0 & 0 & ? & ? & ? \end{array}\right] \left[\begin{array}{cc|ccc} 1 & 0 & 0 & 0 & 0 \\ 0 & 1 & \times & \times & \times \\ \hline 0 & 0 & 1 & \times & \times \\ 0 & 0 & 0 & 1 & 0 \\ 0 & 0 & 0 & 0 & 1 \end{array}\right] = \left[\begin{array}{cc|ccc} \times & \times & ? & ? & ? \\ \times & \times & ? & ? & ? \\ \hline 0 & \times & ? & ? & ? \\ 0 & 0 & \Rightarrow ? & ? & ? \\ 0 & 0 & ? & ? & ? \end{array}\right]$$

$$\qquad \mathbf{S}_3 \qquad\qquad\qquad \mathbf{K}_3^T \qquad\qquad\qquad \mathbf{V}_3$$

where the arrow indicates the next pivot. Write this equation as

$$\begin{bmatrix} \mathbf{S}_{11} & \mathbf{S}_{12} \\ \mathbf{S}_{21} & \mathbf{S}_{22} \end{bmatrix} \begin{bmatrix} \mathbf{K}_{11}{}^T & \mathbf{K}_{21}{}^T \\ \mathbf{0} & \mathbf{K}_{22}{}^T \end{bmatrix} = \begin{bmatrix} \mathbf{V}_{11} & \mathbf{V}_{12} \\ \mathbf{V}_{21} & \mathbf{V}_{22} \end{bmatrix} \tag{1.26}$$

We wish to subtract appropriate multiples of the second row of $\mathbf{V}_{22}$ from the following rows (in this case, only one following row) in order to introduce zeroes in the positions below the pivot. In order to compute the pivot and multipliers we need to compute the first column of $\mathbf{V}_{22}$ but this is readily obtained from (1.22). Specifically, if that equation is written as

$$\left[\begin{array}{cc|ccc} 1 & 0 & 0 & 0 & 0 \\ 0 & 1 & 0 & 0 & 0 \\ \hline 0 & \times & 1 & 0 & 0 \\ 0 & \times & \times & 1 & 0 \\ 0 & \times & \times & 0 & 1 \end{array}\right] \left[\begin{array}{cc|ccc} \times & \times & ? & ? & ? \\ \times & \times & ? & ? & ? \\ \hline 0 & \times & ? & ? & ? \\ 0 & 0 & ? & ? & ? \\ 0 & 0 & ? & ? & ? \end{array}\right] = \left[\begin{array}{cc|ccc} \times & \times & \times & \times & \times \\ \times & \times & \times & \times & \times \\ \hline \times & \times & \times & \times & \times \\ \times & \times & \times & \times & \times \\ \times & \times & \times & \times & \times \end{array}\right]$$

$$\qquad \mathbf{K}_3 \qquad\qquad\qquad \mathbf{V}_3 \qquad\qquad\qquad \mathbf{A}$$

or

$$\begin{bmatrix} \mathbf{K}_{11} & \mathbf{0} \\ \mathbf{K}_{21} & \mathbf{K}_{22} \end{bmatrix} \begin{bmatrix} \mathbf{V}_{11} & \mathbf{V}_{12} \\ \mathbf{V}_{21} & \mathbf{V}_{22} \end{bmatrix} = \begin{bmatrix} \mathbf{A}_{11} & \mathbf{A}_{12} \\ \mathbf{A}_{21} & \mathbf{A}_{22} \end{bmatrix} \tag{1.27}$$

and $\mathbf{v}$ denotes the first column of $\mathbf{V}_{22}$ we have

$$\mathbf{K}_{22}\mathbf{v} = \text{firstcol}(\mathbf{A}_{22}) - \mathbf{K}_{21}.\text{firstcol}(\mathbf{V}_{12}) \tag{1.28}$$

Moreover, from (1.26) we see that

$$\text{firstcol}(\mathbf{V}_{12}) = \mathbf{S}_{11}.\text{firstcol}(\mathbf{K}_{21}{}^T) + \mathbf{S}_{12}.\text{firstcol}(\mathbf{K}_{22}{}^T)$$

so that firstcol$(\mathbf{V}_{12})$ and hence $\mathbf{v}$ may be readily computed. In fact the calculation of firstcol$(\mathbf{V}_{12})$ is particularly simple since $\mathbf{S}_{11}$ is tridiagonal. Once $\mathbf{v}$ has been computed row interchanges may be performed to ensure that the multipliers satisfy the stability condition $|m_{ij}| \le 1$, together with the corresponding column interchanges needed to preserve symmetry. Note that, since the pivot is on the sub-diagonal, these interchanges affect only those columns to the right of the pivotal column of $\mathbf{V}_3$ and thus do not affect the reduction process itself. Once the final form of $\mathbf{v}$ has been established the multipliers and hence $\mathbf{M}_4$ may be computed.

With $\mathbf{v}$ known, the equation $\mathbf{S}_3\mathbf{K}^T_3 = \mathbf{V}_3$ may be written schematically

$$\underset{\mathbf{S}_3}{\left[\begin{array}{ccc|cc} \times & \times & 0 & 0 & 0 \\ \times & \times & \times & 0 & 0 \\ 0 & \times & ? & ? & ? \\ \hline 0 & 0 & ? & ? & ? \\ 0 & 0 & ? & ? & ? \end{array}\right]} \underset{\mathbf{K}_3^T}{\left[\begin{array}{ccc|cc} 1 & 0 & 0 & 0 & 0 \\ 0 & 1 & \times & \times & \times \\ 0 & 0 & 1 & \times & \times \\ \hline 0 & 0 & 0 & 1 & 0 \\ 0 & 0 & 0 & 0 & 1 \end{array}\right]} = \underset{\mathbf{V}_3}{\left[\begin{array}{ccc|cc} \times & \times & \times & ? & ? \\ \times & \times & \times & ? & ? \\ 0 & \times & \times & ? & ? \\ \hline 0 & 0 & \Rightarrow\times & ? & ? \\ 0 & 0 & \times & ? & ? \end{array}\right]}$$

Let the submatrices of (1.26) now refer to the partitions of the above equation. Since $\mathbf{K}_{11}^T$ is unit upper triangular and since only the last columns of $\mathbf{S}_{21}$ and $\mathbf{V}_{21}$ are non-zero it follows that $\mathbf{S}_{21} = \mathbf{V}_{21}$. Now $\mathbf{M}_4$ is chosen so that pre-multiplying $\mathbf{V}_3$ by $\mathbf{M}_4$ has the effect of zeroing all the elements below the pivot while leaving the pivot unchanged. Hence, since $\mathbf{S}_{21} = \mathbf{V}_{21}$, pre-multiplying $\mathbf{S}_3$ by $\mathbf{M}_4$ will have the same effect on $\mathbf{S}_{21}$. It follows from this and from (1.23)-(1.25) that $\mathbf{S}_4\mathbf{K}_4^T = \mathbf{V}_4$ may be written schematically

$$\underset{\mathbf{S}_4}{\left[\begin{array}{ccc|cc} \times & \times & 0 & 0 & 0 \\ \times & \times & \times & 0 & 0 \\ 0 & \times & ? & \times & 0 \\ \hline 0 & 0 & \times & ? & ? \\ 0 & 0 & 0 & ? & ? \end{array}\right]} \underset{\mathbf{K}_4^T}{\left[\begin{array}{ccc|cc} 1 & 0 & 0 & 0 & 0 \\ 0 & 1 & \times & \times & \times \\ 0 & 0 & 1 & \times & \times \\ \hline 0 & 0 & 0 & 1 & \times \\ 0 & 0 & 0 & 0 & 1 \end{array}\right]} = \underset{\mathbf{V}_4}{\left[\begin{array}{ccc|cc} \times & \times & \times & ? & ? \\ \times & \times & \times & ? & ? \\ 0 & \times & \times & ? & ? \\ \hline 0 & 0 & \times & ? & ? \\ 0 & 0 & 0 & ? & ? \end{array}\right]}$$

It only remains to compute the last diagonal element of the new $\mathbf{S}_{11}$ to complete this stage of the reduction and this is done from the expression for the the last diagonal element of $\mathbf{V}_{11}$ in the above equation. Thus, as at the beginning of the elimination stage, $\mathbf{S}_{11}$, $\mathbf{S}_{12}$, $\mathbf{S}_{21}$, and all partitions of $\mathbf{K}$ are known but the submatrices now refer to $\mathbf{S}_4$ and $\mathbf{K}_4$ rather than to $\mathbf{S}_3$ and $\mathbf{K}_3$. Repetition of this procedure n-2 times yields the factorisation $\mathbf{LTL}^T = \mathbf{PAP}^T$ where, as in the case of LU decomposition, the permutation matrix $\mathbf{P}$ is introduced to account for the row and column interchanges caused by pivoting. Once the factorisation has been effected the solution of equation (1.1) is obtained by solving $\mathbf{Ly} = \mathbf{Pb}$ for $\mathbf{y}$, $\mathbf{Tz} = \mathbf{y}$ for $\mathbf{z}$ and $\mathbf{L}^T\mathbf{Px} = \mathbf{z}$ for $\mathbf{Px}$ from which $\mathbf{x}$ is readily recovered. Since $\mathbf{L}$ is unit lower triangular the intermediate results $\mathbf{y}$ and $\mathbf{Px}$ are computed by substitution and $\mathbf{z}$, the other intermediate vector, is computed by Gaussian elimination or LU decomposition. Since, however, $\mathbf{T}$ is tridiagonal the number of operations for this process is $O(n^2)$, the same as for the substitutions, so since μ for the factorisation is $n^3/6$ this is also the value of μ for the process overall. Thus the amount of work required for the method is comparable with that needed when Choleski's method is applied to a positive definite matrix. Aasen [Aa] claims that, provided that the absolute values of the multipliers do not exceed unity, the method is as accurate in practice as Gaussian elimination with row interchanges and this is corroborated by his quoted error bounds.

Other methods of solving equations involving symmetric indefinite matrices have been given by Parlett and Reid [PR], Bunch and Parlett [BP] and Bunch and Kauffman [BK], the last of which is only marginally inferior (see [BG]) to Aasen's method.

1.6 Toeplitz matrices - Levinson's Method

A Toeplitz matrix is a matrix whose elements on a given diagonal are constant. Formally, if $\mathbf{A} = [a_{ij}]$ is a Toeplitz matrix then $a_{i,i+k} = c_k$, $-(n-1) \le k \le n-1$, where c_k is the constant for the diagonal in question. It is straightforward to show that if $\mathbf{A}$ is a Toeplitz matrix then $\mathbf{A}^T = \mathbf{JAJ}$, where $\mathbf{J}$ is the "reverse unit matrix", i.e. the unit matrix with its columns in the reverse order. We here consider, for simplicity, only symmetric and positive definite Toeplitz matrices and from symmetry this equation becomes $\mathbf{A} = \mathbf{JAJ}$ or, since $\mathbf{A}$ is positive definite and thus nonsingular,

$$\mathbf{A}^{-1}\mathbf{J} = \mathbf{J}\mathbf{A}^{-1} \tag{1.29}$$

since $\mathbf{J}^2 = \mathbf{I}$. Another feature of Toeplitz matrices is that all principal submatrices of the same order are identical and Toeplitz.

Now Levinson's method for solving (1.1) when $\mathbf{A}$ is a Toeplitz matrix is based on solving a sequence of equations

$$\mathbf{A}_r\mathbf{x}_r = \mathbf{b}_r \tag{1.30}$$

where $\mathbf{A}_r$ denotes the leading principal submatrix of $\mathbf{A}$ of order r and $\mathbf{b}_r$ denotes the vector consisting of the first r elements of $\mathbf{b}$. Use is also made of a set of auxiliary *r*th order vectors $\{\mathbf{v}_r\}$, where if $\mathbf{A}_{r+1}$ is written as

$$\mathbf{A}_{r+1} = \begin{bmatrix} \mathbf{A}_r & \mathbf{a} \\ \mathbf{a}^T & \alpha \end{bmatrix} \tag{1.31}$$

then $\mathbf{v}_r$ is defined by

$$\mathbf{v}_r = \mathbf{A}_r^{-1}\mathbf{J}\mathbf{a} \tag{1.32}$$

These vectors may be generated recursively and, once known, enable the solutions of (1.30) to be similarly generated. If then

$$\mathbf{A}_{r+2} = \begin{bmatrix} \mathbf{A}_r & \mathbf{a} & \mathbf{a}_1 \\ \mathbf{a}^T & \alpha & \alpha_1 \\ \mathbf{a}_1^T & \alpha_1 & \alpha \end{bmatrix} \tag{1.33}$$

it follows from (1.32) and (1.33) that if $\mathbf{v}_{r+1}^T = [\mathbf{s}^T, \sigma]$ then

$$\begin{bmatrix} \mathbf{A}_r & \mathbf{a} \\ \mathbf{a}^T & \alpha \end{bmatrix}\begin{bmatrix} \mathbf{s} \\ \sigma \end{bmatrix} = \begin{bmatrix} \alpha_1 \\ \mathbf{J}\mathbf{a}_1 \end{bmatrix} = \begin{bmatrix} \mathbf{J}\mathbf{a} \\ \alpha_2 \end{bmatrix}$$

where σ and α_2 are scalars (α_2 is in fact the first element of $\mathbf{a}_1$). Since $\mathbf{A}_r$ is nonsingular (from the hypothesis that $\mathbf{A}$ is positive definite) we have, from (1.29) and (1.32),

$$\mathbf{s} = \mathbf{v}_r - \mathbf{J}\mathbf{v}_r\sigma \tag{1.34a}$$

and

$$\mathbf{a}^T\mathbf{s} + \alpha\sigma = \alpha_2 \tag{1.34b}$$

and since, from (1.31) and (1.32), $\alpha - \mathbf{a}^T\mathbf{J}\mathbf{v}_r$ is a Schur complement of a positive definite matrix it must itself be positive (see appendix). Eliminating $\mathbf{s}$ from these equations then yields

$$\sigma = (\alpha_2 - \mathbf{a}^T\mathbf{v}_r)/(\alpha - \mathbf{a}^T\mathbf{J}\mathbf{v}_r).$$

Let now $\mathbf{x}_{r+1}^T = [\mathbf{z}^T, \zeta]$ and $\mathbf{b}_{r+1}^T = [\mathbf{b}_r^T, \beta]$. Then, from (1.30) and (1.31),

$$\begin{bmatrix} \mathbf{A}_r & \mathbf{a} \\ \mathbf{a}^T & \alpha \end{bmatrix} \begin{bmatrix} \mathbf{z} \\ \zeta \end{bmatrix} = \begin{bmatrix} \mathbf{b}_r \\ \beta \end{bmatrix}$$

so that, from (1.30) and (1.32),

$$\mathbf{z} = \mathbf{x}_r - \mathbf{J}\mathbf{v}_r\zeta$$

and

$$\mathbf{a}^T\mathbf{z} + \alpha\zeta = \beta$$

and eliminating $\mathbf{z}$ yields

$$\zeta = (\beta - \mathbf{a}^T\mathbf{x}_r)/(\alpha - \mathbf{a}^T\mathbf{J}\mathbf{v}_r).$$

Thus, if $\mathbf{x}_r$ and $\mathbf{v}_r$ are known, $\mathbf{x}_{r+1}$ and $\mathbf{v}_{r+1}$ may be readily determined. A further refinement is obtained by observing that the successive denominators of (1.35) may be computed recursively (see [GvL]) and when this is implemented the μ for Levinson's method is $2n^2$. For $\mathbf{A}$ positive definite as assumed above the method is roughly as stable as Choleski's factorisation (see [Cy]).

1.7 Vandermonde matrices - the method of Björck and Pereyra

We describe this method by considering its application to our ubiquitous 5 x 5 example from which the general case may readily be deduced. Let $\mathbf{A}$ be a 5 x 5 Vandermonde matrix given by

$$\mathbf{A} = \begin{bmatrix} 1 & 1 & 1 & 1 & 1 \\ x_1 & x_2 & x_3 & x_4 & x_5 \\ x_1^2 & x_2^2 & x_3^2 & x_4^2 & x_5^2 \\ x_1^3 & x_2^3 & x_3^3 & x_4^3 & x_5^3 \\ x_1^4 & x_2^4 & x_3^4 & x_4^4 & x_5^4 \end{bmatrix}$$

where $x_i \neq x_j$ if $i \neq j$, $\forall$ i,j, and consider the factorisation

$$\mathbf{LA} = \mathbf{U} \tag{1.36}$$

where $\mathbf{U} = [u_{ij}]$ is upper and $\mathbf{L} = [l_{ij}]$ is unit lower triangular. Then the expressions for u_{31} and u_{32} obtained from the left-hand side of the above equation yield

$$u_{31} = l_{31} + l_{32}x_1 + x_1^2 = 0 \tag{1.37}$$

and

$$u_{32} = l_{31} + l_{32}x_2 + x_2^2 = 0 \tag{1.38}$$

If we now define $p_2(x)$ by

$$p_2(x) \equiv l_{31} + l_{32}x + x^2 \tag{1.39}$$

equations (1.37) and (1.38) become $p_2(x_1) = 0$ and $p_2(x_2) = 0$ so that

$$p_2(x) \equiv (x - x_1)(x - x_2) \tag{1.40}$$

Since $u_{3j} = p_2(x_j)$ we have immediately

$$u_{3j} = (x_j - x_1)(x_j - x_2), \qquad j = 3,4,5.$$

and in general this becomes

$$u_{ij} = (x_j - x_1)(x_j - x_2)\ldots(x_j - x_{i-1}),$$
$$1 < i \leq n,\ 1 \leq j \leq n.$$

Trivially, $u_{1j} = 1$ for all j.

To compute $\mathbf{L}$ we use the generalisations of (1.39) and (1.40), and from the second of these we deduce that

$$p_i(x) \equiv (x - x_i)p_{i-1}(x)$$

so that, since $p_{i-1}(x) \equiv l_{i1} + l_{i2}x + \ldots + x^{i-1}$,

$$l_{i+1,1} + l_{i+1,2}x + \ldots + x^i \equiv (x - x_i)(l_{i1} + l_{i2}x + \ldots + x^{i-1}).$$

Equating coefficients of x^j, $j = 0,1,\ldots,i-1$ then yields

$$l_{i+1,1} = -x_i l_{i1} \qquad (j = 0)$$

$$l_{i+1,j+1} = -x_i l_{i,j+1} + l_{ij} \qquad (j = 1,2,\ldots,i-1)$$

where, again trivially, $l_{ii} = 1$ for all i.

Thus **L** may be constructed row-by-row using a process that is essentially the reverse of polynomial deflation, polynomial inflation, perhaps (see [Br]). Unlike forward or backward deflation there is no redundant equation and thus no ambiguity about the form of the process. Once the factors have been obtained the solution of $\mathbf{Ux} = \mathbf{Lb}$ is trivial. According to Björck and Pereyra [BjP] the method gives surprisingly accurate results bearing in mind the inherent ill-conditioning of Vandermonde matrices, and $\mu = n^2$. It has been generalised to deal with confluent Vandermonde systems by Björck and Elfving (see [BE]).

Appendix - Positive Definite Matrices and Schur Complements

Let **A** be symmetric. Then it is also *positive definite* if $\mathbf{x}^T\mathbf{Ax} > 0$ for $\forall\mathbf{x} \neq \mathbf{0}$ and it follows immediately from this definition that positive definite matrices are nonsingular since $\exists$ no $\mathbf{x} \neq \mathbf{0}$ such that $\mathbf{Ax} = \mathbf{0}$.

Let **A** be symmetric and positive definite, and let

$$\mathbf{A} = \begin{bmatrix} \mathbf{A}_{11} & \mathbf{A}_{12} \\ \mathbf{A}_{21} & \mathbf{A}_{22} \end{bmatrix} \quad \text{and} \quad \mathbf{x} = \begin{bmatrix} \mathbf{x}_1 \\ \mathbf{0} \end{bmatrix}$$

Then $\mathbf{x}^T\mathbf{Ax} = \mathbf{x}_1^T\mathbf{A}_{11}\mathbf{x}_1 > 0$ for $\forall\mathbf{x}_1 \neq \mathbf{0}$ so that $\mathbf{A}_{11}$ is also positive definite and hence nonsingular.

If $\mathbf{A}_{11}$ is nonsingular then $\mathbf{A}_{22} - \mathbf{A}_{21}\mathbf{A}_{11}^{-1}\mathbf{A}_{12}$ is called a *Schur complement* of **A** (note: this definition does not require symmetry or definiteness of **A**, only the nonsingularity of $\mathbf{A}_{11}$. It is thus valid for more general matrices than those considered here). Let now **A** be symmetric and positive definite and let

$$\mathbf{x} = \begin{bmatrix} -\mathbf{A}_{11}^{-1}\mathbf{A}_{12}\mathbf{x}_2 \\ \mathbf{x}_2 \end{bmatrix}.$$

Then $\mathbf{x}^T\mathbf{A}\mathbf{x} = \mathbf{x}_2^T(\mathbf{A}_{22} - \mathbf{A}_{21}\mathbf{A}_{11}^{-1}\mathbf{A}_{12})\mathbf{x}_2$, and since $\mathbf{x}^T\mathbf{A}\mathbf{x} > 0$ for $\forall\mathbf{x}_2 \neq 0$ it follows that $\mathbf{A}_{22} - \mathbf{A}_{21}\mathbf{A}_{11}^{-1}\mathbf{A}_{12}$ must be positive definite. Similarly, all other Schur complements of a positive definite matrix are also positive definite.

Part 2. Methods of Conjugate Directions

2.1 The Basis of the Methods

We consider now another class of methods that is specially suited to solving equation (1.1) where **A** is positive definite, large and sparse. Although these methods are technically direct methods (see Part 1 of this survey) their very nature and the inevitable presence of rounding error gives them an iterative character, and there is much interest in using them to achieve an approximate solution in less time than would be needed to carry out a complete implementation. They are similar to purely iterative methods in that the matrix **A** remains unchanged throughout, enabling economies of storage and time to be effected if **A** is sufficiently sparse, and they are based on minimising $\phi(\mathbf{x})$ where

$$\phi(\mathbf{x}) \equiv (\mathbf{x}^T\mathbf{A}\mathbf{x})/2 - \mathbf{b}^T\mathbf{x} \tag{2.1}$$

If $\mathbf{g} = [g_j]$ denotes the gradient of ϕ, i.e. $g_j = \partial\phi/\partial x_j$, then

$$\mathbf{g} = \mathbf{A}\mathbf{x} - \mathbf{b} \tag{2.2}$$

and it is straightforward to show that if

$$\mathbf{e} = \mathbf{x} - \mathbf{A}^{-1}\mathbf{b} \tag{2.3}$$

then $\phi = (\mathbf{e}^T\mathbf{A}\mathbf{e} - \mathbf{b}^T\mathbf{A}^{-1}\mathbf{b})/2$. Since **A** is positive definite this implies that ϕ has a unique minimum at $\mathbf{e} = \mathbf{0}$. Hence, from (2.3), ϕ is minimised when $\mathbf{x} = \mathbf{A}^{-1}\mathbf{b}$ and since this minimum is unique equation (1.1) may be solved by determining the value of $\mathbf{x}$ that minimises ϕ.

In common with purely iterative methods the methods based on conjugate directions generate a sequence $\{\mathbf{x}_i\}$ of approximate solutions of $\mathbf{Ax} = \mathbf{b}$ which should, in the absence of rounding error, terminate at the exact solution after at most n terms have been computed. This contrasts with iterative methods proper which go on indefinitely computing ever more accurate approximations to that solution. There are many ways of considering conjugate direction methods but the one chosen here regards them as minimising ϕ in subspaces of $\mathbb{R}^n$ of ever-increasing dimension so that when that dimension equals n, ϕ is minimised over all $\mathbf{x}$ and thus yields the required solution.

Let now $\mathbf{p}_j$, $j = 1,2,\ldots,n$, be a set of arbitrary nth order vectors with no properties other than that they are assumed to be linearly independent, and define the $n \times j$ matrix $\mathbf{P}_j$, $1 \le j \le n$, by

$$\mathbf{P}_j = [\mathbf{p}_1, \mathbf{p}_2, \ldots, \mathbf{p}_j] \tag{2.4}$$

Let $\mathbf{x}_1$ denote an arbitrary value of $\mathbf{x}$. We consider the minimisation of ϕ subject to the condition that $\mathbf{x}$ is required to satisfy

$$\mathbf{x} = \mathbf{x}_1 + \mathbf{P}_j\mathbf{z} \tag{2.5}$$

where $\mathbf{z} \in \mathbb{R}^j$ is a vector of variables over which ϕ is to be minimised. Note that if $j=1$ this is nothing more than minimising ϕ over a single variable so the minimisation defined above may be regarded as a generalised linesearch. With this restriction on $\mathbf{x}$ it follows from (2.1) that, since $\mathbf{A}$ is symmetric,

$$\phi = (\mathbf{z}^T\mathbf{P}_j^T\mathbf{A}\mathbf{P}_j\mathbf{z})/2 + \mathbf{g}_1^T\mathbf{P}_j\mathbf{z} + k \tag{2.6}$$

where $\mathbf{g}_1$ is the gradient of ϕ evaluated at $\mathbf{x}_1$ and k is some scalar that does not depend on $\mathbf{z}$. Now since $\mathbf{A}$ is positive definite and since, by hypothesis, the columns of $\mathbf{P}_j$ are linearly independent it follows that $\mathbf{P}_j^T\mathbf{A}\mathbf{P}_j$ is also positive definite so by a similar argument to that used to establish that the unconstrained minimum of ϕ occurs at $\mathbf{x} = \mathbf{A}^{-1}\mathbf{b}$ it follows from (2.6) that the value of $\mathbf{z}$ that minimises ϕ when $\mathbf{x}$ satisfies (2.5) is given by

$$\mathbf{z}_j = -(\mathbf{P}_j^T\mathbf{A}\mathbf{P}_j)^{-1}\mathbf{P}_j^T\mathbf{g}_1 \tag{2.7}$$

and from (2.5) the corresponding value of $\mathbf{x}$, $\mathbf{x}_{j+1}$, is given by

$$\mathbf{x}_{j+1} = \mathbf{x}_1 - \mathbf{P}_j(\mathbf{P}_j^T\mathbf{A}\mathbf{P}_j)^{-1}\mathbf{P}_j^T\mathbf{g}_1 \tag{2.8}$$

We note in passing that the minimiser of ϕ is determined only by the subspace and not by the vectors chosen to form the basis of that subspace. Thus if in (2.8) $\mathbf{P}_j$ is replaced by $\mathbf{P}_j\mathbf{M}_j$, where $\mathbf{M}_j$ is any jth order nonsingular matrix, the value of $\mathbf{x}_{j+1}$ is unchanged.

Now we wish to solve (1.1) by minimising ϕ over subspaces of increasing dimension and this is equivalent to determining $\mathbf{x}_{j+1}$ for $j = 1,2,\ldots,n$. Since, however, equation (2.8) involves $(\mathbf{P}_j^T\mathbf{A}\mathbf{P}_j)^{-1}$ this appears to be a quite formidable undertaking if $\mathbf{P}_j$ is chosen completely arbitrarily. If, however, $\mathbf{P}_j$ is chosen to satisfy

$$\mathbf{P}_j^T\mathbf{A}\mathbf{P}_j = \mathbf{D}_j, \quad 1 \le j \le k, \tag{2.9}$$

where $\mathbf{D}_j = \mathrm{diag}(d_i)$ it follows from (2.4), (2.7) and (2.8) that

$$\mathbf{x}_{j+1} = \mathbf{x}_1 - \sum_{i=1}^{j} \mathbf{p}_i(\mathbf{p}_i^T\mathbf{g}_1/d_i), \quad 1 \le j \le k \tag{2.10}$$

so that

$$\mathbf{x}_{j+1} = \mathbf{x}_j - \mathbf{p}_j(\mathbf{p}_j^T\mathbf{g}_1/d_j), \quad 1 \le j \le k \tag{2.11}$$

Now it is always possible to choose $\mathbf{P}_j$, $1 \le j \le n$, to satisfy (2.9) if $\mathbf{A}$ is positive definite (for instance, choose $\mathbf{P}_n$ to be an inverse Choleski factor, see Part 1 above), and such a choice of $\mathbf{P}_j$ not only makes the inversion of $\mathbf{P}_j^T\mathbf{A}\mathbf{P}_j$ trivial but, from (2.11), enables the minimisation to be performed by a sequence of minimisations along the vectors $\mathbf{p}_j$ themselves. Thus the use of conjugate vectors to define the successive subspaces converts a single minimisation over n variables into a sequence of n minimisations, each over a single variable. The problem is thus transformed into the determination of a set of conjugate vectors $\mathbf{p}_j$ and we consider now the possibility that these may be derived from another set of linearly independent vectors $\mathbf{q}_j$, $1 \le j \le n$.

One line of approach would be to let $\mathbf{p}_j$ be a linear combination of the vectors $\mathbf{q}_i$, $1 \le i \le j$, for some linearly independent choice of these vectors. Then, if we define $\mathbf{Q}_j$ by

$$\mathbf{Q}_j = [\mathbf{q}_1, \mathbf{q}_2, \ldots, \mathbf{q}_j], \quad 1 \le j \le k \tag{2.12}$$

it follows that

$$\mathbf{P}_j = \mathbf{Q}_j \mathbf{U}_j, \quad 1 \le j \le k \tag{2.13}$$

where $\mathbf{U}_j$ is a unit upper triangular matrix (chosen thus to ensure that the vectors $\mathbf{p}_j$ are linearly independent). Assume now that $\mathbf{P}_k$ has already been determined and that

$$\mathbf{P}_k^T \mathbf{A} \mathbf{P}_k = \mathbf{D}_k \tag{2.14}$$

Let $\mathbf{p}_{k+1} = \mathbf{q}_{k+1} + \mathbf{Q}_k \mathbf{v}_k$ so that, from (2.13),

$$\mathbf{p}_{k+1} = \mathbf{q}_{k+1} + \mathbf{P}_k \mathbf{w}_k \tag{2.15}$$

where $\mathbf{U}_k \mathbf{w}_k = \mathbf{v}_k$, For conjugacy we require that $\mathbf{P}_k^T \mathbf{A} \mathbf{p}_{k+1} = \mathbf{0}$ so that, from (2.14) and (2.15), $\mathbf{P}_k^T \mathbf{A} \mathbf{q}_{k+1} + \mathbf{D}_k \mathbf{w}_k = \mathbf{0}$. Now it follows from previous assumptions that $\mathbf{D}_k$ is nonsingular so this equation may be solved for $\mathbf{w}_k$, and substituting this value in (2.14) then gives

$$\mathbf{p}_{k+1} = \mathbf{S}_k \mathbf{q}_{k+1} \tag{2.16a}$$

where

$$\mathbf{S}_k = \mathbf{I} - \mathbf{P}_k \mathbf{D}_k^{-1} \mathbf{P}_k^T \mathbf{A} \tag{2.16b}$$

Now, from (2.9), $\mathbf{S}_k^2 = \mathbf{S}_k$ and is thus a nonsymmetric projection matrix so that the computation of $\mathbf{p}_{k+1}$ may be regarded as projecting $\mathbf{q}_{k+1}$ into a particular subspace. However, the evaluation of $\mathbf{p}_{k+1}$ using equations (2.15) and (2.16) would again appear to be computationally burdensome unless some simplifications could be made, and we shall return to this aspect of the problem shortly.

We conclude this section by deriving one of the properties of the gradients at the points generated by this process of successive minimisation. If we denote by $\mathbf{g}_j$ the gradient of ϕ evaluated at $\mathbf{x}_j$ it follows from (2.2), (2.7) and (2.9) that

$$\mathbf{g}_{j+1} = (\mathbf{I} - \mathbf{A}\mathbf{P}_j\mathbf{D}_j^{-1}\mathbf{P}_j^T)\mathbf{g}_1, \quad 1 \le j \le k \tag{2.17}$$

so that, from (2.9),

$$\mathbf{P}_j^T\mathbf{g}_{j+1} = \mathbf{0}, \quad 1 \le j \le k \tag{2.18}$$

and, from (2.13),

$$\mathbf{Q}_j^T\mathbf{g}_{j+1} = \mathbf{0}, \quad 1 \le j \le k \tag{2.19}$$

Thus the gradient at $\mathbf{x}_{j+1}$ is, as would be expected, orthogonal to the subspace over which ϕ was minimised.

2.2 The Method of Conjugate Gradients

We consider now some of the properties of the method for a particular choice of the vectors $\mathbf{q}_j$, namely

$$\mathbf{q}_j = \mathbf{B}\mathbf{g}_j, \quad 1 \le j \le n \tag{2.20}$$

where $\mathbf{B}$ is an arbitrary positive definite matrix. If $\mathbf{G}_j$ is defined by

$$\mathbf{G}_j = [\mathbf{g}_1, \mathbf{g}_2, \ldots, \mathbf{g}_j] \tag{2.21}$$

then equations (2.19) - (2.21) yield

$$\mathbf{G}_j^T\mathbf{B}\mathbf{g}_{j+1} = \mathbf{0}, \quad 1 \le j \le k \tag{2.22}$$

so that the gradients at the successive constrained minima are conjugate with respect to $\mathbf{B}$. Clearly, if $\mathbf{B} = \mathbf{I}$, these gradients are orthogonal. Now equation (2.11) may be written

$$\mathbf{x}_{j+1} = \mathbf{x}_j + \mathbf{p}_j t_j, \quad 1 \le j \le k \tag{2.23}$$

where

$$t_j = -(\mathbf{p}_j^T \mathbf{g}_1)/d_j, \quad 1 \le j \le k$$

so that from (2.2) we have, transposing,

$$\mathbf{g}_{j+1}^T = \mathbf{g}_j^T + t_j \mathbf{p}_j^T \mathbf{A}, \quad 1 \le j \le k \tag{2.24}$$

Now for $j+1 \le k$, $\mathbf{g}_{j+1}^T$ and $\mathbf{g}_j^T$ are rows of $\mathbf{G}_k^T$ so that, from (2.22) with $j=k$, $t_j \mathbf{p}_j^T \mathbf{ABg}_{k+1} = \mathbf{0}$, $1 \le j \le k-1$ so that if $t_j \ne 0$ we have

$$\mathbf{P}_{k-1}^T \mathbf{ABg}_{k+1} = \mathbf{0} \tag{2.25}$$

Now equations (2.15), (2.16) and (2.20) yield

$$\mathbf{p}_{k+1} = (\mathbf{I} - \mathbf{P}_k \mathbf{D}_k^{-1} \mathbf{P}_k^T \mathbf{A}) \mathbf{Bg}_{k+1} \tag{2.26}$$

but since, from (2.4), $\mathbf{P}_k = [\mathbf{P}_{k-1}, \mathbf{p}_k]$ it follows from (2.25) that the first k-1 elements of $\mathbf{P}_k^T \mathbf{ABg}_{k+1}$ are equal to zero. Equation (2.26) thus reduces, since $\mathbf{D}_k$ is diagonal, to

$$\mathbf{p}_{k+1} = \mathbf{Bg}_{k+1} - \mathbf{p}_k (\mathbf{p}_k^T \mathbf{ABg}_{k+1}/d_k) \tag{2.27}$$

so that the formidable expression (2.15) for computing the conjugate vectors reduces to a simple recurrence formula. Setting $\mathbf{B} = \mathbf{I}$ then gives the well-known expression used in the method of conjugate gradients and first described in the classic paper of Hestenes and Stiefel [HS]. Setting $\mathbf{B} = \mathbf{A}$ gives a formula that generates a sequence $\{\mathbf{x}_i\}$ for which the residuals are conjugate with respect to $\mathbf{A}$ (the conjugate-residual formula, see [Hs],[Hg]).

2.3 Alternative expressions

A feature of the conjugate direction methods arising from the various conjugacy properties that obtain is the variety of mathematical expressions available for many of the quantities needed in calculation, and this often enables us when selecting an expression we wish to program to choose one that results in some computational advantage. A case in point occurs with equation (2.11). The use of this equation to compute $\mathbf{x}_{j+1}$ means storing $\mathbf{g}_1$ in addition to $\mathbf{g}_j$ (which must be stored anyway in order to compute $\mathbf{p}_j$ using (2.27) with k = j-1) so it would be desirable if some alternative expression for $\mathbf{x}_{j+1}$ not involving $\mathbf{g}_1$ could be found. This can be achieved by observing that equations (2.9) and (2.24) give

$$\mathbf{g}_{j+1}^T\mathbf{p}_k = \mathbf{g}_j^T\mathbf{p}_k, \quad 1 \le j < k \tag{2.28}$$

so that if we take j = 1, 2, ..., k-1 we get

$$\mathbf{g}_1^T\mathbf{p}_k = \mathbf{g}_2^T\mathbf{p}_k = \ldots = \mathbf{g}_k^T\mathbf{p}_k$$

Putting k = j in this last equation and substituting in (2.11) then gives

$$\mathbf{x}_{j+1} = \mathbf{x}_j - \mathbf{p}_j(\mathbf{p}_j^T\mathbf{g}_j/d_j) \tag{2.29}$$

eliminating $\mathbf{g}_1$ as required. If $\mathbf{B} = \mathbf{I}$, other versions of (2.11) and (2.27) are given by (see e.g. [Fl])

$$\mathbf{x}_{j+1} = \mathbf{x}_j - \mathbf{p}_j(\mathbf{g}_j^T\mathbf{g}_j/d_j) \tag{2.30}$$

and

$$\mathbf{p}_{j+1} = \mathbf{g}_{j+1} - \mathbf{p}_j(\mathbf{g}_{j+1}^T\mathbf{g}_{j+1}/\mathbf{g}_j^T\mathbf{g}_j) \tag{2.31}$$

These versions are particularly useful if for some reason it is important to preserve the positivity of the scalar factors when they are nearly equal to zero and subject to rounding error.

Acknowledgements

The author thanks Mr. W. E. Hart, Dr. J. A. Ford and the University of Essex for their help in preparing this paper.

References, Part 1

[Aa] Aasen, J. O., "On the Reduction of a Symmetric Matrix to Tridiagonal Form," BIT 11, 233-242, 1971.

[BG] Barwell, V. and J. A. George, "A Comparison of Algorithms for Solving Symmetric Indefinite Systems," ACM Trans. Math. Soft. 2, 242-251, 1976.

[BjP] Bjorck, A. and V. Pereyra, "Solution of Vandermonde Systems of Equations," Math. Comp. 24, 893-903, 1970.

[BE] Bjorck, A. and T. Elfving, "Algorithms for Confluent Vandermonde Systems," Numer. Math. 21, 130-137, 1973.

[Br] Broyden, C. G., "Error Propagation in Numerical Processes," J. Inst. Maths. Applics. 14, 131-140, 1974.

[BuP] Bunch, J. R. and B. N. Parlett, "Direct Methods for Solving Symmetric Indefinite Systems of Linear Equations," SIAM J. Numer. Anal. 8, 639-655, 1971.

[BK] Bunch, J. R. and L. Kaufman, "Some Stable Methods for Calculating Inertia and Solving Symmetric Linear Systems," Math. Comp. 31, 162-179, 1977.

[Cy] Cybenko, G., "The Numerical Stability of the Levinson-Durbin Algorithm for Toeplitz Systems of Equations," SIAM J. Sci. & Stat. Comp. 1, 303-310, 1980.

[Di] Dixon, L. C. W., private communication.

[GvL] Golub, Gene H. and Charles van Loan, *Matrix Computations,* Johns Hopkins University Press, Baltimore, 1983.

[Hi] Higham, Nicholas J., "The Accuracy of Solutions to Triangular Systems," SIAM J. Numer. Anal. 26, 1252-1265, 1989.

[Le] Levinson, N., "The Weiner RMS Error Criterion in Filter Design and Prediction," J. Math. Phys. 25, 261-278, 1947.

[MPW] Martin, R. S., G. Peters and J. H. Wilkinson, "Symmetric Decomposition of a Positive Definite Matrix," Numer. Math. 8, 203-216, 1966.

[PR] Parlett, B. N. and J. K. Reid, "On the Solution of a System of Linear Equations whose Matrix is Symmetric but not Definite," BIT 10, 386-397, 1970.

[St] Stewart, G. W., *Introduction to Matrix Computations,* Academic Press, New York, 1973.

[Wi] Wilkinson, J. H., *The Algebraic Eigenvalue Problem,* Clarendon Press, Oxford, 1965.

References, Part 2

[Fl] Fletcher, R., "Conjugate Gradient Methods for Indefinite Systems," in *Proc. Dundee Conference on Numerical Analysis,* Lecture Notes in Mathematics 506, edited by G. A. Watson, Springer, Berlin-Heidelberg, 1976.

[Hs] Hestenes, M. R., "The Conjugate Gradient Method for Solving Linear Systems," in *Proc. Symposia on Applied Mathematics, Vol VI, Numerical Analysis,* Mc.Graw-Hill, New York, 1956.

[Hg] Hegedus, Cs. J., "Generating Conjugate Directions for Arbitrary Matrices by Matrix Equations, Parts 1 and 2, Report No. KFKI-1990-36/M, Hungarian Academy of Sciences, Central Research Institute for Physics, Budapest, 1990.

[HS] Hestenes, M. R. and E. Stiefel, "Methods of Conjugate Gradients for Solving Linear Systems," J. Res. Nat. Bureau of Standards 49, 409-436, 1952.

Complexity of Algorithms for Linear Systems of Equations

Victor Pan†
Department of Mathematics and Computer Science
Lehman College
CUNY, Bronx, NY 10468

and

Department of Computer Science
State University of New York at Albany
Albany, NY 12222

A huge amount of computer resources is spent over the world every day for solving systems of linear equations, which are the backbone of computations in sciences and engineering. Naturally, the solution algorithms are devised so as to decrease the amount of such resources spent, that is, to decrease the estimated computational complexity of the solution.

We are going to survey the state of art in this well developed and thoroughly investigated area, to the extent allowed by the size of this article. For further details, we refer the reader to the cited bibliography and in particular, to the book [BP,a].

Our study and its results decisively depend on the structure of the input coefficient matrix A, and we will study the cases of general (dense and unstructured), sparse, and dense but structured matrices A in sections 1, 2 and 3, respectively. In section 4, we will revisit iterative algorithms for the solution (particularly effective for special linear systems, relatively less covered in sections 1 - 3); in particular, we will recall some additional advantages in applications to partial differential equations (by using multigrid methods and their recent compact versions) and in using symbolic computation. In subsection 3.2 and 3.3 of section 3, we recall some recent research results from [P90a].

In our survey we focus on the asymptotic complexity of the solution and only briefly sketch some important practical aspects of the solution algorithms, for which we refer the reader to [GL], [GeLi], [P87] and, of course, to other articles of this volume.

---

† This work has been supported by NATO Advanced Study Institute, by University of Bergamo, by NSF Grants CCR 8805782 and CCR 9020690 and by PSC CUNY Awards #661340, #668541 and #669290.

NATO ASI Series, Vol. F 77
Computer Algorithms for Solving Linear Algebraic Equations
Edited by E. Spedicato

1. Complexity of Solving a General Linear System

1.1. The Main Requirements to the Algorithms.

1.1. The Main Requirements to the Algorithms. Let us first consider a linear system

$$A\mathbf{x} = \mathbf{b}, \tag{1.1}$$

with a general $n \times n$ nonsingular coefficient matrix A, and recall the three major practical and theoretical requirements to its solution algorithms, that is, their numerical stability, the economization of the memory space they use, and their faster performance, characterized by involving fewer arithmetic operations (hereafter referred to as **ops**). Note that numerically unstable algorithms, however, may be stabilized and have some (more limited) practical value, in the context of symbolic computation, by means of reduction of the intermediate results modulo primes or prime powers (see section 4.3 below).

The most popular solution algorithms rely on Gaussian elimination with (partial) pivoting; they are numerically stable, require near optimum storage space of $n^2 + O(n)$ words of memory and involve about $(2/3)\ n^3$ ops [as well as about $(1/2)\ n^2$ comparisons of pairs of real numbers], ([GL], [P87]).

1.2. Asymptotic Acceleration by Reduction to Matrix Multiplication. May $o(n^3)$ ops suffice? The negative answer, that is, a lower bound an^3, for a positive constant a, was ruled out in 1968, when V. Strassen presented a solution using less than $4.7n^{2.808}$ ops ([St69]). The basis for this solution was a new simple but surprising algorithm that reduced the 2×2 block matrix multiplication to 7 (rather than to 8) block multiplications and to 18 block additions/subtractions. Adding/subtracting every pair of $k \times k$ blocks in k^2 ops and recursively applying the same algorithm to block multiplications, we may multiply a pair of $H \times H$ matrices, for $H = 2^h$, by using 7^h multiplications and $4 * 7^h + 18 * 4^h$ additions/subtractions, that is, by using less than $4.7\ H^{\beta}$ ops, $\beta \leq \log_2 7 = 2.807\ldots$

The constant 4.7 can be decreased to 4.54 by applying Winograd's improved algorithm that uses 7 block multiplications and 15 block additions/subtractions for 2×2 block matrix multiplications. Specifically (compare [BM]), denote

$$A = \begin{bmatrix} a_{11} & a_{12} \\ a_{21} & a_{22} \end{bmatrix}, \quad B = \begin{bmatrix} b_{11} & b_{12} \\ b_{21} & b_{22} \end{bmatrix}$$

and compute

$$S_1 = a_{21} + a_{22}, \qquad S_2 = S_1 - a_{11}, \qquad S_3 = a_{11} - a_{21}, \qquad S_4 = a_{12} - S_2,$$

$S_5 = b_{12} - b_{11}$, $S_6 = b_{22} - S_5$, $S_7 = b_{22} - b_{12}$, $S_8 = S_6 - b_{21}$;

$P_1 = S_2\, S_6$, $P_2 = a_{11}\, b_{11}$, $P_3 = a_{12}\, b_{21}$, $P_4 = S_3\, S_7$,

$P_5 = S_1\, S_5$, $P_6 = S_4\, b_{22}$, $P_7 = a_{22}\, S_8$;

$S_9 = P_3 + P_2$, $S_{10} = P_1 + P_2$, $S_{11} = S_{10} + P_4$, $S_{12} = S_{10} + P_5$,

$S_{13} = S_{12} + P_6$, $S_{14} = S_{11} - P_7$, $S_{15} = S_{11} + P_5$.

Then

$$AB = \begin{bmatrix} S_9 & S_{13} \\ S_{14} & S_{15} \end{bmatrix}.$$

To extend the above asymptotic estimate to matrix inversion [and then, of course, to computing the solution vector $\mathbf{x} = A^{-1}\mathbf{b}$ to the system (1.1)], let us recall the well-known (Gauss-Jordan) factorization of a 2×2 block matrix:

$$A = \begin{bmatrix} B & C \\ D & E \end{bmatrix} = \begin{bmatrix} I & O \\ DB^{-1} & I \end{bmatrix} \begin{bmatrix} B & O \\ O & S \end{bmatrix} \begin{bmatrix} I & B^{-1}C \\ O & I \end{bmatrix}, \tag{1.2}$$

$$A^{-1} = \begin{bmatrix} I & -B^{-1}C \\ O & I \end{bmatrix} \begin{bmatrix} B^{-1} & O \\ O & S^{-1} \end{bmatrix} \begin{bmatrix} I & O \\ -DB^{-1} & I \end{bmatrix}, \tag{1.3}$$

where $S = E - DB^{-1}C$ is the Schur complement of B in A, and O and I denote the null and the identity matrices, respectively.

The factorization (1.2), (1.3) is well-defined and is numerically well-conditioned if the matrix A is diagonally dominant and/or positive definite, which may be ensured by means of symmetrization, since $A^{-1} = (A^T A)^{-1} A^T$ (see [AHU], sections 6.4-6.5, on a modification of (1.2), (1.3) with using pivoting as a good practical alternative to the symmetrization).

Now, for simplicity, let $n = 2^h$, h be an integer, and let M(k) and I(k) denote the minimum numbers of ops required for $k \times k$ matrix multiplication and inversion, respectively. Then (1.3) implies that $I(n) \le 2\,I(n/2) + 6\,M(n/2) + 4\,(n/2)^2$,

and recursively,

$$I(n) \le 4\,I(n/4) + 12\,M(n/4) + 8\,(n/4)^2 +$$

$$6\,M \quad (n/2) + 4\,(n/2)^2 \le$$

$$\cdots\cdots\cdots\cdots$$

$$n\,I(1) + 3 \sum_{i=1}^{n} 2^i\, M(n/2^i) + 2 \sum_{i=1}^{h} 2^i\,(n/2^i)^2,$$

so that

$$I(n) < n + 3cn^\beta / (1 - 2^{1-\beta}) + 2n^2$$

$$\text{if } M(n) \le cn^\beta, \quad \beta \ge 2.$$

1.3. Even Faster Algorithms. Besides matrix inversion, numerous other algebraic and combinatorial computations can be ultimately reduced to matrix multiplication and thus asymptotically accelerated to the level $O(n^\beta)$, $\beta \le 2.808$ ([AHU], [P84], [P84a]). Furthermore, Strassen's recursive construction can be immediately extended to define the bound

$$M(n) = O(n^\beta), \quad \beta \le \log \mu / \log n, \tag{1.4}$$

on the exponent β of matrix multiplication (and of all the computations reducible to it), as soon as we are given a *bilinear algorithm* that, for a fixed natural h, multiplies $h \times h$ block matrices $X = [x_{ij}]$ and $Y = [Y_{jk}]$ as follows:

$$(XY)_{ik} = \sum_j x_{ij}\, y_{jk} = \sum_{m=1}^{\mu} c_{ki}^{(m)}\, L_m(X)\, L_m^*(Y), \tag{1.5}$$

where

$$L_m(X) = \sum_{i,j} a_{ij}^{(m)} X_{ij},\ L_m^*(Y) = \sum_{j,k} b_{jk}^{(m)} Y_{jk},\ i, j, k = 1, \ldots, h;\ m = 1, \ldots, \mu;$$

$a_{ij}^{(m)}$, $b_{jk}^{(m)}$ and $c_{ki}^{(m)}$ are scalar constants, x_{ij}, y_{jk}, and thus also $L_m(X)$, $L_m^*(Y)$ are blocks, whose multiplication can be reduced to recursive application of the same bilinear algorithms. [Indeed, add and multiply by a constant the $k \times k$ blocks in k^2 ops and multiply such blocks together by recursively using (1.5).]

Strassen's and classical algorithms give us two examples of such bilinear algorithms where $\mu = \mu(2) = 7$ $(h=2)$ and $\mu = \mu(h) = h^3$ (for any h), respectively.

The observation (1.4) and Strassen's success instigated numerous attempts of devising new bilinear algorithms, first with $\mu = \mu(2) \le 6$ and then with $\mu = \mu(3) \le 21$, such that (1.4) would imply the bounds $\beta < \log_2 7$. This turned out to be impossible for $\mu(2) \le 6$ and too hard (if possible) even for $\mu(3) \le 23$, so that the actual decrease of the exponent $\log_2 7$ (first to $\beta \le 2.795\ldots$) only occurred in 1978 and was based on a new bilinear algorithm, with $\mu(70) = 143640$, devised by using then novel techniques of *trilinear aggregating* ([P78], [P80]).

Further extension of these techniques and their combination with several other nontrivial and novel ideas successively decreased the exponent to its current record level of $\beta < 2.376$ ([CW], see also [P84] and simpler expositions in [P84a] and [BP,a]).

1.4. Some Practical Aspects. Typically, each asymptotic acceleration of matrix multiplication relied on a nontrivial basis linear algorithm for computing a matrix product, 2 or 3 disjoint matrix products, or some related bilinear forms. Such an algorithm was then recursively applied, so that it improved the straightforward algorithm when the problem size grew sufficiently large. To support the exponents β below 2.77, however, all the known algorithms require nontrivial multistage recursive construction, so that they only beat the straightforward bound of $2n^3 - n^2$ ops for immense n.

Thus, only few known algorithms may be considered for practice, that is, ones of [St69], of Winograd (see [BM], pp. 45-46), (both for $\beta \le \log_2 7$) and of [P84], Part III, and [LPS] (for $\beta < 2.78$).

Weak numerical stability has been proven in [BL80], for all the bilinear algorithms known for fast matrix multiplication, and even much stronger stability has been observed in numerous experiments.

Strassen's and Winograd's algorithms (with $\beta \le \log_2 7$) have been successfully implemented on CRAY and IBM computers, and their popularity, as well as users' interest in the reduction of solving linear systems to matrix multiplication, grows with their interest to block algorithms in linear algebra.

1.5. Nonasymptotic Acceleration. Possible asymptotic acceleration should not, of course, preclude using other improvements. In particular, we may effectively multiply matrices filled with zeros and ones or even with any bounded ("short") integers by applying the binary segmentation techniques. These techniques, for instance, enable us to recover the outer product of two vectors filled with zeros and ones (or by "short" integers) from the product of two longer integers associated with the two input vectors, thus reducing the computation to a single integer multiplication (see [P84]), [BP,a]). To see a simple example of binary segmentation, recover "short" integer coefficients p_i of a polynomial $p(x) = \sum_i p_i x^i$ from its value $p(2^h)$, for a sufficiently large h; then apply this technique to arrive at the product of two polynomials with "short" integer coefficients by multiplying a single pair of their values at $x = 2^h$, for a large integer h.

Clearly, it suffices to choose any $h > \log_2 n$ if the output vectors are filled with zeros and ones and any $h > \log_2(n M^2)$ if they are filled with integers from 0 to M. If the input vectors are filled with the integers from $-M$ to M, we choose $h > \log_2(2n M^2)$ and recover the coefficients of p(x) from the value $p(2^h) + \sum_{i=0}^{h-1} 2^{hi+h-1}$.

Now, if we need to compute the outer product of the two vectors, that is, to compute the $n \times n$ matrix $[u_i v_j]$, $i,j=0,\ldots,n-1$, where the input entries u_i, v_j are filled with integers from $-M$ to M, we choose $h = \lceil 1+2 \log M \rceil$, compute $p(2^h) = (\sum_{i=0}^{n-1} 2^{inh} u_i)(\sum_{j=0}^{n-1} 2^{jh} v_j) = \sum_{i,j=0}^{n-1} u_i v_j \, 2^{(in+j)h}$, and recover all the products $u_i v_j$ by means of binary segmentation.

In yet another example, for the same input, we let $h = \lceil 1 + \log(n M^2) \rceil$ and compute the product of two integers

$$(\sum_{i=0}^{n-1} 2^{ih} u_i)(\sum_{j=0}^{n-1} 2^{(n-1-j)h} v_j) = \sum_{k=0}^{2n-2} z_k 2^{kh},$$

and then recover the inner product: $z_{n-1} = \sum_i u_i v_i$. In these examples we assume that the values $n^2 h$ (for the outer product) and $(2n-1)h$ (for the inner product) fit the computer precision, and then a single multiplication suffices. Even if the computer precision is less than this but greater than sh, for a reasonably large s, then still, the above techniques by roughly s times accelerate the computations, which is a dramatic acceleration, for larger s, and which directly applies to matrix multiplication and to several other computations in linear algebra.

Less dramatic but still substantial an improvement can be obtained for multiplication of matrices, filled with complex numbers. Indeed, in this case, the matrix $(X + iY)(U + iV) = XU - YV + i(XV + YU)$ (where $i = \sqrt{-1}$, X, Y, U, V are matrices) can be computed by using 3 (rather than 4) multiplications of real matrices, since

$$XU - YV = X(U + V) - (X + Y)V,$$

$$XV + YU = Y(U - V) + (X + Y)V.$$

In another instance, recall that we may replace about 1/2 of all multiplications by additions when we compute any $n \times n$ matrix product (see [Kn], [P84], Part III).

1.6. Parallel Algorithms. A Model and Some Basic Estimates. Parallel architecture is a very powerful tool for acceleration of linear algebra computations. To specify parallel complexity estimates, let us hereafter assume the customary Parallel Random Access Machine (PRAM) models ([Quinn], [EG88], [KR]), under which in each step each processor performs at most one arithmetic operation. The models are not realistic (since they assume access of all the processors to their shared global memory) but can be simulated on quite realistic models ([Val90]). The reader, however, should note that we will omit the estimates for the communication and synchronization complexity, which would require to use distinct tools and concepts; we will, however, bound the number of processors used, which will decrease the impact of this omission. Furthermore, by following R. Brent (see [KR]), we will assume that processors can be saved if we slow down the computations, so that our estimate of the form $O_A(t, p)$ will mean O(st) arithmetic steps and p/s processors for any s such that $1 \le s \le p$.

Here is a table of the complexity estimates for some basic computations (FFT and Toeplitz matrices will be used in the next sections).

Table 1.1

Parallel Computation for	Parallel time (t_{par})	Processors (p)	Source
1. FFT at n points (n is a power of 2)	$3(1+\log_2 n)$	2n	[Pease]
2. Summation of n numbers	$2\lceil \log_2 n \rceil$	$\lceil n/\log_2 n \rceil$	[Quinn]
3. Inner product	$2\lceil \log_2 n \rceil + 1$	$\lceil n/\log_2 n \rceil$	(reduce to summation)
4. n×n matrix times a vector	$2\lceil \log_2 n \rceil + 1$	$n\lceil n/\log_2 n \rceil$	(n inner products)
5. Product of n×n matrices	$2\lceil \log_2 n \rceil + 1$	$n^2\lceil n/\log_2 n \rceil$	(n^2 inner products)
6. n×n Toeplitz matrix by a vector (n is a power of 2)	$1+9(2+\log_2 n)$	4n	(reduce to 3FFTs at 2n points, [AHU])

1.7. Parallel Complexity of Solving Linear Systems. Parallel acceleration of Gaussian elimination supports the estimates $O_A(n, n^2)$, which are quite satisfactory for practical solution of the general linear systems (1.1).

Theoretically, it is interesting to reach parallel time $O(\log^h n, n^g)$ for two constants g and h, and this has been done for h=2 in several ways (see [Cs], [PS], [GP], [P87b]), with the smallest g slightly below $\beta + 0.5$, $\beta < 2.376$ (see [GP]). Moreover, the near optimum probabilistic bounds $O_A(\log^2 n, n^\beta)$ have been recently derived in [P90a] (see also the end of section 3.2 below), by using some random parameters and, which is more painful, the coefficients of the minimum polynomial of the input matrix A. The latter defect causes numerical stability problems and thus limits the application of this algorithm to the area of algebraic and symbolic computation.

Even from the practical point of view, however, it is desirable to improve parallel Gaussian elimination by reducing the solution of (1.1) to few matrix multiplications, which are quite effective operations with respect to their communication complexity. Indeed, let each of them be performed by means of parallelization of the straightforward algorithm. Then $2n^3 - n^2$ ops are performed per $2n^2$ data fetches (of the $2n^2$ input entries), thus reaching the ratio of about n ops per fetch, which is much higher than, for instance, in the matrix-by-vector multiplication, where not more than 2 ops are performed per data fetch.

This observation draws our attention to Newton's iteration,

$$X_0 = \alpha_0 A^T, \; X_{k+1} = \alpha_{k+1} X_k \; (2I - AX_k), \; k = 0, 1, \ldots, \tag{1.6}$$

which, already for $\alpha_k = 1$ and for all $k \geq 1$, converges to A^{-1} or, if A is singular, to the Moore-Penrose generalized (pseudo) inverse, A^+, of A, thus defining the least-squares solution $\mathbf{x} = A^+\mathbf{b}$ to the system (1.1). (Note that $A^+ = A^{-1}$ if A is nonsingular and see [GL], [BP,a] on some other properties of A^+.) For each k, the iteration step (1.6) is reduced to 2 matrix multiplications, and squares the residual matrix already for $\alpha_k = 1$:

$$I - AX_k = (I - AX_{k-1})^2 = (I - AX_0)^K, \; K=2^k, \; k = 1, 2, \ldots \tag{1.7}$$

The latter property implies local quadratic convergence and self-correcting of the approximations by X_k to A^{-1} (see [PS88] on the numerical stability of such an approximation for singular A). For an appropriate choice of α_k, for $k = 1, 2, \ldots$, the convergence in roughly $\log_2 \kappa(A)$ steps (1.6) [$\kappa(A)$ being the maximum ratio of two singular values of A] has been ensured in [PS88], with a further acceleration by twice for a Hermitian (real symmetric) A. [A modified version of (1.6), proposed in [PS88], converges to the matrix $A^+(e)$ obtained from the matrix A^+ by setting to zero the singular values less than e in the SVD of A^+, for any fixed positive e; this has some further important applications.] Thus, the parallel complexity of such an approach is $O_A(\log n \log \kappa(A), M(n))$, where M(n) denotes the number of ops sufficient for $n \times n$ matrix multiplication, $M(n) \leq \text{minimum}\ \{2n^3 - n^2, 4.7\, n^{\log_2 7}, Cn^{2.376}\}$ for a constant C. Generally, $\kappa(A)$ is not easy to estimate, but $\kappa(A)$ is not large for many classes of linear systems, arising, for instance, in the solution of integral equations.

2. Sparse Linear Systems

2.1. General Review. The solution of a linear system (1.1) with a sparse $n \times n$ matrix A having $\gamma(A)$ nonzero entries, $\gamma(A) = o(n^2)$, usually involves fewer computational resources of time and storage space provided that the sparsity patterns are sufficiently regular, so that we may use appropriate data structures and avoid storing zero entries of A and operating with them. In particular, the Lanczos and conjugate gradient algorithms only require few (say, 1 or 2) multiplications of A or A^T by vectors at each iteration and converge to the solution in at most n iterations (if performed with no errors) and in O(n) iterations [and actually in o(n) iterations] in practice of finite precision computations. (Symmetrization needed in these algorithms may be implicit and may rely on the matrix equations: $A^{-1} = (A^T A)^{-1} A^T$ or

$$\begin{bmatrix} O & A \\ A^T & O \end{bmatrix}^{-1} = \begin{bmatrix} O & A^{-T} \\ A^{-1} & O \end{bmatrix}, \quad \left[\text{with } A^{-T} = (A^{-1})^T\right];$$

in the latter case, the matrix is indefinite but better conditioned than $A^T A$, compare [GL]). Thus $O(n\,\gamma(A))$ ops and about $\gamma(A)$ words of memory shall always suffice.

Note that $\gamma(A) < (w+1)n$, in the important special case where A is banded with bandwidth w. In this case the bounds of about $\gamma(A) \le w\,n$ words and $O(n\,\gamma(A)) = O(w\,n^2)$ ops are implied by the above methods, but also by some alternate methods ([GL], sect. 5.3). Furthermore, if w is small and if we seek a faster parallel algorithm, it can be most effective (unless numerical stability problems arise) to treat A as an $(n/w) \times (n/w)$ tridiagonal block matrix with (generally dense) $w \times w$ blocks and to apply the block cyclic reduction algorithm, which solves the system (1.1) in $\lceil \log_2 (n/w) \rceil$ cyclic reduction steps, each step essentially amounting to $O(n/w)$ concurrent inversions and multiplications of the $w \times w$ blocks ([GL], sect. 5.5).

Many special sparse linear systems may be solved by using $o(n\,\gamma(A))$ ops. In particular, O(n) ops suffice where the system (1.1) is obtained by discretization of a partial differential equation (PDE) and is solved by a multigrid algorithm (see [HT82], [McC87]), whereas $O(n^{1.5})$ ops suffice where A is associated with a graph having an $O(\sqrt{n})$-separator family ([GeLi], [LRT]).

The latter approach is usually applied to symmetric and positive definite matrices A, and the algorithm, called a (generalized) nested dissection algorithm, computes the Choleski's factorization $PAP^T = L^T L$, L being a lower triangular matrix and P being a permutation matrix, chosen so as to decrease the size of the *fill-in* , that is, of the set of the nonzeros of L and L^T that appear on the places of the zeros of A.

The most widely used and simple policy for choosing P is to proceed with the symmetric version of Gaussian elimination, where in each step the pivotal entry (i, j) is chosen so as to minimize the value $(r(i)-1)(c(j)-1)$, r(i) and c(j) denoting the numbers of nonzeros in the i–th row and in the j–th column of the current coefficient matrix, respectively. To see the power of this policy (called *the Markowitz rule,* or, in the symmetric case, *the minimum degree rule*), consider a matrix filled with zeros except for its diagonal and its first row and column; then, without pivoting, the first step of Gaussian elimination turns the original sparse matrix into dense one, but no fill-in appears at all in this case if the Markowitz rule is applied.

2.2. (Generalized) Nested Dissection. For the general sparse matrix A, computing a permutation matrix P that minimizes the fill-in is an NP- *complete* problem ([AHU]), that is, we cannot hope to solve it even for moderate n (say, for n = 100). For a large and important class of linear systems (1.1), however, an effective method, called *(generalized) nested dissection* ([GeLi], [GT], [LRT]) enables us to bound the fill-in and, consequently, the time and memory space involved in the computation. Typically, the bounds are $O(n^{3/2})$ ops, O(n) words of memory space, and O(n) elements of the fill-in, for the (generalized) nested dissection algorithm that computes the Choleski factorization of a sparse and appropriately structured matrix A.

With this factorization available, O(n) ops and O(n) words of memory suffice to solve the system (1.1) for any vector **b** on the right side, which is particularly attractive when we need to solve several such systems for a fixed A and variable **b**.

As the first major step in the study of the nested dissection, we replace the nonzero entries of A by ones and thus arrive at the adjacency matrix of the associated graph G = (V,E). The edges of E correspond to the nonzero entries of A, and the n vertices of V are identified with n unknown variables of the system (1.1). For simplicity, we will assume that the matrix A is symmetric, so that the graph G is undirected.

The elimination of a variable v corresponds to the deletion of a vertex v, together with all the edges adjacent to it, followed by adding the edges (u, w) to the graph as soon as its two edges (v, u) and (v, w) have been deleted. For a large and important class of linear systems (1.1), particularly, for many linear systems arising in the solution of ordinary and partial differential equations, the associated graphs G = (V, E) have small separator families, that is, can be recursively partitioned, so that removing a set S of $O(\sqrt{n})$ vertices from G (together with all edges adjacent to S) partitions G into two subgraphs $G_1 = (V_1, E_1)$ and $G_2 = (V_2, E_2)$ that have no edges remaining in G and connecting them and such that each of G_1 and G_2 has at most a fixed fraction (say, 2/3) of the vertices of G; moreover, similar partitions are recursively defined for the subgraphs G_1 and G_2 ([GT]) or, in an alternate definition of [LRT], [PR88], for the graphs $\tilde{G}_1$ and $\tilde{G}_2$ obtained from G_1 and G_2, respectively, by adjoining to each of them the separator set S, together with all the edges of G adjacent to S. The recursive process ends where G_i (for i=1 and/or for i=2) has less than n_o vertices, for a fixed (smaller) constant n_o. To better understand this definition, the reader may examine the important special case where G is a grid graph (say, for the two-dimensional 7×7 grid), and to define the separators recursively, by removing the vertical and horizontal edges in the middle of the grid. In particular, by setting $n_o = 1$ and by using the definition of [GT] for the separators, we will finally arrive at the singletons. Computing the Choleski factorization (which we will identify with the elimination) proceeds in the order opposite to the partition (and again the reader may examine this process for the 7×7 grid graph G, as an exercise). In particular, at the last, d–th, step of the recursive partition, we arrive at the graphs $G_{h,\,d}$, $h = 1, 2, ..., k(d)$, with the vertex sets of cardinalities less than n_o, whose vertices are eliminated first. This corresponds to the Choleski factorization of the block diagonal matrix of the northwestern part of $A_o = PAP^T$. The remaining graph corresponds to the matrix A_1 of the southeastern block of A. Then, the remaining vertices of the graphs $G_{j,\,d-1}$, $j = 1, 2, ..., k(d-1)$, obtained at the $(d-1)$–st partition step, are eliminated; they correspond to the blocks of the block diagonal submatrix of A_1 lying in its northwestern corner; then the process is recursively repeated for the southeastern block A_2 of A_1, and so on.

Due to the bound on the size of the vertex set involved in each step of the recursive partition of G, the block diagonal matrices eliminated in the elimination stage g, $g \le d \le \log_{(3/2)} n$, have blocks of the size $O(\sqrt{(2/3)^{d-g} n})$, and this implies a small upper bound on the complexity of the elimination, even though the (small) eliminated blocks are generally dense. We refer the reader to [GT] and [LRT] for many details involved in the most general case, where the graphs G are given with their families of small separators (the important special cases of grid graphs and of finite element graphs are relatively simple).

Finally, we would like to mention the option of replacing the Choleski factorization by computing the recursive factorization (1.2), by using which we may still solve (1.1) in $O(n)^2$ ops (see [PR88]). This version (unlike the other version) can be performed over semirings (where divisions and subtractions are not allowed) and thus supports the extension of the generalized nested dissection algorithm to many highly important computations of paths in planar graphs ([PR,a]); both versions allow parallel implementation with parallel time of the order of $\log^3 n$ and $O(n^{3/2})$ processors (see [PR88] and [Haf]).

3. Dense Structured Linear Systems

3.1. General Comments. If the coefficient matrix A of (1.1) is dense but well structured, the solution of (1.1) can be much simpler than in the general case. Typically, with a structured $n \times n$ matrix A, O(n) words of memory and from $O(n \log n)$ to $O(n \log^2 n)$ ops suffice (versus the orders of n^2 words and of $n^{2.8}$ to n^3 ops required in the general case). Numerical stability of the fast algorithms is easily ensured in the cases of symmetric positive definite or diagonally dominant input matrices A, as well as of input matrices A of smaller sizes, but otherwise can be a severe problem ([Bun]), restricting the practical applications of the algorithms to the area of symbolic computations.

3.2. Toeplitz and Hankel Linear Systems. Particularly important are the classes of linear systems with Toeplitz and Hankel matrices, $T = [t_{ij}]$ and $H = [h_{ij}]$, such that $t_{ij} = t_{i-j}$ and $h_{ij} = h_{i+j}$, for all entries (i, j), respectively. Each such an $n \times n$ matrix has at most $2n-1$ distinct entries, and is completely defined by its two columns or by its two rows (the first and the last), which reduces to n entries of the first column in the symmetric case. For any Hankel matrix H and for the reflection matrix J (zero everywhere except for its antidiagonal, filled with ones), JH and HJ are Toeplitz matrices, so that we will just study Toeplitz linear systems (1.1), and this will actually cover the Hankel case too.

A powerful method that facilitates operations with $n \times n$ Toeplitz matrices exploits their embedding in larger $N \times N$ circulant matrices $C = [c_{ij}]$, which constitute a subclass of the class of N×N Toeplitz matrices, $c_{ij} = c_{i-j \bmod N}$, $N = O(n)$. Computing $C\mathbf{b}$, given a vector $\mathbf{b}$ and an N×N circulant matrix C, is immediately reduced to 3 fast Fourier transforms (FFTs) at N points performed for the overall cost $O_A(\log N, N)$, ([Da74]), and similarly for computing both $C^{-1}\mathbf{b}$ if C is nonsingular (note that C^{-1} is a circulant matrix too) and $T\mathbf{b}$ where T is any Toeplitz matrix of size $n \times n$, $2n \le N$, for we surely may embed T in an $N \times N$ circulant matrix.

The latter observation suggests using conjugate gradient algorithms for Toeplitz systems (1.1), where every iteration step reduces to few FFTs and thus costs $O_A(\log n, n)$. Indeed, the power of such algorithms and their fast convergence have been proven for a large and important subclass of all Toeplitz systems (1.1) ([Strang,a], [BdB]).

For other Toeplitz systems (1.1), for which the convergence of the latter algorithms is slow, the most customary solution algorithms are Levinson's, Schur's (both using the order of n^2 ops), and modified Schur's [using $O(n \log^2 n)$ ops], (see [CB], [AG,b], [Kai] and, for some earlier solutions in $O(n \log^2 n)$ ops, [BGY], [BA], [Mus], [dH]). These are direct algorithms (they compute the solution exactly if performed with infinite precision), and in their parallel implementation, at least n parallel arithmetic steps are required.

In some applications, a greater parallel acceleration is needed. For instance, in some signal processing applications, the solution to a Toeplitz linear system must be updated in real time, together with the (slowly) updated input Toeplitz matrix of the coefficients, so that the solution computed at the previous time-step can be used as a good initial approximation to the solution at the current time-step.

In such cases, Newton's iteration of section 1.7 is a better choice (a good alternative is also given by various methods of the steepest descent). Specifically, each Newton's iteration step essentially amounts to a pair of matrix multiplications. In the case of a Toeplitz input matrix A, we may modify such a step and perform it for the low cost of $O_A(\log n, n)$ ([P89b], [P90a]).

In particular, this approach may employ the Gohberg-Semencul formulae (theorem 3.1 below) or their further extensions ([Io], [T90]).

Definition 3.1. Z is the matrix of lower shift (displacement), filled with zeros except for its first subdiagonal, filled with ones; L(**v**) is the lower triangular Toeplitz matrix defined by its first column vector **v**. J is the reflection matrix (see above).

Theorem 3.1 ([T64], [GSe]). *Let the matrices* Z, L(**v**) *and* J *be defined in definition 3.1,* T *be a nonsingular Toeplitz matrix,* **x** *and* **y** *be the two columns of* T^{-1}, *the first and the last, respectively,* x_o *be the first component of* **x**. *Then* $x_o T^{-1} = L(\mathbf{x}) L^T (J\mathbf{y}) - L(Z\mathbf{y}) L^T (Z^T J\mathbf{x})$.

Now, for an arbitrary matrix W, let **u**, **v** and x_o denote its first column, its last column and the first component of **v**, respectively, and let $x_o \neq 0$. Then we denote

$$X(W) = (1/x_o)(L(\mathbf{u}) L^T (J\mathbf{v}) - L(Z\mathbf{v}) L^T (Z^T J\mathbf{u})). \tag{3.1}$$

Hereafter, let $|x_o|$ always be large enough so as not to let the division by x_o cause any numerical stability problem in (3.1); otherwise, we could just shift to some modified versions of (3.1) and of theorem 3.1 (see [Io] or [T90]).

Now, we define the following modification of Newton's iteration (1.6):

$$X_{k+1} = X(Y_{k+1}),\ Y_{k+1} = \alpha_{k+1} X_k (2I - AX_k),\ k = 0, 1, \ldots, \tag{3.2}$$

where $X(Y_{k+1})$ is defined by (3.1), for $W = Y_{k+1}$, and actually we only need to compute the two columns (the first and the last), for each of the two matrices Y_{k+1} and X_{k+1} in the k–th iteration step (3.2), whose cost is thus bounded by $O_A(\log n, n)$.

On the other hand, due to (1.7), a rapid convergence of the matrices X_k to A^{-1} is ensured if some (vector induced) matrix norm $\| I - AX_o \|$ is substantially less than 1 (say, if it is less than 0.8), and the latter assumption is satisfied, for instance, in the cited real time computations, where we may thus effectively apply the modification (3.2) of Newton's iteration, so as to ensure that $\|I - AX_k\| < 2^{-s}$ already for $k = O(\log s)$, that is, for the cost $O_A(\log s \log n, n)$. Similar comments apply if, say, the entries of A rapidly decrease their magnitude as they move away from the diagonal; then we have a good approximation to A and A^{-1} by a band Toeplitz matrix X_0 and by its inverse, respectively; X_0 is obtained from A by zeroing its smaller entries that lie farther from the diagonal of A.

The complexity of the solution of the system (1.1) decreases almost to $O_A(\log n, n)$ if A is a triangular Toeplitz or circulant matrix and decreases even more if A is a banded Toeplitz matrix ([BP,a]).

Now, suppose, no good initial approximation to A^{-1} is available; we may still extend the above approach (see [P89b], [P90a]). The estimated parallel complexity grows to $O_A(\log^2 n, n^2/\log n)$, and moreover, the algorithms involve computing the coefficients of the characteristic polynomial of A, and consequently may lead to some problems of numerical stability.

The algorithm of [P90a], however, seems to be a practical algorithm for computing a minimum norm least-squares solution to (1.1), that is, $\mathbf{x} = A^{+}\mathbf{b}$, since for this problem, both its parallel cost bounds $O_A(\log^2 n, n^2/\log n)$ and its sequential time bound $O_A(n^2 \log n)$ are the record bounds.

Generally, the precision of these computations is prohibitively high for numerical computing (this, however, applies to any method for computing $A^{+}\mathbf{b}$, if, say, A is a rank deficient matrix), but we will avoid such problems by using modular arithmetic.

Let us next outline the algorithm. (This outline, as well as the exposition of a large part of the next subsection, rely on the very recent results of [P90a] and reproduce some novel techniques from that paper. The results assume using modular arithmetic to avoid numerical instability. Those readers who are not interested in these results and techniques may now go to section 4.) We first reduce our problem to computing the coefficients of the characteristic polynomial of A,

$$\det(\lambda I - A) = \sum_{i=0}^{n} p_i \lambda^i, \; p_n = 1, \; p_0 = (-1)^n \det A;$$

having such coefficients available, we then compute $A^{-1}\mathbf{b} = -\sum_{i=1}^{n} (p_i/p_0) A^{i-1}\mathbf{b}$ (if A is nonsingular) or

$$A^{+}\mathbf{b} = (1/p_{n-r}) \sum_{i=n-r+1}^{n-1} ((p_{n-r+1}/p_{n-1})\, p_i - p_{i+1})\, A^{i-n+r}\, \mathbf{b} + (p_{n-r+1}/p_{n-r}^2)\, A^r\, \mathbf{b} \qquad (3.3)$$

(if A is a symmetric matrix of rank r, so that $p_0 = p_1 = \ldots = p_{n-r-1} = 0 \neq p_{n-r}$). These expressions follow from the Cayley-Hamilton theorem (see [P90]). Symmetrization of the coefficient matrix of (1.1) may rely on the matrix identity:

$$\begin{bmatrix} O & A \\ A^T & O \end{bmatrix}^{+} = \begin{bmatrix} O & (A^{+})^T \\ A^{+} & O \end{bmatrix}.$$

Thus, the evaluation of $A^{+}\mathbf{b}$ has been reduced to computing:

a) the coefficients $p_0, \ldots, p_{n-1}$ and

b) the products of the matrix powers A^k by $\mathbf{b}$, for $k=1,\ldots,n-1$.

Stage a) in turn reduces [for the cost $O_A(\log^2 n, n/\log n)$] to computing the traces of A^k, $s_k = \text{trace}(A^k) = \sum_i a_{ii}^{(k)}$, $A^k = [a_{ij}^{(k)}]$, for $k=1,\ldots,n$ (see [P90]), so that we only need to know the traces of the matrices A^k and the products $A^k\mathbf{b}$.

We introduce an auxiliary parameter λ and the matrix $B = I - \lambda A$. Then $B^{-1} = \sum_{k=0}^{\infty} (\lambda A)^k$, and it suffices to compute the coefficients of the following polynomials in λ:

$$\text{trace}(B^{-1}) \bmod \lambda^{n+1} = \sum_{k=0}^{n} \lambda^k \,\text{trace}(A^k),$$

and

$$B^{-1}\mathbf{b} \bmod \lambda^{n+1} = \sum_{k=0}^{n} \lambda^k \, B^k\mathbf{b}.$$

To invert B modulo λ^{n+1}, we apply Newton's iteration (1.6) starting with $X_0 = B^{-1} \bmod \lambda = I$. Due to (1.7), we have $X_k = B^{-1} \bmod \lambda^K$, $K=2^k$, $k=0,1,\ldots$, and therefore, $d = \lceil \log_2(n+1) \rceil$ Newton's steps suffice.

Since A is a Toeplitz matrix, so are the matrices B and X_k, for all k. We may apply theorem 3.1 [note that the entry (1,1) equals 1 modulo λ, so it does not vanish]. Thus, we only need to deal with two columns of X_k (the first and the last), for each k, and Newton's step k is reduced to 4 multiplications modulo λ^{2K}, $K=2^k$, of Toeplitz matrix polynomials by vector polynomials (that is, of Toeplitz matrices filled with polynomials by vectors filled with polynomials). Every such an operation can be reduced to multiplication of two bivariate polynomials and can be performed for the cost $O_A(\log n, n\, 2^k)$, which (for $k=0,1,\ldots,n$) gives us the overall cost bound $O_A(\log^2 n, n^2/\log n)$. (Here, we slow down Newton's steps that otherwise would require more processors.) Having the first and the last columns of $X_d = B^{-1} \bmod \lambda^D$, $D=2^d$, and thus also of $B^{-1} \bmod \lambda^{n+1}$, we then apply theorem 3.1 and compute $B^{-1}\mathbf{b} \bmod \lambda^{n+1}$ and $\text{trace}(B^{-1}) \bmod \lambda^{n+1}$, remaining within the desired cost bounds $O_A(\log^2 n, n^2/\log n)$. □

These results have several further applications, in particular, to computing the least common multiple and the greatest common divisor of two polynomials (see Appendix) and to supporting the record asymptotic estimates $O_A(\log^2 n, n^\beta)$, $2 \le \beta < 2.376$, for the parallel complexity of probabilistic algorithms for computing the solution or a least-squares solution to a general linear system $A\mathbf{x} = \mathbf{b}$ with $n\times n$ matrix A. In [Kal], [Wied], computing $\mathbf{x} = A^{-1}\mathbf{b}$ and $\det A$ has been reduced to solving a Toeplitz system of linear equations as follows:

First compute the coefficients of the minimum polynomial of A, $m(\lambda) = \sum_{i=0}^{k} m_i \lambda^i$, $k \le n$, $m_k = 1$, such that $m(A) = O$, and then obtain

$\mathbf{x} = A^{-1}\mathbf{b} = -\sum_{i=1}^{k} (m_i / m_0) A^{i-1}\, \mathbf{b}$ or $\mathbf{x} = A^{+}\mathbf{b}$ [relying on the extension of (3.3)] in three steps (compare [Cs], [P87b]):

a) compute the powers $A^2, A^4, A^8, \ldots, A^H$, $H=2^h$, $2^{h-1} < k \le 2^h$;

b) compute the Krylov sequence of the vectors $A\mathbf{b}, A^2\mathbf{b}, \ldots, A^{k-1}\mathbf{b}$:

$$A\mathbf{b} = A\mathbf{b},$$

$$A^2[A\mathbf{b}, \mathbf{b}] = [A^3\mathbf{b}, A^2\mathbf{b}],$$

$$A^4[A^3\mathbf{b}, A^2\mathbf{b}, A\mathbf{b}, \mathbf{b}] = [A^7\mathbf{b}, A^6\mathbf{b}, A^5\mathbf{b}, A^4\mathbf{b}]$$

....

(compare [BM], p. 128, and [K-G]).

c) compute the linear combination $-\sum_i (m_i / m_0)\, \mathbf{v}_i$, $\mathbf{v}_i = A^{i-1}\mathbf{b}$.

To obtain the coefficients $m_0, m_1, \ldots, m_k$, follow [Wied] and first compute the sequence of 2n scalars $s_i = \mathbf{u}A^i\mathbf{v}$, $i=0,1,\ldots,2n-1$, for two random real vectors $\mathbf{u}$ and $\mathbf{v}$. [It is easy to verify that the randomized parallel complexity of all these computations has the bounds $O_A(\log^2 n, n^\beta)$]. The scalars s_i form a linear recurrence,

$$s_{i+1} = c_h s_i + c_{h-1} s_{i-1} + \ldots + c_0\, s_{i-h},$$

for some scalars $c_0, c_1, \ldots, c_h$ and for $h < k \le n$. For a random choice of the real vectors $\mathbf{u}$ and $\mathbf{v}$, the latter scalars, with the probability 1, give us the coefficients of $m(\lambda)$, that is, $-m_0 = c_0, -m_1 = c_1, \ldots, -m_{k-1} = c_h$, $k-1 = h$, provided that $m_k = 1$.

On the other hand, the problem of computing $c_0, \ldots, c_h$ (given $s_0, s_1, \ldots, s_{2h}$), called the Berlekamp-Massey linear span problem or the shift register problem, is equivalent to computing the (h,h) Padé approximation for a given analytic function, which is in turn equivalent to solving a consistent (though perhaps singular) Toeplitz system of h linear equations with h unknowns ([BGY]). With this stage included, the overall randomized parallel complexity of the entire computation still remains bounded by $O_A(\log^2 n, n^\beta)$ since $\beta \ge 2$, and since we know how to solve such Toeplitz linear systems of n equations with n unknowns, for the cost $O_A(\log^2 n, n^2/\log n)$. (Actually, if the coefficient matrix of this system has rank $r < h$, then it is known that its $r \times r$ northwestern, that is, leading principal, submatrix is nonsingular, and this enables us to solve the system for the above cost.)

3.3. Toeplitz-like and Hankel-like Linear systems. Toeplitz and Hankel matrices constitute an important subclass of a more general class of dense structured matrices characterized by a lower rank of their images in the application of certain operators of displacement and/or scaling ([B83], [KVM], [KKM], [CKL-A], [P90a]). In particular, for the displacement matrix Z of definition 3.1, the four displacement operators

$$F_-(A) = A - Z^T AZ, \tag{3.4}$$

$$F_+(A) = A - ZAZ^T, \tag{3.5}$$

$$F^-(A) = AZ^T - Z^T A, \tag{3.6}$$

$$F^+(A) = AZ - ZA \tag{3.7}$$

are naturally associated with Toeplitz and Toeplitz-like matrices, for which the images of such operators have lower ranks, in particular, 2, for every Toeplitz matrix, and $m + n$, for every $m \times n$ block matrix with Toeplitz blocks. Representing an $n \times n$ matrix Y of a lower rank r as the product $Y = GH^T$, where G and H are $r \times n$ matrices, we will call the pair (G, H) a *generator of length* r for Y. If $Y = F(A) = GH^T$, for a displacement operator F(A) of (3.4)-(3.7), that is, if F(A) is $F_-(A)$, $F_+(A)$, $F^-(A)$ or $F^+(A)$, then this pair of matrices (G,H) will also be called a *displacement generator* (or an F- *generator* , for a specific operator F) *of length r for A.*

Let us recall some fundamental properties of displacement generators (see [KKM], [BA], [P90a], [BP,a]).

Theorem 3.2 ([KKM]). *Let*

$$G = [\mathbf{g}_1, \dots, \mathbf{g}_r], \; H = [\mathbf{h}_1, \dots, \mathbf{h}_r].$$

Then

$$A = \sum_{i=1}^{r} L(\mathbf{g}_i) L^T(\mathbf{h}_i) \text{ if } A - ZAZ^T = GH^T, \tag{3.8}$$

$$A = \sum_{i=1}^{r} L^T(J\mathbf{g}_i) L(J\mathbf{h}_i) \text{ if } A - Z^T AZ = GH^T. \tag{3.9}$$

Thus, an $n \times n$ matrix A can be defined by the 2nr entries of its displacement generator of length r, rather than by the n^2 entries of A. For the same matrix A, its L^TL- and LL^T–representations (3.8) and (3.9) are closely related to each other (see [BA], [BP,a]), so that an F_+ (respectively, F_-)-generator of length r for a matrix A immediately define its F_- (respectively, F_+)-generator of length at most $r + 2$. Indeed, it is easy to verify ([BA], [P90a], [BP,a]) the following result:

Proposition 3.1. *For any pair of vectors* **g** *and* **h** *of the same dimension,*

$$L^T(\mathbf{g})\, L(\mathbf{h}) = L(\mathbf{u}) + L^T(\mathbf{v}) - L(ZJ\mathbf{g})\, L^T(ZJ\mathbf{h}),$$

$$L(\mathbf{g})\, L^T(\mathbf{h}) = L^T(\mathbf{w}) + L(\mathbf{y}) - L^T(ZJ\mathbf{g})\, L(ZJ\mathbf{h})$$

where $\mathbf{u}$ *is the first column and* $\mathbf{v}^T$ *is the first row of* $L^T(\mathbf{g})\,L(\mathbf{h})$, $J\mathbf{w}$ *is the last column and* $\mathbf{y}^T J$ *is the last row of* $L(\mathbf{g})\,L^T(\mathbf{h})$.

The operators F_-, F_+, F^- and F^+ are closely related to each other:

Proposition 3.2. *Let* $\mathbf{i}_1 = [1,0,\ldots,0]^T$, $\mathbf{i}_n = [0,\ldots,0,1]^T$. *Then, for any matrix* A,

$$F^-(A)Z = F_-(A) - A\mathbf{i}_n\mathbf{i}_n^T,\ F_-(A)Z^T = F^-(A) + Z^T A\mathbf{i}_1\mathbf{i}_1^T,$$

$$F^+(A)Z^T = F_+(A) - A\mathbf{i}_1\mathbf{i}_1^T,\ F_+(A)Z = F^+(A) + ZA\mathbf{i}_n\mathbf{i}_n^T,$$

$$ZF^-(A) = \mathbf{i}_1\mathbf{i}_1^T A - F_+(A),\ ZF_-(A) = \mathbf{i}_1\mathbf{i}_1^T AZ - F^+(A),$$

$$Z^T F^+(A) = \mathbf{i}_n\mathbf{i}_n^T A - F_-(A),\ Z^T F_+(A) = \mathbf{i}_n\mathbf{i}_n^T AZ^T - F^-(A).$$

Proof. Pre- and postmultiply the matrix equations (3.4) - (3.7) by Z and Z^T and combine the resulting equations with (3.4) - (3.7) and with the matrix equations $Z^T Z = I - \mathbf{i}_n\mathbf{i}_n^T$, $ZZ^T = I - \mathbf{i}_1\mathbf{i}_1^T$. □

We will call A a *Toeplitz-like matrix* wherever $r = \text{rank}(F(A))$ is small, for a displacement operator F of (3.4) - (3.7) (where, say, $r = O(1)$ as $n \to \infty$). We may define a *Hankel-like* matrix A as such that JA is a Toeplitz-like matrix, or we may define Hankel-like matrices by introducing associated displacement operators such as $F(A) = A - ZAZ$ or $F(A) = Z^T A - AZ$. We may operate with the displacement generators of Toeplitz-like matrices rather than with the matrices themselves (and similarly with Hankel-like matrices) by relying on the following result:

Theorem 3.3. *Given a displacement operator* F *of (3.3) - (3.6) and a pair of F-generators of lengths a and b for a pair of* $n \times n$ *matrices A and B, we may immediately obtain an F-generator of length at most* $a + b$ *for* $A + \alpha B$ *(for any fixed scalar* α*); furthermore, we may also compute [for the cost of* $O_A(\log n, abn)$*)] F-generators of lengths at most* $a + b + \delta(F)$, *for* AB *and* $A^H A$ *where* $\delta(F) = 0$ *if* $F = F^-$ *or* $F = F^+$ *and* $\delta(F) = 1$ *if* $F = F_-$ *or* $F = F_+$ (see [CKL-A], [P89a], [P90a]).

Proof. The result for $A + \alpha B$ is obvious. For AB (and similarly for $A^H A$), it suffices to consider the cases of $F = F^-$ and $F = F_-$. Let $F = F^-$. Then the result follows since

$$F^-(AB) = ABZ^T - Z^T AB = A(BZ^T - Z^T B) + (AZ^T - Z^T A)B =$$

$$A\,F^-(B) + F^-(A)\,B = AG_B^-(H_B^-)^T + G_A^-(H_A^-)^T B = G_{AB}^-(H_B^-)^T + G_A^-(H_{AB}^-)^T,$$

provided that

$$F^-(A) = G_A^-(H_A^-)^T,\ F^-(B) = G_B^-(H_B^-)^T,\ G_{AB}^- = AG_B^-,\ (H_{AB}^-)^T = (H_A^-)^T B.$$

If $F = F_-$, theorem 3.3 follows since

$F_-(AB) = AB - Z^T AIBZ =- AB - (Z^T AZ)\,(Z^T BZ) +$ $Z^T A\mathbf{i}_1\mathbf{i}_1^T BZ = (A - Z^T AZ)\,B +$

$Z^T AZ(B - Z^T BZ) + \mathbf{uv}^T =$ $F_-(A)B + Z^T AZ F_-(B) + \mathbf{uv}^T =$ $G_-(A)(H_-^T(A)B) + (Z^T AZ G_-(B))H_-^T(B) + \mathbf{uv}^T$ provided that $F_-(A) = G_-(A)H_-^T(A)$, $F_-(B) = G_-(B)H_-^T(B)$, $\mathbf{u} = Z^T A\mathbf{i}_1$, $\mathbf{v}^T = \mathbf{i}_1^T BZ$. The proof is similar for $F = F^+$ and $F = F_+$. □

The above machinery enables us to extend our algorithm of section 3.2 to solving Toeplitz-like linear systems. The problem is to ensure a lower cost of Newton's iteration, which was done by using (3.1), (3.2) in section 3.2 in the Toeplitz case. For the Toeplitz-like input matrix polynomial $B = I - \lambda A$, given with its F-generator of length r, where, say, $F = F^-$, we will compute F-generators of length at most r, for the matrices $X_k = B^{-1} \bmod \lambda^K$, $K=2^k$, for all k, k=0,1,..., by induction on k. Due to theorem 3.2, we may then reduce the cost of Newton's step k to $O_A(\log n, r^2 2^k n)$ (see corollary 3.1 below).

The basis for the above induction on k is trivial since $X_0 = I$, but the proof of the inductive hypothesis about the low cost transition from a generator of length r, for B and $X_k \bmod \lambda^K$, $K=2^k$, to a similar one, for $X_{k+1} \bmod \lambda^{2K}$, $K=2^k$, requires some auxiliary results and techniques (compare [KKM], [B83] and [P90a]).

Theorem 3.4. *Let A be a nonsingular matrix polynomial. Then* $\text{rank}\, F_-(A^{-1}) = \text{rank}\, F_+(A)$, $\text{rank}\, F^-(A^{-1}) = \text{rank}\, F^-(A)$, $\text{rank}\, F^+(A^{-1}) = \text{rank}\, F^+(A)$.

Proof. The last two equations are immediately obtained by pre- and postmultiplying by A^{-1} the matrix equations (3.6) and (3.7). To deduce the first equation, observe that $\text{rank}\, F_-(A^{-1}) = \text{rank}(A^{-1} - Z^T A^{-1} Z) =$ $\text{rank}(I - A Z^T A^{-1} Z) = \text{rank}(I - Z A Z^T A^{-1}) =$ $\text{rank}(A - ZAZ^T)$. The second and the last equations are deduced by pre- and post multiplication of the parenthesized expressions by A. The third equation follows since $\text{rank}(I - BZ) = \text{rank}(I - ZB) = 1 + \text{rank}(I_{n-1} - B_{n,1})$, for any n×n matrix B (in particular, for $B = AZ^T A^{-1}$) and for its (n−1)×(n−1) submatrix $B_{n,1}$ obtained by deleting the first column and the last row of B. □

Combining theorems 3.2 and 3.3 and propositions 3.1 and 3.2, we arrive at

Corollary 3.1. *An* F- *generator of length at most* $3r + c(F)$ *for the matrix polynomial* $X_{i+1} \bmod \lambda^s$, $s = 2^{i+1}$, *can be computed for the cost* $O_A(\log n, r^2 sn)$ *given an* F- *generator of length* r *for* $X_i \bmod \lambda^{s/2}$ *and an* F^*- *generator of length at most* r *for* B, *provided that* $X_{i+1} = X_i(2I - BX_i)$ *and that one of the four cases occurs:*

a) $F = F_-$, $F^* = F_+$, $c(F) = 5$,

b) $F = F_+$, $F^* = F_-$, $c(F) = 5$,

c) $F = F^* = F^-$, $c(F) = 1$,

d) $F = F^* = F^+$, $c(F) = 1$.

Remark 3.1. A more careful examination enables us to decrease c(F) to 2 in the cases a) and b) and to 0 in the cases c) and d) (see [P90a]).

Finally, we will need the following result from [P90a]:

Proposition 3.3. *Let us be given a displacement operator* F *of (3.4) – (3.7); four integers* n, s, r and R, $1 \le r < R \le n$, $s \ge 1$; *an* n×n *matrix polynomial* $W = W(\lambda) \bmod \lambda^s$ *having* F*-rank* r *and its* F*-generator* (G_R, H_R) *of length* R. *Then an* F*-generator* (G_r, H_r) *of length* r *for* $W \bmod \lambda^s$ *can be computed for the cost* $O_A(R \log s,\ n\,Rs)$.

Outline of the proof (see the full proof in [P90a]). First apply Gaussian elimination with pivoting to the matrix polynomial G_R and H_R and represent them as $G_R = PLU$, $H_R = P^*L^*U^*$ where P and P^* are permutation matrices, L and L^* are n×R unit triangular (or more precisely, unit trapezoidal) matrix polynomials having ones on their diagonal and zeros above it, and U and U^* are R×R upper triangular matrix polynomials. (The entries with the lowest degree nonzero terms among all the diagonal and subdiagonal entries of the pivot columns are used as the pivots in the elimination process.) Then reduce the R×R matrix polynomial $U(U^*)^T$ to Smith's normal form, $U(U^*)^T = K \sum C$ where K and C are unimodular matrices (that is, det K and det C neither vanish nor depend on λ) and $\sum$ is a diagonal matrix polynomial. It follows that $r = \text{rank}\, F(W) = \text{rank}(U(U^*)^T) = \text{rank} \sum$, so that there are only r nonzero entries in $\sum$.

We now have the representation $F(W) = \hat{G} \sum \hat{H}^T$ where $\hat{G} = PLK$, $\hat{H} = P^*L^*C^T$ are n×R matrices. Finally, delete the columns of $\hat{G}$, $\sum$ and $\hat{H}$ and the rows of $\sum$, corresponding to the vanishing diagonal entries of $\sum$ and obtain the decomposition $F(W) = \tilde{G} \tilde{\sum} H^T = GH^T$, where $G = \tilde{G} \tilde{\sum}$, $\tilde{G}$ and H are n×r matrix polynomial. The inspection shows that the cost of these computations is within the required bounds. □

Now, we may fill the remaining gap in our extension of the algorithm of the previous subsection to the Toeplitz-like case, that is, define a desired extension of Newton's iteration (3.2) to the Toeplitz-like case.

Specifically, we just need to show how to compute an F^--generator of length r for $X_{k+1} = X_k(2I - AX_k) \bmod \lambda^{2K}$, $K=2^k$, given F^--generators of lengths at most r for A and $X_k \bmod \lambda^K$, $K=2^k$. We first apply corollary 3.1 to compute an F^--generator of length at most $R = 3r + 1$ for $X_{k+1} \bmod \lambda^{2K}$ and then compute a desired F^--generator G_r, H_r of length at most r, for $X_{k+1} \bmod \lambda^{2K}$, $K=2^k$, by applying proposition 3.3. This can be done since $\text{rank}\, F^-(X_{k+1} \bmod \lambda^{2K}) = \text{rank}\, F^-(B^{-1} \bmod \lambda^{2K})$ [see (1.7)] and since $\text{rank}\, F^-(B^{-1} \bmod \lambda^{2K}) = \text{rank}\, F^-(B \bmod \lambda^{2K}) = r$ (due to theorem 3.4). At the final stage $k = \lceil \log_2(n+1) \rceil$, we apply theorem 3.2 as a substitution for theorem 3.1 used earlier, but for this we need to have a short F_- or F_+-generator. We obtain it from G_r, H_r by using proposition 3.2. Thus we arrive at the desired modification of the expression (3.1) and of the iteration (3.2) in the Toeplitz-like case (see the details in [P90a]), and we may similarly solve the Hankel-like linear systems (1.1) or we may reduce them to Toeplitz-like linear systems.

Remark 3.2. Actually, the above algorithm can be further simplified based on the identities $F(A^{-1}) = -A^{-1}F(A)A^{-1}$ for a nonsingular matrix A and for an operator F of (3.6) or (3.7); since $X_k = B^{-1} \bmod \lambda^K$, $K = 2^k$, this immediately defines an F-generator of length r for X_{k+1}, given F-generators of length r for A and X_k (see [BP,a]).

3.4. Some Other Classes of Dense Structured Linear-Systems.

Two other important classes of dense structured matrices are represented by Vandermonde matrices,

$$V = \left[v_{ik}\right], \; v_{ik} = v_i^k, \; i, k = 0, 1, \ldots, n-1, \tag{3.10}$$

and generalized Hilbert matrices,

$$B = \left[b_{ij}\right], \; b_{ij} = \frac{1}{s_i - t_j}, \; i, j = 0, 1, \ldots, n-1, \tag{3.11}$$

for some scalars v_i, s_i, t_j, such that $s_i \neq t_j$ for all i and j. Clearly, we may well define every matrix V of (3.10) by its n entries, and every matrix B of (3.11) is well defined by its $2n-1$ entries. Multiplication of such a matrix by a vector, as well as the solution of each linear system, $V\mathbf{x} = \mathbf{b}$ or $B\mathbf{x} = \mathbf{b}$, amounts [for (3.10)] or can be reduced [for (3.11)] to the multipoint polynomial evaluation and interpolation and thus can be performed for the cost $O_A(\log^2 n, n)$ (see, for instance, [BP,a]). This computation may lead to numerical stability problems, but they can be avoided by using some alternate computations, with a certain increase of the computational cost bound in the Vandermonde case ([GL]) and with shifting from the exact to an approximate solution in the generalized Hilbert case ([BP,a], chapter 3).

On the other hand, for symbolic computations (with infinite precision), the complexity estimates of $O_A(\log^2 n, n^2)$ and of $O(n \log^2 n)$ ops, for Hankel-like and Toeplitz-like linear systems (1.1), can be extended to the case of Vandermonde-like and Hilbert-like linear systems (1.1).

For demonstration, let $A = V^T = \left[a_{ij}\right]$, $a_{ij} = v_j^i$, be the transpose of a Vandermonde matrix. Then $H = AA^T = V^TV = \left[h_{ij}\right]$, $h_{ij} = \sum_k v_k^{\,i+j}$, is a Hankel matrix, and we may solve the system (1.1) as follows ([CKL89]):

1. Compute the coefficients of the polynomial $p(x) = \sum_k (x - v_k)$.
2. Compute the power sums of the zeros of p(x), $\sum_k v_k^{\,s}$, $s = 1,2,\ldots,2n$.
3. Solve the Hankel linear system $H\mathbf{y} = \mathbf{b}$, $H = V^T V$.
4. Compute the solution $\mathbf{x} = V\mathbf{y} = V^{-T}\mathbf{b}$ to the original linear system $V^T\mathbf{x} = \mathbf{b}$ [where $V^{-T} = (V^T)^{-1}$].

The well-known techniques of polynomial computation ([BP,a]) enable us to perform each stage 1. - 4., and thus also the entire algorithm, for the cost $O_A(\log^2 n, n)$.

The same complexity estimates for solving the system $V^T\mathbf{x} = \mathbf{b}$ can be similarly obtained by means of an alternate reduction ([P89a]) of the original linear system to the linear system $V^T V^- \mathbf{y} = \mathbf{b}$, where $V^- = \left[v_{j\bar{k}}\right]$, $v_{j\bar{k}} = v_j^{-k}$, is a Vandermonde matrix, and $V^T V^{-1} = \left[t_{ij}\right]$, $t_{ij} = \sum_k v_k^{i-j}$, is a Toeplitz matrix, so that the desired solution $\mathbf{x} = V^{-T}\mathbf{b}$ to (1.1) is obtained as $\mathbf{x} = V^-\mathbf{y}$.

These two ad hoc approaches to the solution of (1.1), for $A = V^T$, have been extended to the general reduction of solving Vandermonde-like and Hilbert-like linear systems to Toeplitz-like and/or Hankel-like linear systems (see [P89a]). The reduction relies on generalizing the concepts of the displacement ranks and the displacement generators (by allowing the operators of scaling instead of or in addition to the displacement operators and on a respective extension of theorem 3.2 above) and on the respective extension of theorem 3.3, which ensures a low cost evaluation of an F-generator of length at most $r_1 + r_2$ or $r_1 + r_2 + 1$ for $A_1 A_2$ given F_i-generators of lengths r_i for A_i, i=1,2, where the operator F is defined by the operators F_1 and F_2, based on displacement and scaling, and where r_1 and r_2 are small.

These techniques can be also further extended to some other classes of dense structured linear systems by using appropriate linear operators.

4. Iterative Solution

4.1. Iterative Algorithms (General Outline). We have already considered Newton's iteration and cited the classes of conjugate gradient and Lanczos algorithms. There are several other customary classes of iterative algorithms for linear systems (1.1), such as Jacobi's, Gauss-Seidel's, SOR, SSOR, and various methods of residual minimization in the Krylov space formed by vectors $\mathbf{v}$, $A\mathbf{v}$, $A^2\mathbf{v},\ldots,A^m\mathbf{v}$, for fixed vector $\mathbf{v}$ and natural m. Typically, the complexity of each iteration step (except for the steps of Newton's iteration) is dominated by the cost of 1 or 2 multiplications of the input matrix A, its transpose and/or their submatrices by vectors, which requires small storage space and relatively few ops for linear systems (1.1) with *special matrices* A (sparse and/or well structured), so that such iterative algorithms are effective and are customarily recommended for such linear systems, provided that their sufficiently fast convergence can be ensured. The convergence rate depends on the eigenvalues or singular values of A or CA where C is a preconditioning matrix chosen depending on the structure of A. The choice of such a matrix C is a crucial means of the convergence acceleration, but it usually relies on some empirical rules and experimental tests. In the next subsection, we will review an important special class of linear systems (1.1) effectively solved by iterative algorithms.

4.2. Multigrid Methods. An important example of effective framework for various iterative methods is the multigrid/multilevel approach to the solution of PDEs discretized over some region. In particular, on a grid with n points, the associated linear system (1.1) with n×n matrix A can be solved by using O(n) ops and O(n) memory space (up to the level of discretization error). Moreover, a recent version of the *compact multigrid* enables us to solve and to store the solution to the system (1.1) arising from a (piecewise) constant coefficient linear PDE by using O(n) bit-operations and O(n) bits of memory ([PR90]), the order of log n time improvement of the previous approach.

A typical scheme is to discretize a given partial differential equation (PDE) over a sequence of grids of points starting with the coarsest grid G_0 of few points $N_0 = |G_0|$ and successively refined to finer grids $G_1, G_2, \ldots, G_k$ containing, say, about
$N_1 = |G_1| = 2^d N_0$, $N_2 = |G_2| = 2^d N_1 = 2^{2d} N_0, \ldots, N_k = |G_k| = 2^d N_{k-1} = 2^{kd} N_0 = N$
points, respectively, where d is the dimension of the grids (the number of the variables of the PDE). An approximation to the solution to the (piecewise) linear PDE on the superimposed grids $G_0, G_1, \ldots, G_k$ is defined as the solutions of the associated systems of N_j linear equations with N_j unknowns,

$$D_j \mathbf{u}_j = \mathbf{b}_j,\ j=0,1,\ldots,k, \tag{4.1}$$

where $\mathbf{u}_j$ and $\mathbf{b}_j$, for each j, are N_j-dimensional vectors, $\mathbf{b}_j$ is given, $\mathbf{u}_j$ is unknown, and D_j is a given matrix, defined by the PDE, sparse and usually having a special structure. In particular, the number of nonzeros per row of D_j is bounded by a constant (independent of j), and for a large class of PDEs, iterative algorithms are available such that in each iteration step the norm of the approximation error vector

$$\mathbf{e}_j^{(h)} = \mathbf{u}_j^{(h)} - \mathbf{u}_j$$

decreases at least by a fixed constant factor $\Theta < 1$, $\mathbf{u}_j^{(h)}$ being an approximation to $\mathbf{u}_j$ computed in h iteration steps ([HT82], [McC87]).

The function $u(\mathbf{x})$ satisfying the PDE is usually smooth, so that its approximation on the grid G_j within the discretization error E_j usually has the order of $O(1/N_j^c)$ for a fixed positive constant c (say, c=2 or c=3). Surely, we only need to approximate to $\mathbf{u}_j$ within the error of the order of E_j and shall end the iteration when this is done.

Therefore, in the transition from the approximation $\mathbf{u}_{j-1}$ to $\mathbf{u}_j$, we only need to decrease the approximation error norm by the factor of 2^{cd}, which means a constant number [of the order of $\log(2^{cd})/\log(1/\Theta) = cd/\log(1/\Theta)$] of iteration steps. Each step amounts to O(1) multiplications of D_j (or D_j^T) and/or their submatrices by vectors, and this involves $O(N_j)$ ops, since D_j only has O(1) nonzeros per row. Summarizing, we only need $\sum_j O(N_j) = O(\sum_j N_j) = O(\sum_{j=0}^{k} 2^{-dj}N) = O(N)$ ops.

Furthermore, for a large class of linear PDEs with piecewise constant coefficients, every arithmetic operation in this computation only involves operands represented with a small constant number of binary bits: shift from (4.1) to the linear system

$$D_j \mathbf{e}_j = \mathbf{r}_j$$

(where $\mathbf{r} = \mathbf{b}_j - D_j \mathbf{u}_j^*$, $\mathbf{u}_j^*$ is the current approximation to $\mathbf{u}_j^*$) and observe that the entries of D_j and of $\mathbf{e}_j$ are O(1)-bit binary numbers.

There are various ways of taking advantage of operating only with lower precision numbers. Most important, we may decrease the storage space if, instead of storing the order of $\log_2 N_j$ bits of each component of the vector $\mathbf{u}_j$, we will only store O(1) bits of each component of $\mathbf{u}_0$ and of $\mathbf{e}_j$ for $j=0,1,\ldots,k$, $k = O(\log N)$. Thus, we decrease by the factor of $\log_2 N$ the overall storage space used, and similarly we save by the factor of $\log_2 N$ or more ops if we may perform ks ops with O(1)-bit operands as fast as or faster than s ops with k-bit operands, and this can be ensured either by using a specialized computer (such as MASPAR) or special algorithms (such as binary segmentation of [P84], sect. 40, recalled above, in section 1, and in [BP,a]).

The above observation about a possible decrease (by the factor of $\log_2 N$) of the time and, particularly, the space bit-complexity of multigrid computations (due to [PR90]) can be extended to various other types of the discretization sets (say, to grids with step sizes variable over the points of the grids, to nonrectangular, say, triangular, grids, and so on), to linear PDEs with variable coefficients (then we may approximate the PDE by the matrix D_j whose coefficients are defined with the precision increasing with j), to nonlinear PDEs, still reduced to solving linear equations (see [PR91]). Note the practical significance of the factor of $\log_2 N$ reduction. Indeed, $\log_2 N > 10$ if $N > 1024$, and this is just a timid lower bound on N for customary multigrid computations.

4.3. P-adic (Newton-Hensel's) Lifting. In this section we will recall an effective method from the area of symbolic computation that outputs the solution **x** to (1.1) reduced modulo a prime power p^k, that is, the vector $A^{-1}\mathbf{b} \bmod p^k$, given a matrix A and a vector **b** with integer entries. The vector **x** can be easily recovered from $\mathbf{x} \bmod p^k$ (with k of the order of $n \log \|A\|_2 / \log p$) by means of the continued fraction approximation algorithm ([HW]).

This iterative method (called p-adic, Hensel's and Newton-Hensel's) reduces modulo p or p^2 most of the order of n^2 ops involved in each iteration, so that only O(n) ops per iteration are reduced mod p^i for larger i. Since p is relatively small, the computations modulo p and p^2 can be performed with lower precision, with the resulting advantages.

Here is a desired effective solution ([MC]).

Input: Prime p, natural k, vector **b**, matrix A, such that $\det A \neq 0 \bmod p$.

Output: $A^{-1}\mathbf{b} \bmod p^k$.

Step 0 (initialization). Compute $S(0) = A^{-1} \bmod p$, $\mathbf{x}_1 = S(0)\mathbf{b} \bmod p$, $\mathbf{v}_1 = A\,\mathbf{x}_1 \bmod p^2$.

Step j, j=1,...,k−1. Successively compute the vectors:

$$\mathbf{r}_i = \mathbf{b} - \mathbf{v}_i \bmod p^{i+1},$$

$$\mathbf{z}_i = (S(0)\,\mathbf{w}_i / p^i) \bmod p^2,$$

$$\mathbf{v}_{i+1} = \mathbf{v}_i + p^i A \mathbf{y}_i \bmod p^{i+2},$$

$$\mathbf{x}_{i+1} = \mathbf{x}_i + p^i \mathbf{y}_i \bmod p^{i+1}.$$

Note that only two matrix-by-vector multiplications are needed in each step i, that is, of S(0) by $(\mathbf{w}_i / p^i) \bmod p^2$ and of A by $\mathbf{y}_i$, and both these vectors $\mathbf{w}_i / p^i$ and $\mathbf{y}_i$ have to be reduced modulo p^2, whereas all other computations of step i involve O(n) ops.

Appendix

Let us show an application of our algorithms for Toeplitz computations to computing $m(x) = \mathrm{lcm}(p_1(x), p_2(x))$, the least common multiple (lcm) of two polynomials $p_1(x)$, $p_2(x)$. This also gives us $d(x) = \gcd(p_1(x), p_2(x))$, the greatest common divisor (gcd) of these two polynomials, since $d(x) = p_1(x)\,p_2(x)/\,m(x)$. Conversely, $m(x) = p_1(x)\,p_2(x)/\,d(x)$.

Computing m(x), we assume (with no loss of generality) that the x-free terms of both polynomials $p_1(x)$, and $p_2(x)$ equal 1. Let $n_i = \deg p_i(x)$, i=1,2; $m = \sum_{i=1}^{2} n_i$, $n = \deg \sum_{i=1}^{2} p_i(x)$, $N = m + n + 1$, and apply the following algorithm:

Algorithm A.1, computing polynomial lcms.

1. Compute the first N Taylor's coefficients of the analytic function $a(x) = \left(\sum_{i=1}^{2} \frac{1}{p_i(x)}\right)^{-1} = \sum_{j=0}^{+\infty} a_j x^j$, that is, compute the coefficients of the polynomial $\left(\sum_{i=1}^{2} (1/p_i(x))\right)^{-1} \bmod x^N = \sum_{j=0}^{N-1} a_j x^j$.
2. Compute the rank r of an n×n Toeplitz matrix T with the first row $[a_m, a_{m+1}, \ldots, a_{m+n-1}]$ and with the first column $[a_m, a_{m-1}, \ldots, a_{m-n+1}]^T$, where $a_s = 0$ for $s < 0$. (The r×r leading principal submatrix of T must be nonsingular.)
3. Compute the (m-r,n-r) Padé approximation [u(x), v(x)] to the function a(x) and output $u(x) = \mathrm{lcm}(p_1(x), p_2(x))$.

The correctness of this algorithm immediately follows from the part d) of theorem 2 of [BGY] (reproduced from [G]).

The complexity of this algorithm is upper bounded by the complexity of computing the rank in stage 2 and the (m–r,n–r) Padé approximation in stage 3, which is in turn reduced to solving a consistent linear system of equations with an n×n Toeplitz matrix of coefficients and to multiplying an m×n Toeplitz matrix by a vector ([BGY], sect. 2). Applying our algorithms here, we dramatically improve the known parallel solutions for computing polynomial lcms and gcds.

Algorithm A.1 can be modified in order to directly compute $\gcd(p_1(x), p_2(x))$ by setting $a(x) = p_1(x)/p_2(x)$, $m = \deg p_1(x) \geq n = \deg p_2(x)$, $u(x) = \gcd(p_1(x), p_2(x))$, and can be extended to randomize evaluation of $\mathrm{lcm}(p_1(x), \ldots, p_k(x))$, $k \geq 2$, by choosing $a(x) = \sum_{i=1}^{k} (c_i/p_i(x))^{-1}$ for random scalars $c_1 \ldots, c_k$.

References

[AHU] A.V. Aho, J.E. Hopcroft, and J.D. Ullman, *The Design and Analysis of Computer Algorithms,* Addison-Wesley, 1976.

[AG,b] G.S. Ammar and W.G. Gragg, "Superfast Solution of Real Positive Definite Toeplitz Systems," *SIAM J. on Matrix Analysis and Applications*, vol. 9,1, pp. 61-76, 1988.

[B83] D. Bini, "On the Class of Matrices Related to Toeplitz Matrices," Tech. Report 83-5, *Computer Science Dept., SUNYA,* Albany, NY, 1983.

[BdB] D. Bini and F. di Benedetto, "A New Preconditioner for the Parallel Solution of Positive Definite Toeplitz Systems," *Proc. 2-nd Ann. ACM Symp. on Parallel Algorithms and Architecture (SPAA 90)*, pp. 220-223. 1990; accepted to *SIAM J. on Scient. and Statis. Computing.*

[BL80] D. Bini and G. Lotti, Stability of Fast Algorithms for Matrix Multiplication, *Numerische Math.*, vol. 36, pp. 63-72, 1980.

[BP,a] D. Bini and V. Pan, *Numerical and Algebraic Computations with Matrices and Polynomials*, Birkhäuser, Boston, 1991.

[BA] R.R. Bitmead and B.D.O. Anderson, "Asymptotically Fast Solution of Toeplitz and Related Systems of Linear Equations," *Linear Algebra and Its Applics.*, vol. 34, pp. 103-116, 1980.

[BM] A. Borodin and I. Munro, *The Computational Complexity of Algebraic and Numeric Problems,* American Elsevier, New York, 1975.

[BGY] R.P. Brent, F.G. Gustavson, and D.Y.Y. Yun, "Fast Solution of Toeplitz Systems of Equations and Computation of Padé Approximations," *J. of Algorithms*, vol. 1, pp. 259-295, 1980.

[Bun] J.R. Bunch, "Stability of Methods for Solving Toeplitz Systems of Equations," *SIAM J. on Scientific and Statistical Computing*, vol. 6,2, pp. 349-364, 1985.

[CKL89] J.F. Canny, E. Kaltofen, and Y. Lakshman, "Solving Systems of Non-Linear Polynomial Equations Faster," *Proc. ACM-SIGSAM Int. Symp. on Symb. and Alg. Comp.*, pp. 121-128, 1989.

[CKL-A] J. Chun, T. Kailath, and H. Lev-Ari, "Fast Parallel Algorithm for QR-factorization of Structured Matrices," *SIAM J. on Scientific and Statistical Computing*, vol. 8,6, pp. 899-913, 1987.

[CW] D. Coppersmith and S. Winograd, "Matrix Multiplication via Arithmetic Progressions," *J. of Symbolic Computations* , vol. 9,3, 1990 (short version in *Proc. 19th Ann. ACM Symp. on Theory of Computing*, pp. 1-6, 1987).

[Cs] L. Csanky, "Fast Parallel Matrix Inversion Algorithm," *SIAM J. Computing*, vol. 5,4, pp. 618-623, 1976.

[CB] G. Cybenko and M. Berry, "Hyperbolic Householder Algorithms for Factoring Structured Matrices," Tech. Report, Center for Supercomputing Research and Development, University of Illinois, Urbana, IL , 1989, to appear in *SIAM J. on Matrix Analysis*..

[Da74] P. Davis, *Circulant Matrices,* Wiley, New York, 1974.

[dH] F.R. deHoog, "On the Solution of Toeplitz Systems," *Linear Algebra and Its Applics.*, vol. 88/89, pp. 123-138, 1987.

[EG88] D. Eppstein and Z. Galil, "Parallel Algorithmic Techniques for Combinatorial Computation," *Annual Review of Computer Science*, vol. 3, pp. 233-283, 1988.

[GP] Z. Galil and V. Pan, "Parallel Evaluation of the Determinant and of the Inverse of a Matrix," *Inf. Proc. Letters*, vol. 30, pp. 41-45, 1989.

[GeLi] J.A. George and J.W. Liu, *Computer Solution of Large Sparse Positive Definite Systems,* Prentice-Hall, New Jersey, 1981.

[GT] J.R. Gilbert and R.E. Tarjan, "The Analysis of a Nested Dissection Algorithm," *Numer. Math*, vol. 50, pp. 377-404, 1987.

[GSe] I.C. Gohberg and A.A. Semencul, "On the Inversion of Finite Toeplitz Matrices and Their Continuous Analogs," *Mat. Issled.*, vol. 2, pp. 201-233 (in Russian), 1972.

[GL] G.H. Golub and C.F. van Loan, *Matrix Computations,* Johns Hopkins Univ. Press, Baltimore, Maryland, 1989.

[G] W.B. Gragg, "The Padé Table and Its Relation to Certain Algorithms of Numerical Analysis," *SIAM Review*, vol. 14,1, pp. 1-62, 1972.

[HT82] W. Hackbusch and U. Trottenberg (eds.), "*Multigrid Methods*," Springer's Lecture Notes in Math., vol. 960, 1982.

[Haf] H. Hafsteinsson, "Parallel Sparse Choleski Factorization," Ph.D. Thesis and Tech. Report, TR 88-940, *Computer Science Dept., Cornell Univ.*, 1988.

[HW] G.H. Hardy and E.M. Wright, *An Introduction to the Theory of Numbers,* Clarendon Press, Oxford, 1962.

[Io] I.S. Iohvidov, *Hankel and Toeplitz Matrices and Forms,* Birkhäuser, Boston, MA, 1982.

[Kai] T. Kailath, "Signal Processing Applications of Some Moment Problems," *Proc. AMS Symp. in Applied Math.*, vol. 37, pp. 71-100, 1987.

[KKM] T. Kailath, S.-Y. Kung, and M. Morf, "Displacement Ranks of Matrices and Linear Equations," *J. Math. Anal. Appl.*, vol. 68,2, pp. 395-407, 1979.

[KVM] T. Kailath, A. Viera, and M. Morf, "Inverses of Toeplitz Operators, Innovations, and Orthogonal Polynomials," *SIAM Review*, vol. 20,1, pp. 106-119, 1978.

[Kal] E. Kaltofen, "Processor-Efficient Parallel Computation of Polynomial Greatest Common Divisors," Tech. Report, *Dept. of Computer Sci., RPI*, Troy, NY, 1990.

[KR] R. Karp and V. Ramachandran, "A Survey of Parallel Algorithms for Shared Memory Machines," *Handbook of Theoretical Computer Science*, North-Holland, Amsterdam, 1991.

[K-G] W. Keller-Gehrig, "Fast Algorithms for Characteristic Polynomial," *Theoretical Computer Science*, vol. 36, pp. 309-317, 1985.

[Kn] D.E. Knuth, *The Art of Computer Programming: Seminumerical Algorithms,* vol. 2, Addison-Wesley, 1981.

[LPS] J. Laderman, V. Pan, and X.-H. Sha, "On Practical Acceleration of Matrix Multiplication," Techical Report, TR 90-14, *Computer Science Dept., SUNYA,* Albany, NY, 1990.

[LRT] R.J. Lipton, D. Rose, and R.E. Tarjan, "Generalized Nested Dissection," *SIAM J. on Numerical Analysis*, vol. 16,2, pp. 346-358, 1979.

[McC87] S. McCormick, editor, *Multigrid Methods,* SIAM, Philadelphia, 1987.

[MC] R.T. Moenck and J.H. Carter, "Approximate Algorithms to Derive Exact Solutions to Systems of Linear Equation," *Proc. EUROSAM, Lecture Notes in Computer Science*, vol. 72, pp. 63-73, Springer, 1979.

[Mus] B.R. Musicus, "Levinson and Fast Choleski Algorithms for Toeplitz and Almost Toeplitz Matrices," Internal Report, *Lab. of Electronics, M.I.T.*, 1981.

[P78] V. Pan, "Strassen's Algorithm Is Not Optimal. Trilinear Technique of Aggregating, Uniting and Cancelling for Constructing Fast Algorithms for Matrix Multiplication," *Proceedings of the 19th Annual IEEE Symposium on Foundations of Computer Science*, pp. 166-176, 1978.

[P80] V. Pan, "New Fast Algorithms for Matrix Operations," *SIAM J. on Computing*, vol. 9,2, pp. 321-342, 1980.

[P84a] V. Pan, "How Can We Speed Up Matrix Multiplication?," *SIAM Review*, vol. 26, 3, pp. 393-415, 1984.

[P84] V. Pan, *How to Multiply Matrices Faster*, Lecture Notes in Computer Science, 179, Springer Verlag, 1984.

[P87b] V. Pan, "Complexity of Parallel Matrix Computations," *Theoretical Computer Science*, vol. 54, pp. 65-85, 1987.

[P89b] V. Pan, "Parallel Inversion of Toeplitz and Block Toeplitz Matrices," *Operator Theory: Advances and Applications*, vol. 40, pp. 359-389, Birkhauser, Basel, 1989.

[P89a] V. Pan, "On Some Computations with Dense Structured Matrices," *Proc. ACM-SIGSAM Intern. Symp. on Symbolic and Alg. Comp.*, pp. 34-42, 1989 and *Math. of Comp.*, vol. 55, 191, pp. 179-190, 1990.

[P90] V. Pan, "Parallel Least-Squares Solution of General and Toeplitz-like Linear Systems," *Proc. 2nd Ann. ACM Symp. on Parallel Algorithms and Architecture*, pp. 244-253, 1990.

[P87] V. Pan, "Linear Systems of Algebraic Equations," in *Encyclopedia of Physical Sciences and Technology*, ed. Marvin Yelles, vol. 7, pp. 304-329, Academic Press, NY (second edition), 1991.

[P90a] V. Pan, "Parametrization of Newton's Iteration for Computations with Structured Matrices and Applications," Techn. Report CUCS-032-90, *Columbia University, Computer Science Dept., NY, 1990,* accepted by *Computers and Mathematics (with Applications).*

[PR88] V. Pan and J. Reif, "Fast and Efficient Parallel Solution of Sparse Linear Systems," Technical Report 88-18, Computer Science Dept., SUNYA, 1988.

[PR,a] V. Pan and J. Reif, "Fast and Efficient Solution of Path Algebra Problems," *J. Computer and System Sciences*, vol. 38, pp. 494-510, 1989.

[PR91] V. Pan and J. Reif, "Generalized Compact Multigrid," Technical Report, *International Computer Science Institute* , Berkeley, CA, 1991.

[PR90] V. Pan and J. Reif, "Compact Multigrid," Tech. Report, 90-18, *Computer Science Department, SUNYA*, 1990 (accepted by *SIAM J. on Scientific and Statistical Computing*).

[PS88] V. Pan and R. Schreiber, "An Improved Newton Iteration for the Generalized Inverse of a Matrix, with Applications," Technical Report 88-35, Computer Science Dept., SUNYA, 1988 (to appear in *SIAM J. Sci. Stat. Comp.*).

[Pease] M. Pease, "An Adaptation of Fast Fourier Transform for Parallel Processing," , vol. 15, pp. 252-284, J. ACM, 1968.

[PS] F.P. Preparata and D.V. Sarwate, "An Improved Parallel Processor Bound in Fast Matrix Inversion," *Inform. Proc. Letters*, vol. 7,3, pp. 148-149, J. ACM, 1978.

[Quinn] M.J. Quinn, *Designing Efficient Algorithms for Parallel Computers,* 7,3, pp. 148-149, McGraw-Hill, New York, 1987.

[Strang,a] G. Strang, "A Proposal for Toeplitz Matrix Calculations," *Stud. Appl. Math.*, vol. 74, pp. 171-176, McGraw-Hill, New York, 1986.

[St69] V. Strassen, "Gaussian Elimination is Not Optimal," *Numerische Math.*, vol. 13, pp. 354-356, McGraw-Hill, New York, 1969.

[T64] W.F. Trench, "An Algorithm for Inversion of Finite Toeplitz Matrices," *J. of SIAM*, vol. 12,3, pp. 515-522, McGraw-Hill, New York, 1964.

[T64] W.F. Trench, "An Algorithm for Inversion of Finite Toeplitz Matrices," *J. of SIAM*, vol. 12,3, pp. 515-522, McGraw-Hill, New York, 1964.

[T90] W.F. Trench, "A Note on a Toeplitz Inversion Formula," *Linear Algebra and its Applics*, vol. 129, pp. 55-61, McGraw-Hill, New York, 1990.

[Val90] L. Valiant, "General Purpose Parallel Architectures," in *Handbook of Theoretical Computer Science*, ed. J. vanLeeuven, vol. 129, pp. 55-61, North Holland, Amsterdam, 1990.

[Wied] D.H. Wiedemann, "Solving Sparse Linear Equations Over Finite Fields," *IEEE Trans. on Inf. Theory*, vol. IT-32,1, pp. 54-62, North Holland, Amsterdam, 1986.

Algorithms for Linear Least Squares Problems

Å. Björck
Department of Mathematics
Linköping University
S-581 83 Linköping
Sweden

Abstract

A survey is first given of the singular value decomposition (SVD) and its use for analyzing and solving linear least squares problems. Refined perturbation bounds based on componentwise perturbations in the data are given. The SVD is expensive to compute, and for large sparse problems is not a practical alternative. We discuss two recent algorithms for numerically rank deficient problems based instead on the QR factorization.

New backward stable methods based on the modified Gram-Schmidt (MGS) factorization are given for linear least squares problems and underdetermined linear systems. For large sparse problems, where Q is discarded to save storage, the method of seminormal equations is often used. It is shown that provided a correction step is added this method is, under mild conditions, as accurate as those using the full QR factorization.

Iterative methods applied to the factored normal equations $A^T(b - Ax) = 0$ and preconditioners are surveyed. In least squares problems arising from multidimensional models the matrices often possess a block structure, perhaps at several levels, which reflects a "local connection" structure in the underlying physical problem. For such problems parallel block preconditioners are developed.

1 THE LINEAR LEAST SQUARES PROBLEM

1.1 Introduction

The linear least squares problem is a computational problem of primary importance in many applications. Assume, for example, that one wants to fit a linear mathematical model to given data. In order to reduce the influence of errors in the data one can then use a greater number of measurements than the number of unknowns. The resulting problem is to "solve" an overdetermined linear system i.e. to find a vector $x \in R^n$ such that Ax is the "best" approximation to the known vector $b \in R^m$, where $A \in R^{m,n}$ and $m > n$.

There are many possible ways of defining the "best" solution. A choice which can often be motivated for statistical reasons (see below) and also leads to a simple computational problem is to let x be a solution to the minimization problem

$$\min_x \|Ax - b\|_2, \quad A \in R^{m,n}, \quad b \in R^m, \tag{1}$$

NATO ASI Series, Vol. F 77
Computer Algorithms for Solving Linear Algebraic Equations
Edited by E. Spedicato

where $\|\cdot\|_2$ denotes the Euclidean vector norm. We call this a **linear least squares problem** and x a linear least squares solution of the system $Ax = b$.

One important source of least squares problems is linear statistical models. Here one assumes that vector of observations b is related to the unknown vector of parameters x by

$$Ax = b + \epsilon,$$

where $\epsilon = (\epsilon_1, \epsilon_2, \ldots, \epsilon_m)^T$ and $\epsilon_i, i = 1, \ldots, m$ are random errors with zero mean and covariance matrix $\sigma^2 I$. More generally, if the covariance matrix is $V(\epsilon) = W$, where $W \in \mathbf{R}^{m,m}$ is a positive definite symmetric matrix, then the appropriate minimization problem is

$$\min_x (Ax - b)^T W^{-1} (Ax - b).$$

The set of least squares solutions to a system $Ax = b$ is characterized by the property that $A^T(b - Ax) = 0$. It follows that a least squares solution satisfies the **normal equations**

$$A^T Ax = A^T b. \tag{2}$$

The matrix $A^T A$ is symmetric and nonnegative definite and the normal equations are always consistent. If rank $(A) = n$, then there is a unique least squares solution given by $x = (A^T A)^{-1} A^T b$. The corresponding residual $r = b - Ax$ can be written

$$r = (I - P_{\mathcal{R}(A)})b, \quad P_{\mathcal{R}(A)} = A(A^T A)^{-1} A^T, \tag{3}$$

where $P_{\mathcal{R}(A)}$ is the orthogonal projection onto Range(A).

The **minimum norm solution** to an underdetermined consistent linear system, solves the problem

$$\min_y \|y\|_2, \text{ subject to } A^T y = c. \tag{4}$$

This problem can be considered as a dual problem to (1). Similarly, if rank $(A) = n$, the solution to (2) is given by

$$-A^T Ax = c, \qquad y = -Ax. \tag{5}$$

Both problems (1) and (4) are special cases of the augmented system of $m+n$ equations

$$\begin{pmatrix} I & A \\ A^T & 0 \end{pmatrix} \begin{pmatrix} y \\ x \end{pmatrix} = \begin{pmatrix} b \\ c \end{pmatrix}. \tag{6}$$

The system matrix in (6) is square and symmetric but indefinite if $A \neq 0$. If we put $y = r = b - Ax$, and $c = 0$ we get the normal equations (2) and the defining equations for the residual. The equations (5) for the minimum norm solution are easily seen to be equivalent to the augmented system with $b = 0$.

1.2 The Singular Value Decomposition

The singular value decomposition (SVD) of a matrix $A \in \mathbf{R}^{m,n}$ is of great theoretical and practical importance for least squares problems. It provides a diagonal form of A under an orthogonal equivalence transformation. The history of this matrix decomposition goes back more than a century, but only recently has it been as much used as it should be. Because applications exist also for complex matrices we consider here matrices $A \in \mathbf{C}^{m,n}$.

Theorem 1.1 *(Singular Value Decomposition) Let $A \in C^{m,n}$ be a matrix of rank r. Then there exist unitary matrices $U \in C^{m,m}$ and $V \in C^{n,n}$ such that*

$$A = U\Sigma V^*, \quad \Sigma = \begin{pmatrix} \Sigma_r & 0 \\ 0 & 0 \end{pmatrix}, \tag{7}$$

where $\Sigma \in R^{m,n}$, $\quad \Sigma_r = \mathrm{diag}(\sigma_1, \sigma_2, \ldots, \sigma_r)$ and

$$\sigma_1 \geq \sigma_2 \geq \ldots \geq \sigma_r > 0.$$

The σ_i are called the singular values of A and if we write

$$U = [u_1, \ldots, u_m], \quad V = [v_1, \ldots, v_n] \tag{8}$$

the u_i and v_i are, respectively, the left and right singular vectors corresponding to σ_i, $i = 1, \ldots, r$.

From (1.7) it follows that

$$A^*A = V\Sigma^*\Sigma V^*, \qquad AA^* = U\Sigma\Sigma^*U^* \tag{9}$$

Thus σ_j^2, $j = 1, \ldots, r$ are the nonzero eigenvalues of the Hermitian and positive semidefinite matrices A^*A and AA^*, and v_j and u_j are the corresponding eigenvectors. Hence, in principle the SVD can be reduced to the eigenvalue decomposition for Hermitian matrices. However, this is not a suitable way to compute the SVD.

Consider, for example, the case $n = 2$ and take

$$A = (a_1, a_2) \in \mathbf{R}^{m,2}, \quad a_1^T a_2 = \cos\gamma$$

and $\|a_1\|_2 = \|a_2\|_2 = 1$. Here γ is the angle between the vectors a_1 and a_2. The matrix

$$A^TA = \begin{pmatrix} 1 & \cos\gamma \\ \cos\gamma & 1 \end{pmatrix}$$

has eigenvalues $\lambda_1 = 2\cos^2(\gamma/2), \lambda_2 = 2\sin^2(\gamma/2)$ and so

$$\sigma_1 = \sqrt{2} \cdot \cos\frac{\gamma}{2}, \quad \sigma_2 = \sqrt{2}\sin\frac{\gamma}{2}.$$

Now assume that γ is less than the square root of the floating point precision. Then the computed values of the elements $\cos\gamma$ in A^TA will equal 1. Thus, the computed matrix A^TA will be singular with eigenvalues 2 and 0, and it is not possible to retrieve the small singular value $\sigma_2 \approx \gamma/\sqrt{2}$. This illustrates that ***information may be lost in computing A^TA unless sufficient precision is used.***

If A is square and $A = U\Sigma V^*$, then

$$C = \begin{pmatrix} 0 & A^* \\ A & 0 \end{pmatrix} = \frac{1}{\sqrt{2}}\begin{pmatrix} V & V \\ U & -U \end{pmatrix}\begin{pmatrix} \Sigma & 0 \\ 0 & -\Sigma \end{pmatrix}\frac{1}{\sqrt{2}}\begin{pmatrix} V & V \\ U & -U \end{pmatrix}^* \tag{10}$$

For a general matrix $A \in \mathbf{C}^{m,n}$ it can be shown that the eigenvalues of C are $\pm\sigma_1, \ldots, \pm\sigma_r$ and 0 repeated $(m+n-2r)$ times. (Note that any rectangular matrix can be made into a square matrix by adding zero rows or columns, without changing the nonzero singular values.) This gives another useful relation between the SVD and the Hermitian eigenvalue problem.

The singular values of A are unique. The singular vector v_j, $j \leq r$ will be unique only when σ_j^2 is a simple eigenvalue of A^*A. For multiple singular values, the corresponding singular vectors can be chosen as any orthonormal basis for the unique subspace that they span. Once the singular vectors v_j, $1 \leq j \leq r$ have been chosen, the vectors u_j, $1 \leq j \leq r$ are uniquely determined from

$$u_j = \sigma_j^{-1}Av_j, \qquad j = 1, \ldots, r. \tag{11}$$

Similarly, given u_j,$1 \leq j \leq r$, the vectors v_j, $1 \leq j \leq r$ are uniquely determined from

$$v_j = \sigma_j^{-1}A^*u_j, \qquad j = 1, \ldots, r. \tag{12}$$

The SVD is a powerful tool to solve the linear least squares problem. The reason for this is that the unitary matrices that transform A to diagonal form do not change the l_2-norm of vectors.

Consider the linear least squares problem (1), where $A \in R^{m,n}$ and rank $(A) = r \leq \min(m,n)$. If $A = U\Sigma V^T$, we put

$$z = V^Tx = \begin{pmatrix} z_1 \\ z_2 \end{pmatrix}, \qquad c = U^Tb = \begin{pmatrix} c_1 \\ c_2 \end{pmatrix}, \tag{13}$$

where $z_1, c_1 \in R^r$, then

$$\|b - Ax\|_2 = \|U^T(b - AVV^Tx)\|_2 = \left\| \begin{pmatrix} c_1 - \Sigma_r z_1 \\ c_2 \end{pmatrix} \right\|_2 \tag{14}$$

Thus, the residual norm will be minimized for $z_1 = \Sigma_r^{-1}c_1$, and z_2 arbitrary. Obviously the choice $z_2 = 0$ minimizes $\|z\|_2$ and therefore also $\|x\|_2 = \|Vz\|_2$. Thus the least squares solution of minimum norm is $x = A^\dagger b$, where

$$A^\dagger = V\begin{pmatrix} \Sigma_r^{-1} & 0 \\ 0 & 0 \end{pmatrix}U^T \in R^{n,m} \tag{15}$$

is the unique least squares solution for which $\|x\|_2$ is minimum. $A^\dagger$ is called the **pseudoinverse** of A, and $x = A^\dagger b$ the pseudoinverse solution.

A geometric characterization of the pseudoinverse solution is

$$x \perp \mathcal{N}(A), \qquad Ax = P_{\mathcal{R}(A)}b,$$

where $P_{\mathcal{R}(A)}$ is the orthogonal projection onto $\mathcal{R}(A)$,

The SVD gives complete information about the four fundamental subspaces associated with A and A^*. We obtain simple expressions for the orthogonal projections onto these four subspaces:

$$\begin{aligned} P_{\mathcal{R}(A)} &= AA^\dagger = U_1U_1^T, \qquad P_{\mathcal{N}(A^T)} = I - AA^\dagger = U_2U_2^T, \\ P_{\mathcal{R}(A^T)} &= A^\dagger A = V_1V_1^T, \qquad P_{\mathcal{N}(A)} = I - A^\dagger A = V_2V_2^T, \end{aligned} \tag{16}$$

where U and V have been partitioned as

$$U = (U_1, U_2), \qquad V = (V_1, V_2), \qquad U_1 \in R^{m,r}, \qquad V_1 \in R^{n,r}.$$

The singular value decomposition also plays an important role in a number of matrix approximation problems. In the theorem below we consider the approximation of one matrix by another of lower rank.

Theorem 1.2 *(Mirsky [49]) Let $\mathcal{M}_k^{m\times n}$ be the set of matrices in $\mathbf{C}^{m\times n}$ of rank k. Assume that $A \in \mathcal{M}_r^{m\times n}$ and let $B \in \mathcal{M}_k^{m\times n}, k < r$ be such that*

$$\|A - B\|_2 \le \|A - X\|_2 \text{ for all } X \in \mathcal{M}_k^{m\times n}.$$

Then if $A = U\Sigma V^* = \sum_{i=1}^r \sigma_i u_i v_i^*$ is the SVD of A we have

$$B = \sum_{i=1}^k \sigma_i u_i v_i^*, \qquad \|A - B\|_2 = \sigma_{k+1}.$$

The theorem was originally proved for the Frobenius norm by Eckhard and Young [21]. For this norm the minimum distance is $\|A - B\|_F = (\sigma_{k+1}^2 + \ldots + \sigma_r^2)^{1/2}$, and the solution is unique.

Like the eigenvalues of a real Hermitian matrix, the singular values of a general matrix have a minmax characterization (cf. the Courant-Fischer theorem for the symmetric eigenvalue problem.)

Theorem 1.3 *Let $A \in \mathbf{C}^{m,n}$ have singular values*

$$\sigma_1 \ge \sigma_2 \ge \ldots \ge \sigma_p \ge 0, \qquad p = \min(m, n).$$

and S be a linear subspace of $\mathbf{C}^n$ of dimension $\dim(S)$. *Then*

$$\sigma_i = \min_{dim(S)=n-i+1} \max_{\substack{x\in S \\ x\neq 0}} \|Ax\|_2/\|x\|_2. \tag{17}$$

The minmax characterization of the singular values may be used to establish a relation between the singular values of two matrices A and B.

Theorem 1.4 *Let $A, B \in \mathbf{C}^{m,n}$ have singular values $\sigma_1 \geq \sigma_2 \geq \ldots \geq \sigma_p$ and $\tau_1 \geq \tau_2 \geq \ldots \geq \tau_p$ respectively, where $p = \min(m,n)$. Then*

$$|\sigma_i - \tau_i| \leq \|A - B\|_2, \qquad \sum_i |\sigma_i - \tau_i|^2 \leq \|A - B\|_F^2.$$

These result shows that the singular values of a matrix A are insensitive to perturbations of A, which is of great importance for the use of the SVD to determine the "numerical rank" of a matrix. Perturbations of the elements of a matrix produce perturbations of the same, or smaller, magnitude in the singular values.

1.3 The Sensitivity of Least Squares Solutions

In this section we give results on the sensitivity of pseudoinverses and least squares solutions to perturbations in A and b. We mainly follow the exposition in Wedin [62]. For a survey of perturbation theory see also Stewart and Sun [59]. We introduce the condition number of the matrix $A \in R^{m,n}$

$$\kappa(A) = \|A\|_2 \|A^\dagger\|_2 = \sigma_1/\sigma_r, \qquad \text{rank}(A) = r.$$

The last equality follows from the relations $\|A\|_2 = \sigma_1$, $\|A^\dagger\|_2 = \sigma_r^{-1}$.

We now give a perturbation bound for the pseudoinverse. The pseudoinverse $A^\dagger$ is not in general a continuous function of the elements of A. If $\text{rank}(A) < n$, then $B^\dagger$ can be very different from $A^\dagger$ even when $B - A$ is small. Indeed it can be shown that if $\text{rank}(B) \neq \text{rank}(A)$ then

$$\|B^\dagger - A^\dagger\|_2 \geq 1/\|B - A\|_2.$$

On the other hand, if $\text{rank}(B) = \text{rank}(A)$, then (see Wedin [62])

$$\|B^\dagger - A^\dagger\|_2 \leq \frac{1+\sqrt{5}}{2} \|B^\dagger\|_2 \|A^\dagger\|_2 \|B - A\|_2$$

We now consider the effect of perturbations of A and b upon the pseudoinverse solution $x = A^\dagger b$. We denote the perturbed data $\tilde{A} = A + \delta A$, $\tilde{b} = b + \delta b$, the perturbed solution $\tilde{x} = x + \delta x = \tilde{A}^\dagger \tilde{b}$, and $\tilde{r} = b - \tilde{A}\tilde{x}$.

Theorem 1.5 *Assume that $\text{rank}(A) = n$, and that δA and δb satisfy*

$$\epsilon_A = \|\delta A\|_2 / \|A\|_2, \quad \epsilon_b = \|\delta b\|_2 / \|b\|_2. \tag{18}$$

and that $\eta = \kappa \epsilon_A < 1$. Then $\text{rank}(A + \delta A) = n$, and the perturbations δx and δr in the least squares solution x and residual $r = b - Ax$ satisfy

$$\|\delta x\|_2 \leq \frac{\kappa}{1-\eta} \left(\epsilon_A \|x\|_2 + \epsilon_b \frac{\|b\|_2}{\|A\|_2} + \epsilon_A \kappa \frac{\|r\|_2}{\|A\|_2} \right), \tag{19}$$

$$\|\delta r\|_2 \leq \epsilon_A \|x\|_2 \|A\|_2 + \epsilon_b \|b\|_2 + \epsilon_A \kappa \|r\|_2. \tag{20}$$

Note that if the system is compatible, e.g. if rank$(A) = m$, then $r = 0$ and the term involving κ^2 in (19) vanishes. By considering first order approximations of the terms it can be shown that there are perturbations δA and δb such that the estimates in Theorem 1.5 can almost be attained for an arbitrary matrix A and vector b.

It should be stressed that the perturbation analysis above, based on the class of perturbations defined by (18), is relevant only if the components of the errors δA and δb are of equal magnitude. A much better estimate can often be obtained by scaling the matrix $A = (a_1, a_2, \ldots, a_n)$ so that all columns have equal length

$$Ax = (AD^{-1})(Dx) = \bar{A}\bar{x}, \qquad D = \mathrm{diag}(\|a_1\|_2, \ldots, \|a_n\|_2).$$

Van der Sluis [58] has shown that this scaling approximately minimizes $\kappa(AD^{-1})$ over all $D > 0$. If $\kappa(\bar{A}) \ll \kappa(A)$, we get a much smaller bound for the perturbation. Note, however, that scaling the columns changes the norm in which the error in x is measured.

Similarly, if the rows in A differ widely in norm, then (19) and (20) may considerably overestimate the perturbations. In the next section we give a more refined analysis based on component-wise bounds on the perturbations.

1.4 Component-Wise Perturbation Analysis

Here we derive perturbation results corresponding to component-wise errors in A and b for the least squares problem (1.1). The case when A is square and nonsingular has been studied by Bauer [6] and Skeel [56]. The basic identity for this perturbation analysis is

$$|\delta x| \leq |(I + A^{-1}\delta A)^{-1}||A^{-1}|(|\delta b| + |\delta A||x|). \tag{21}$$

Here, and in the following, inequalities are to be interpreted component-wise. Provided that the spectral radius $\rho(|A^{-1}||\delta A|) < 1$, we obtain the estimate

$$|\delta x| \leq (I - |A^{-1}||\delta A|)^{-1}|A^{-1}|(|\delta b| + |\delta A||x|). \tag{22}$$

To extend the Bauer–Skeel analysis to the full rank least squares problem we apply this to the augmented system

$$Mz = c \equiv \begin{pmatrix} I & A \\ A^T & 0 \end{pmatrix} \begin{pmatrix} r \\ x \end{pmatrix} = \begin{pmatrix} b \\ 0 \end{pmatrix}.$$

The perturbations δM and δc have the special form

$$\delta M = \begin{pmatrix} 0 & \delta A \\ \delta A^T & 0 \end{pmatrix}, \qquad \delta c = \begin{pmatrix} \delta b \\ 0 \end{pmatrix}.$$

Using (22) and an explicit expression for M^{-1} we obtain

$$\begin{pmatrix} |\delta r| \\ |\delta x| \end{pmatrix} \leq (I - |M^{-1}||\delta M|)^{-1} \begin{pmatrix} |P_{\mathcal{N}(A^T)}| & |(A^\dagger)|^T \\ |A^\dagger| & |(A^TA)^{-1}| \end{pmatrix} \begin{pmatrix} |\delta b| + |\delta A||x| \\ |\delta A|^T|r| \end{pmatrix}$$

provided that $\rho(|M^{-1}||\delta M|) < 1$.

Assume now that the perturbations satisfy the component-wise bounds

$$|\delta A| \leq \omega E, \qquad |\delta b| \leq \omega f, \tag{23}$$

where $E \geq 0$ and $f \geq 0$ are given, we obtain the first order bound

$$\begin{pmatrix} |\delta r| \\ |\delta x| \end{pmatrix} \leq \omega \begin{pmatrix} |P_{\mathcal{N}(A^T)}| & |(A^\dagger)|^T \\ |A^\dagger| & |(A^T A)^{-1}| \end{pmatrix} \begin{pmatrix} f + E|x| \\ E^T|r| \end{pmatrix} + O(\omega^2), \tag{24}$$

and using the infinity norm in the solution space

$$\|\delta x\|_\infty \leq \omega \Big(\| \, |A^\dagger| (E\,|x| + f) \, \|_\infty + \| \, |(A^T A)^{-1} E^T |r| \, \|_\infty \Big) + O(\omega^2). \tag{25}$$

For a consistent system, $r = 0$, the second term is zero. Then, taking $E = |A|$ and $f = |b|$ and using $|b| \leq |A||x|$ it follows that

$$\|\delta x\|_\infty \leq 2\omega \Big(\| \, |A^\dagger| \, |A| \, |x| \, \|_\infty \Big) \leq 2\omega \; \mathrm{cond}(A) \|x\|_\infty, \tag{26}$$

where $\mathrm{cond}(A) = \| \, |A^\dagger| \, |A| \, \|_\infty$. In the special case when A is square and nonsingular, $A^\dagger = A^{-1}$ and $\mathrm{cond}(A) = \| \, |A^{-1}| \, |A| \, \|_\infty$ is the condition number of the matrix A introduced by Bauer [6] and Skeel [56]. In this case it holds that $\mathrm{cond}(A) \leq \kappa_\infty(A)$, i.e. the matrix is never worse conditioned with respect to component-wise relative error norm. Indeed, it is often true that $\mathrm{cond}(A) \ll \kappa_\infty(A)$.

It is possible to compute a good approximation of the upper bound in (25) using the method of Hager [34] for which FORTRAN codes are given in Higham [39]. This method estimates the ∞-norm (or 1-norm) of a matrix B using only matrix-vector products of the form Bx and $B^T y$. We note that the terms in (25)) are of the form $\| |B| \, g \|_\infty$, where $g > 0$. Arioli, Demmel and Duff [1] have shown that using the equality

$$\| \, |B| \, g \|_\infty = \|BG\|_\infty, \qquad G = \mathrm{diag}(g).$$

Hager's method can be used to estimate $\| \, |B| \, g \, \|_\infty$ provided only that matrix vector products BGx and $G^T B^T y$ can be computed. Now both terms in (25) have the form $\| \, |B_i| \, g_i \, \|_\infty$, $i = 1, 2$, where

$$B_1 = A^\dagger, \quad g_1 = E|x| + f, \qquad B_2 = (A^T A)^{-1}, \quad g_2 = E^T |r|.$$

It follows that in order to apply Hager's algorithm it suffices that we can compute the matrix-vector products $A^\dagger x$, $(A^\dagger)^T y$, and $(A^T A)^{-1} x$. This can be done cheaply if, e.g., the QR-factorization of A is known.

In the square nonsingular case it holds that $|(AD)^{-1}| = |A^{-1}||D^{-1}|$ for $D > 0$, and hence $\mathrm{cond}(A)$ *independent of the row scaling of* A. In the general case when $m > n$ $\mathrm{cond}(A)$ the error bound (26) above is insensitive to the row scaling D, in a sense that is more complicated to define. In particular for stiff problems, where some rows are scaled with a large weight w and $\kappa(A)$ grows linearly with w, the bound (25) can be much smaller than (19).

1.5 A Posteriori Estimation of Error

Given a computed least squares solution $\bar{x}$ the component-wise relative backward error is defined as the smallest $\omega \geq 0$ in Eq. (23) such that and $\bar{x}$ is the *exact* solution of

$$\min_x \|(A + \delta A)x - (b + \delta b)\|_2,$$

For a consistent linear system $Ax = b$ Oettli and Prager [51] showed that

$$\omega = \max_{1 \leq i \leq n} \frac{|A\bar{x} - b|_i}{(E|\bar{x}| + f)_i}, \tag{27}$$

where $0/0$ should be interpreted as 0. Together with the perturbation result (25) this can be used to compute an a posteriori bound on the error in $\bar{x}$.

To extend this result to linear least squares problems, we apply it to the augmented system (6). We do not allow any perturbations in the diagonal block of zeros in the augmented matrix and neither in the last zero vector in the augmented right hand side. However, as the perturbation analysis above shows, we can allow *different* perturbations in the blocks A and A^T respectively without increasing the *forward error bound* (25). Also, by a slight modification of the perturbation analysis in the previous section we can accommodate a perturbation to the first diagonal block. Hence, for a posteriori error analysis it makes sense to take the relative backward error of a computed solution $\bar{r}$ and $\bar{x}$ to be the smallest nonnegative number ω such that

$$|\delta I| \leq \omega I, \quad |\delta A_i| \leq \omega E, \quad i = 1, 2, \qquad |\delta b| \leq \omega f,$$

and

$$\begin{pmatrix} I + \delta I & A + \delta A_1 \\ A^T + \delta A_2 & 0 \end{pmatrix} \begin{pmatrix} \bar{r} \\ \bar{x} \end{pmatrix} = \begin{pmatrix} b + \delta b \\ 0 \end{pmatrix}.$$

Using the result by Oettli and Prager $\omega(\bar{r}, \bar{x}) = \max(\omega_1, \omega_2)$, where

$$\omega_1 = \omega_1(\bar{r}, \bar{x}) = \max_{1 \leq i \leq m} \frac{|\bar{r} + A\bar{x} - b|_i}{(|\bar{r}| + E|\bar{x}| + f)_i}, \qquad \omega_2 = \omega_2(\bar{r}) = \max_{1 \leq i \leq n} \frac{|A^T\bar{r}|_i}{(E^T|\bar{r}|)_i}, \tag{28}$$

gives the backward error for a computed solution $\bar{r}$ *and* $\bar{x}$.

If we only have a computed $\bar{x}$ it may be feasible to put $\bar{r} = b - A\bar{x}$ and apply the result above. With this choice we have $\omega_1 = 0$ (exactly) and hence

$$\omega(\bar{r}, \bar{x}) = \omega_2 = \max_{1 \leq i \leq n} \frac{|A^T(b - A\bar{x})|_i}{(E^T|b - A\bar{x}|)_i}.$$

However, in case of a nearly consistent least squares problem $fl(b - A\bar{x})$ will mainly consist of roundoff, and will not be accurately orthogonal to the range of A. Hence, although $\bar{x}$ may have a small relative backward error ω_2 will not be small. This illustrates a fundamental problem in computing the backward error: For $\bar{x}$ to have a small backward error it is sufficient that *either* $(b - A\bar{x})$ *or* $A^T(b - A\bar{x})$ is small, but neither of these conditions is *necessary*.

One reason for the interest in component-wise relative backward error is a theorem of Skeel [Theorem 5.1][57]. This states that for a square linear system $Ax = b$, if

A is not too ill-conditioned and if the components of the vector $|A||x|$ do not vary too much, then Gaussian elimination with one step of iterative refinement will have a component-wise relative backward error on the order of machine precision. This is true even if the residuals are computed in single precision. For sparse problems, the assumptions on the components may be violated, and the sparsity structure of δb must be sacrificed by occasionally permitting $f_i > b_i$, see Arioli, Demmel and Duff [1].

For the least squares problem iterative refinement can be applied to the augmented system. Corrections to $\bar{x}$ and $\bar{r}$ are computed from

$$\begin{pmatrix} I & A \\ A^T & 0 \end{pmatrix} \begin{pmatrix} \delta\bar{r} \\ \delta\bar{x} \end{pmatrix} = \begin{pmatrix} fl(b - A\bar{x} - \bar{r}) \\ fl(-A^T\bar{r}) \end{pmatrix},$$

and this process is repeated until convergence. For problems where A and possibly b are sparse, the sparsity structure in the perturbation to b may again have to be sacrificed, see Arioli, Duff and de Rijk [2], who apply this scheme using a symmetric factorization $M = LDL^T$ of the augmented system matrix, where D is block diagonal with 1×1 and 2×2 pivots. The same process for dense problems, using a QR factorization of A has been studied by Higham [40].

2 RANK DEFICIENT PROBLEMS

2.1 Rank Deficient Problems and the SVD

In solving least squares problems it is important to detect ill-conditioning and possible rank deficiency in A. Failure to do so will lead to meaningless solutions of very large norm, or even to failure of the algorithm. When A is rank deficient, $\text{rank}(A) = r < n$, the minimum norm least squares solution can in principle be computed from the SVD of A. However, inaccuracy of data and rounding errors made during the computation usually mean that the matrix A is not exactly known. In this situation the mathematical notion of rank is not appropriate. For example, suppose that A is a matrix that originally was of rank $r < n$, but whose elements have been perturbed by rounding errors. Then it is most likely that the perturbed matrix has full rank n. However, it will be very close to a rank deficient matrix, and it should be considered as *numerically rank deficient.*

From the above considerations it follows that the numerical rank assigned to a matrix should depend on a tolerance which reflects the error level in the data. We will say that a matrix A has numerical δ-rank equal to r if

$$r = \min\{\text{rank}(B) \mid \|A - B\|_2 \leq \delta\} \tag{29}$$

By Theorem 1.2 we have for $r < n$

$$\inf_{\text{rank}(B)\leq r} \|A - B\|_2 = \sigma_{r+1}, \tag{30}$$

where $\sigma_i, i = 1, 2, \ldots, \min(m, n)$ are the singular values of A. This infimum is actually attained for the matrix $B = \sum_{i=1}^{r} \sigma_i u_i v_i^T$. Hence a matrix A has numerical δ-rank r if and only if

$$\sigma_1 \geq \ldots \geq \sigma_r > \delta \geq \sigma_{r+1}, \geq \ldots \geq \sigma_n. \tag{31}$$

The definition (11.3) is satisfactory only when there is a well defined gap between σ_{r+1} and σ_r. If the exact matrix A is rank deficient but well conditioned, then this should be the case. However, there may not exist a gap for any r, e.g., suppose $\sigma_{i+1} = 0.9\sigma_i, i = 1, 2, \ldots, n-1$. In such a case the rank obviously is not well defined. For such difficult problems additional information usually has to be supplied to define a meaningful solution.

The choice of the parameter δ in (31) is not always an easy matter. Also the definition (31) will not be meaningful unless the matrix A is reasonably scaled. Let the error in the matrix A be $E = (e_{ij})$, and assume that the magnitudes of the elements e_{ij} in the error matrix are about the same, and that $|e_{ij}| \leq \epsilon$, for all i, j. Then, since

$$\|E\|_2 \leq \|E\|_F \leq (mn)^{1/2}\epsilon$$

a reasonable choice in (11.3) is to take $\delta = (mn)^{1/2}\epsilon$. If the magnitudes of the elements e_{ij} are not about the same, one could try to scale rows and columns of A so that they become nearly equal. (Note that any diagonal scaling $D_1 A D_2$ will induce the same scaling $D_1 E D_2$ of the error matrix.) However, scaling the rows of A is normally not allowed, since it would change the solution to the least squares problem.

2.2 Rank Deficient Problems and QR Factorization

The singular value factorization is generally regarded as the most reliable method to determine the numerical rank of a matrix. However, in practice the QR factorization with column pivoting often works as well, and requires substantially less work. The QR factorization of $A\Pi$, with $r = \text{rank}(A)$, has the form

$$A\Pi = Q \begin{pmatrix} R_{11} & R_{12} \\ 0 & 0 \end{pmatrix}, \quad R_{11} \in \mathbf{R}^{r\times r}. \tag{32}$$

where Π is a permutation matrix, and R_{11} is upper triangular and nonsingular.

The standard pivoting strategy in both the Householder and modified Gram-Schmidt algorithms can be viewed as choosing a remaining column of largest distance to the subspace spanned by the previously chosen columns. The overhead of the above column pivoting strategy is only $0(mn)$ operations, and will at each step maximize the diagonal element r_{kk}. Hence the diagonal elements in R will form a non-increasing sequence; indeed the elements in R satisfy the stronger inequalities

$$r_{kk}^2 \geq \sum_{i=k}^{j} r_{ij}^2, \quad j = k+1, \ldots, n. \tag{33}$$

In particular this implies that if $r_{kk} = 0$, then $r_{ij} = 0$, $i, j \geq k$.

The orthogonal matrix Q is usually computed as a product of Householder transformations,

$$Q^T = P_m \cdots P_2 P_1, \qquad P_k = I - 2u_k u_k^T,$$

where P_k is chosen to zero elements in the kth column of A. We have after k steps computed

$$A^{(k)} = (P_k \cdots P_1) A (\Pi_1 \cdots \Pi_k) = \begin{pmatrix} R_{11} & R_{12} \\ 0 & A_{22} \end{pmatrix}, \tag{34}$$

where $R_{11} \in \mathbf{R}^{k\times k}$. Because of rounding errors the submatrix A_{22} will usually not be exactly zero for $k = \text{rank}(A)$. From Theorem 1.2 it follows that

$$\sigma_{k+1} \leq \|A_{22}\|_2 \leq (n-k)^{1/2}|r_{k+1,k+1}|.$$

Hence, if for some tolerance $\epsilon > 0$ we have $|r_{k+1,k+1}| \leq \epsilon$, then A has numerical rank at most equal to k. This criterion can be used to terminate both the Householder or modified Gram-Schmidt algorithm.

Unfortunately it is not true that if A is close to a matrix of rank $k < n$, then the diagonal element $r_{k+1,k+1}$ must be small. Hence, the near singularity of A may not be revealed by a small element in R. For an upper triangular matrix, whose elements satisfy (33), the best lower bound is $\sigma_n(R) \geq 2^{1-n}|r_{nn}|$, and this can almost be attained.

Although the condition estimate

$$\kappa(A) \geq |r_{11}/r_{nn}| \tag{35}$$

may considerably underestimate the condition number, extensive numerical testing indicates that, provided that column pivoting is used, (35) is in practice a fairly reliable estimate of $\kappa(A)$. Usually κ is underestimated only by a factor of 2 - 3 and very seldom by more than 10.

It may not always be practical to perform column pivoting as outlined above. This is the case, e.g., in the QR factorization of sparse matrices, where one wants the column order to be determined from sparsity considerations. Another important case is when the Householder (or Gram-Schmidt) algorithm is to be implemented on a multiprocessor machine. Then the standard pivoting strategy makes pipelining of the algorithm impossible. Extensive testing by Foster [27] on certain random matrices has shown that the probability that $\min_i |r_{ii}|$ is much bigger than σ_n is significant when no pivoting is done. Hence, using $\min_i |r_{ii}|$ to detect rank deficiency is far less satisfactory if column pivoting is not used.

Two alternative pivoting techniques, which have recently been developed, are discussed below. In the first approach a QR factorization without pivoting is first computed. Then redundant columns are identified and permuted to the last positions, and the QR factorization is updated. The second approach uses a local pivoting criterion based on an incremental condition estimator.

2.3 The Rank-Revealing QR Factorization

In this section we describe an a posteriori pivoting technique, developed by Foster [26] and Chan [18] for computing a rank revealing QR factorization, i.e., a factorization of the form (34) with $k = \text{rank}(A)$, and A_{22} small.

We first consider the case when the numerical rank of A equals $r = n-1$, and show that we can always find a column permutation of A such that the resulting R factor has a small element r_{nn}. Such a permutation can be found by inspecting the elements of the right singular vector $v = v_n$ of R corresponding to the singular value σ_n. (Note that A and R have the same singular values and right singular vectors.) We have $Av = \sigma_n u$, where $\|v\|_2 = \|u\|_2 = 1$, and hence $\|Av\|_2 = \sigma_n$. Let Π be a permutation such that if $\Pi^T v = w$, then $|w_n| = \|w\|_\infty$. Then if $A\Pi = QR$ is the QR factorization of A, it holds (see Chan [18]) $|r_{nn}| \le n^{1/2}\sigma_n$. Hence, we have the upper bound

$$\sigma_n \ge n^{-1/2}|r_{nn}|.$$

Chan suggests that an approximation to v_n is determined by one or two steps of inverse iteration, from which the permutation can be determined. The LINPACK condition number estimator (see next section) could be used to select the initial vector. This leads to a two step procedure: first *any* QR factorization of A is determined, and a condition estimator is used to determine an elementary permutation matrix Π; then the QR factorization $R\Pi = \bar{Q}\bar{R}$ is computed, from which we derive

$$A\Pi = Q\begin{pmatrix} R\Pi \\ 0 \end{pmatrix} = Q\text{diag}(\bar{Q}, I_{m-n})\begin{pmatrix} \bar{R} \\ 0 \end{pmatrix},$$

the QR factorization of $A\Pi$. The QR factorization of $R\Pi$ can be computed in less than $2n^2$ flops using updating techniques, see, e.g., subroutines in LINPACK.

The above procedure can be extended to the case when A is nearly rank deficient with rank $r < n - 1$, by repeatedly applying the one dimensional algorithm to smaller and smaller leading blocks of R. Chan [18] callsf the resulting algorithm RRQR, and shows that it is guaranteed to work for matrices of small rank deficiency, and that it is very likely to work also for large rank deficiency. Algorithm RRQR is closely related to algorithms given by Foster [26]. The main difference is that RRQR produces a QR factorization of the original matrix whereas Fosters algorithm, which is primarily designed for the sparse case, only produces a factorization for a subset of columns of the original matrix. The availability of the full QR factorization has the advantages that it allows the computation from RRQR of a posteriori upper and lower bounds on the singular values of A.

2.4 Condition Estimators

Methods for estimating the condition number can be based on *inverse iteration*. Given a vector d one computes y and z from

$$R^T x = d, \quad Rz = x.$$

Since $z = (R^T R)^{-1} d = (A^T A)^{-1} d$ this is equivalent to one step of inverse iteration with $A^T A$. Let $R = U\Sigma V^T$ be the singular value factorization of R, and write $d = \sum_{i=1}^n \alpha_i v_i$. Then it is easily seen that $z = \sum_{i=1}^n (\alpha_i/\sigma_i^2) v_i$, and the vector z is likely to be dominated by its component on v_n. Hence,

$$\sigma_n \le \min\{\|d\|_2/\|x\|_2, \|x\|_2/\|z\|_2\} \tag{36}$$

will usually be a good estimate of σ_n.

In the LINPACK condition estimator uses the vector $d = (\pm 1, \pm 1, \ldots, \pm 1)^T$, and the 1-norm. The algorithm can also be used for the 2-norm or ∞-norm. To enhance the probability of obtaining a good estimate the signs are chosen adaptively as the solution of $R^T x = d$ proceeds, by a "look-ahead" strategy. Although counter-examples have been constructed, the LINPACK condition estimator has proven to be very reliable in practice. The large accumulated experience with it suggests that it produces good order-of-magnitude estimates. It will usually detect near rank deficiency even in the case when this is not revealed by a small diagonal element in R.

Recently Bischof [8] has developed a new technique for estimating the smallest singular value of an upper triangular matrix *as it is generated one column at a time.* Let $R \in R^{k,k}$ be an upper triangular matrix and x an approximate *left* singular vector such that $R^T x = d$, $\|d\|_2 = 1$, and $\sigma_{min}(R) \approx 1/\|x\|_2$. Given a new column of R,

$$\hat{R} = \begin{pmatrix} R & u \\ 0 & \rho \end{pmatrix} \in R^{(k+1),(k+1)},$$

one generates a new vector $\hat{x}$ such that $\sigma_{min}(\hat{R}) \approx 1/\|\hat{x}\|_2$, *without accessing R again.*

As in the LINPACK estimator one makes use of the implication that if $Rx = d$, $\|d\|_2 = 1$, then $\sigma_{min}(R) \leq 1/\|x\|_2$. The new vector $\hat{x}$ is taken to be

$$\hat{x} = \begin{pmatrix} sx \\ (c - s\alpha)/\rho \end{pmatrix}, \quad s = \sin\phi, \quad c = \cos\phi, \quad \alpha = u^T x, \tag{37}$$

where $(s, c)^T$ is chosen to maximize $\|\hat{x}\|_2$. In case $\alpha \neq 0$ we obtain $s = \mu/\sqrt{\mu^2 + 1}$, $c = -1/\sqrt{\mu^2 + 1}$, where

$$\mu = \eta + \operatorname{sign}(\alpha)\sqrt{\eta^2 + 1} \qquad \eta = \frac{\beta}{2\alpha}, \qquad \beta = \rho^2\|x\|_2^2 + \alpha^2 - 1.$$

In case $\alpha = 0$ we take $s = 1$ if $|\rho|\ \|x\|_2 > 1$, and $c = 1$ otherwise. We can now calculate $\hat{x}$ from Eq. (37), and further we have $\|\hat{x}\|_2 = \sqrt{\alpha\mu + 1}/|\rho|$. Note that we only have to save the current approximative singular vector x, and that the only vector operations needed are a scalar product and a scaling. The total cost of including this incremental condition estimator in a QR factorization routine is therefore about n^2 flops.

Bischof shows that his incremental condition estimator can underestimate the condition number when $|u^T x| \ll \|x\|_2\|u\|_2$. However, numerical testing indicates that the scheme is quite reliable, and generally produces good estimates. In particular, the scheme works well even when no pivoting is used in the QR factorization.

3 THE MODIFIED GRAM-SCHMIDT METHOD

3.1 Introduction

In the modified Gram-Schmidt method we compute a sequence of matrices, $A = A^{(1)}$, $A^{(2)}, \ldots, A^{(n)} = Q \in \mathbf{R}^{m\times n}$, where

$$A^{(k)} = (q_1, \ldots, q_{k-1}, a_k^{(k)}, \ldots, a_n^{(k)}).$$

Here the first $(k-1)$ columns are final columns in Q and $a_k^{(k)}, \ldots, a_n^{(k)}$ have been made orthogonal to $q_1, \ldots, q_{k-1}$. In the kth step we take

$$r_{kk} = \|a_k^{(k)}\|_2^{1/2}, \quad q_k = a_k^{(k)}/r_{kk}, \tag{38}$$

and orthogonalize $a_{k+1}^{(k)}, \ldots, a_n^{(k)}$ against q_k

$$a_j^{(k+1)} = P_k a_j^{(k)} = a_j^{(k)} - r_{kj} q_k, \quad j = 2, \ldots, n. \tag{39}$$

where $P_k = I - q_k q_k^T$, is the orthogonal projection onto q_k. After n steps we have obtained the factorization $A = QR$, where the columns of $Q = (q_1, q_2, \ldots, q_n)$ are orthonormal by construction. The algorithm requires mn^2 flops, which is slightly more than the $mn^2 - n^3/3$ flops required for the Householder QR factorization. However, the modified Gram-Schmidt algorithm operates on vectors of equal length, which may be an advantage on some vector computers.

The numerical properties of the MGS factorization are analyzed in Björck [9]. Let $\bar{R}$ and $\bar{Q}_1$ denote the *computed* factors, using a square root free version of the algorithm, but scaled so that the columns of $\bar{Q}_1$ have *exactly* unit length. Then if inner products are accumulated in double precision it holds that

$$A + E_1 = \bar{Q}_1 \bar{R}, \qquad \|E_1\|_2 \le \bar{c}_1 u \|A\|_2, \quad \bar{c}_1 = 1.5 n^{3/2}. \tag{40}$$

where u is the machine precision. Hence $\bar{Q}_1 \bar{R}$ very accurately represents A. However, due to round-off, there will in general be a gradual loss of orthogonality in the computed Q. The departure from orthogonality can be bounded in terms of κu,

$$\|I - \bar{Q}_1^T \bar{Q}_1\|_2 \le \bar{c}_2 \kappa u + O(u^2), \qquad \bar{c}_2 = 2n^2. \tag{41}$$

(If inner products are not accumulated in double precision the constants $\bar{c}_i$ in the above bounds will approximately increase by a factor of m.) Because of the lack of orthogonality in Q_1 it seems that algorithms based on the modified Gram-Schmidt method could give much worse errors than algorithms based on Householder's method. We will see that this is *not* so, and in particular the R obtained from MGS is numerically just as good as that obtained from Householder or Givens transformations.

In the following we show how the loss of orthogonality in Q_1 from MGS, can effectively be regained without altering the MGS algorithm or reorthogonalizing the columns of Q_1.

3.2 Modified Gram-Schmidt as a Householder Method

The analysis of MGS hinges on an interesting observation, apparently by Charles Sheffield and communicated to the author by Gene Golub, that MGS is *numerically* equivalent to Householder's method applied to the matrix A *augmented with an n by n square matrix of zero elements on top*. The Householder method computes the factorization

$$P^T \begin{pmatrix} 0 \\ A \end{pmatrix} = \begin{pmatrix} R \\ 0 \end{pmatrix}, \qquad P^T = \tilde{P}_n \cdots \tilde{P}_2 \tilde{P}_1, \tag{42}$$

where $\tilde{P}_k = I - 2\hat{v}_k \hat{v}_k^T / \|\hat{v}_k\|_2^2$, $k = 1, 2, \ldots, n$, are Householder transformations. Because of the special structure of the augmented matrix the vectors v_k have the form

$$\hat{v}_k = (-r_{kk} e_k, \hat{q}_k)^T, \qquad r_{kk} = \|\hat{q}_k\|_2.$$

where e_k is the kth unit vector, and we have chosen the sign so that R has a positive diagonal. It follows that

$$\tilde{P}_k = I - v_k v_k^T, \qquad v_k = (-e_k, q_k)^T, \qquad \|v_k\|_2^2 = 2, \tag{43}$$

where $q_k = \hat{q}_k / r_{kk}$, and since the first n rows are initially zero, the scalar products of the vector v_k with later columns will only involve q_k. In particular, after the first step

$$\tilde{P}_1 \begin{pmatrix} 0 \\ A^{(1)} \end{pmatrix} = \begin{pmatrix} r_{11} & r_{12} & \cdots & r_{1n} \\ 0 & 0 & \cdots & 0 \\ \vdots & \vdots & \vdots & \vdots \\ 0 & 0 & \cdots & 0 \\ 0 & a_2^{(2)} & \cdots & a_n^{(2)} \end{pmatrix},$$

where these values are clearly *numerically* the same as in the first step of MGS on A. We see the next Householder transformation step produces the second row of R and $a_3^{(3)}, \ldots, a_n^{(3)}$, just as in MGS, and carrying on this way we see that the computed quantities r_{kj} and q_k are *numerically* equivalent to the same quantities in the modified Gram-Schmidt method.

Because of the numerical equivalence outlined above the backward error analysis for the Householder QR factorization can be applied to the modified Gram-Schmidt algorithm. It follows, see Wilkinson [63], that there exists an *exactly* orthogonal matrix $\tilde{P} \in R^{(m+n),(m+n)}$ such that for the *computed* matrix $\bar{R}$ it holds

$$\begin{pmatrix} \tilde{E}_1 \\ A + \tilde{E}_2 \end{pmatrix} = \tilde{P} \begin{pmatrix} \bar{R} \\ 0 \end{pmatrix}, \qquad \|\tilde{E}_i\|_2 \le \tilde{c}_i u \|A\|_2, \quad i = 1, 2. \tag{44}$$

where $\tilde{c}_i(m,n)$ are constants depending on the m, n and the details of the arithmetic. Using this result it can be shown, see Björck and Paige [16], that when $\bar{R}$ is computed by MGS, there exist an *exactly* orthonormal matrix $\tilde{Q}_1$ such that

$$A + E = \tilde{Q}_1 \bar{R}, \qquad \tilde{Q}_1^T \tilde{Q}_1 = I, \qquad \|E\| \le (c_1 + c_2)\|A\|_2. \tag{45}$$

Thus $\bar{R}$ from MGS is not only the same as $\bar{R}$ from the Householder QR factorization applied to A augmented by a square block of zeros, but (45) shows it is comparable in accuracy to the upper triangular matrix from the Householder or Givens QR factorization applied to A alone.

The key point is that because of (41), we cannot use $\bar{Q}_1$ as our set of orthonormal vectors, but we *can* use the computed $\bar{P}$, as the theoretical product of the computed $\bar{P}_k = I - \bar{v}_k \bar{v}_k^T$, since we have all the necessary information in $\bar{Q}_1$, that is $\bar{v}_k^T = (-e_k^T, \bar{q}_k^T)$. Thus we can solve problems as accurately using MGS as using Householder or Givens QR factorizations, if instead of using the computed $\bar{Q}_1$ directly, we formulate the problems in terms of (42) and use the $\bar{q}_k$ to define the computed $\bar{P}$. In the next section we show some important uses of this idea.

3.3 The Augmented System Formulation

The class of problems we consider is that of solving the symmetric indefinite linear system involving A

$$\begin{pmatrix} I & A \\ A^T & 0 \end{pmatrix} \begin{pmatrix} y \\ x \end{pmatrix} = \begin{pmatrix} b \\ c \end{pmatrix}. \tag{46}$$

When $c = 0$ this has been called the augmented system formulation of the linear least squares problem, and in general we can call it the Augmented System Formulation (ASF) of the following two problems, since it represents the conditions for their solution.

$$\min \|b - y\|_2, \qquad A^T y = c, \tag{47}$$

$$\min \|b - Ax\|_2 + 2c^T x. \tag{48}$$

The ASF is obtained by differentiating the Lagrangian $\|b - y\|_2^2 + 2x^T(A^T y - c)$, and equating to zero. Here x is the vector of Lagrange multipliers. The ASF can also be obtained by differentiating (48) to give $A^T(b - Ax) = c$, and setting y to be the least squares residual $y = b - Ax$.

Setting $b = 0$ in (47), and so in (46), gives the problem of finding the minimum 2-norm solution of a Linear Underdetermined System (LUS), which occurs for example as a subproblem in optimization algorithms. Setting $c = 0$ in (48) gives the Linear Least Squares problem (LLS). Thus the ASF covers two important special cases. It also occurs in its full form (46) in the iterative refinement of least squares solutions, see Björck [10], and has been studied extensively in its form (47) by Arioli and Laratta [3, 4].

Using the singular value decomposition A we find the general solution to (46) is

$$x = A^\dagger b - (A^T A)^\dagger c + x_{\mathcal{N}(A)}, \qquad y = P_{\mathcal{N}(A^T)} b + (A^\dagger)^T c, \tag{49}$$

where $x_{\mathcal{N}(A)} \in \mathcal{N}(A)$ is arbitrary.

The ASF can be rephrased as the *normal equations*

$$A^T A x = A^T b - c, \qquad y = b - Ax, \tag{50}$$

but this method will not give good numerical accuracy when A is ill-conditioned. Here we restrict ourselves to the case with $\text{rank}(A) = n$, and concentrate solely on methods which use some QR factorization of A. If we replace $A^T A$ by $R^T R$ and A by $Q_1 R$ in (50) we get $Rx = Q_1^T b - R^{-T} c$, $y = b - Q_1 Rx$, so one method is

$$z = R^{-T} c, \qquad d = Q_1^T b, \qquad x = R^{-1}(d - z), \qquad y = b - Q_1(d - z). \tag{51}$$

Using instead a QR factorization with a square orthogonal factor Q we have

$$z = R^{-T} c, \qquad \begin{pmatrix} d \\ f \end{pmatrix} = Q^T b, \qquad x = R^{-1}(d - z), \qquad y = Q \begin{pmatrix} z \\ f \end{pmatrix}. \tag{52}$$

3.4 Stability of Algorithms

The algorithms we will deal with exhibit different forms of numerical stability, and we will use the following notions. A *backward stable* algorithm is one which is guaranteed to give the exact answer to a nearby problem. This is the concept developed and used so effectively by Wilkinson, and for which the term *stable* is used in Bunch [17].

Following [17] we will say an algorithm for solving linear equations is *strongly stable* for a class of matrices $\mathcal{M}$ if for each $M \in \mathcal{M}$, and for each g, the computed solution $\bar{z}$ to $Mz = g$ satisfies $\bar{M}\bar{z} = \bar{g}$, where $\bar{M} \in \mathcal{M}$, and $\bar{M}$ and $\bar{g}$ are close to M and g respectively. For example a strongly stable algorithm for solving (46) would give the exact answer to a problem

$$\begin{pmatrix} I & A+\delta A \\ (A+\delta A)^T & 0 \end{pmatrix} \begin{pmatrix} \bar{y} \\ \bar{x} \end{pmatrix} = \begin{pmatrix} b+\delta b \\ c+\delta c \end{pmatrix}, \tag{53}$$

with satisfactory bounds on the sizes of δA, δb, δc. Note that backward stable algorithms which only compute y for (47), or x for (48), can be considered *strongly* stable for y (respectively x) in (46), in that there must in addition exist $\bar{y}$ (respectively $\bar{x}$) so that (53) holds. Put another way, a method for (53) *must* be strongly stable in order to be backward stable in the sense mentioned above for (47) or (48), otherwise the solution of either of these last two cannot be the *exact* solution to any nearby problem.

We define an *acceptable-error stable* algorithm to be one which gives a solution whose error size is never significantly worse than the error bound obtained from a tight perturbation analysis of a backward stable algorithm for the same problem. That is, the error (but not necessarily the residual) in an acceptable-error stable algorithm will not be significantly worse in magnitude than the worst error that could result from a backward stable algorithm applied to a problem having the same condition and magnitude of data.

Björck [10] showed that (52) is backward stable for (46) using the Householder QR factorization. A similar analysis is given in by Arioli and Laratta [4] for (47) only, however that paper also contains a perturbation analysis to bound the error in computed $\bar{x}$ in (47) for such backward stable algorithms. It can be shown, see Björck and Paige [16] that (52) is not only backward stable, but also gives the exact answer for a nearby problem having exactly the same structure as (46).

Björck showed MGS can be used as the basis for an acceptable-error stable algorithm for solving LLS, that is (48) with $c = 0$. Two variants of (51)–(52) using MGS are also shown to be acceptable-error stable for (47) in Arioli and Laratta [3].

3.5 Backward Stable Solution of the ASF using MGS

We saw that (52) with Householder or Givens QR factorizations was backward stable for the ASF, but the same cannot be said when we use (51) with MGS. In this section we develop a new backward stable algorithm for (46) using Q_1 and R from MGS, and show how it specializes to LUS and LLS and solution of square nonsingular systems.

Despite the lack of orthogonality in Q_1, Björck [9] showed the standard method using $\bar{Q}_1$ from MGS in LLS is acceptable-error stable. The same is true for the more general problem (46), but since there is a backward stable algorithm for the ASF based on MGS, we will give this first. This simplifies the proof of acceptable-error stability for a standard method using MGS. This new algorithm also specializes to backward stable algorithms using MGS for LLS, LUS and the square nonsingular systems $Ay = b$ and $A^T x = c$.

Our original ASF (46) is equivalent to the augmented system

$$\begin{pmatrix} I & 0 & 0 \\ 0 & I & A \\ 0 & A^T & 0 \end{pmatrix} \begin{pmatrix} w \\ y \\ x \end{pmatrix} = \begin{pmatrix} 0 \\ b \\ c \end{pmatrix}, \tag{54}$$

so applying Householder transformations gives the augmented version of

$$z = R^{-T} c, \quad \begin{pmatrix} d \\ h \end{pmatrix} = P^T \begin{pmatrix} 0 \\ b \end{pmatrix}, \quad \begin{pmatrix} w \\ y \end{pmatrix} = P \begin{pmatrix} z \\ h \end{pmatrix}, \quad x = R^{-1}(d - z), \tag{55}$$

which is strongly stable for (54). But MGS is *numerically* equivalent to Householder transformations applied to $\begin{pmatrix} 0 \\ A \end{pmatrix}$, and so if we use the q_k from MGS to produce

$$\tilde{P}_k = I - v_k v_k^T, \qquad v_k^T = (-e_k^T, q_k^T),$$

and use $P^T = \tilde{P}_n \ldots \tilde{P}_2 \tilde{P}_1$ in (55), we have a strongly stable algorithm for (54) based on MGS. It can be shown, see Björck and Paige [16], that a method which is strongly stable for (54) is also strongly stable for (46), and so in particular our MGS method is.

We now show how to take advantage of the structure of the P_k, then we will summarize this numerically stable use of MGS for the ASF. To compute d and h in (55) note that $P^T = \tilde{P}_n \ldots \tilde{P}_1$, and define

$$\begin{pmatrix} d^{(1)} \\ h^{(1)} \end{pmatrix} = \begin{pmatrix} 0 \\ b \end{pmatrix}, \qquad \begin{pmatrix} d^{(k+1)} \\ h^{(k+1)} \end{pmatrix} = \tilde{P}_k \ldots \tilde{P}_1 \begin{pmatrix} 0 \\ b \end{pmatrix} = \tilde{P}_k \begin{pmatrix} d^{(k)} \\ h^{(k)} \end{pmatrix}.$$

Now using induction we see $d^{(k)}$ has all but its first $k-1$ elements zero, and

$$\begin{pmatrix} d^{(k+1)} \\ h^{k+1)} \end{pmatrix} = \begin{pmatrix} d^{(k)} \\ h^{(k)} \end{pmatrix} - \begin{pmatrix} -e_k \\ q_k \end{pmatrix} \begin{pmatrix} -e_k^T & q_k^T \end{pmatrix} \begin{pmatrix} d^{(k)} \\ h^{(k)} \end{pmatrix} = \begin{pmatrix} d^{(k)} + e_k(q_k^T h^{(k)}) \\ h^{(k)} - q_k(q_k^T h^{(k)}) \end{pmatrix},$$

giving the computation $h^{(1)} = b$,

$$\textbf{for } k = 1, \ldots, n \ \textbf{do}\{\delta_k := q_k^T h^{(k)}; \quad h^{(k+1)} := h^{(k)} - q_k \delta_k\},$$

so $h = h^{(n+1)}, d = d^{(n+1)} = (\delta_1, \ldots, \delta_n)^T$.

This costs $2mn$ flops (1 flop = 1 multiplication and 1 addition in floating point arithmetic), compared with the mn flops required to form $d = Q_1^T b$ in (51). The computation for d and h is exactly the same as if the n MGS steps in had been applied to (A, b) instead of A, so that h is theoretically the component of b orthogonal to the columns of A, but now d has elements $q_k^T h^{(k)}$ instead of $q_k^T b$.

To compute y in (55), define

$$\begin{pmatrix} w^{(n)} \\ y^{(n)} \end{pmatrix} = \begin{pmatrix} z \\ h \end{pmatrix}, \qquad \begin{pmatrix} w^{(k-1)} \\ y^{(k-1)} \end{pmatrix} = \tilde{P}_k \ldots \tilde{P}_n \begin{pmatrix} z \\ h \end{pmatrix} = \tilde{P}_k \begin{pmatrix} w^{(k)} \\ y^{(k)} \end{pmatrix},$$

so that

$$\begin{pmatrix} w^{(k-1)} \\ y^{(k-1)} \end{pmatrix} = \begin{pmatrix} w^{(k)} \\ y^{(k)} \end{pmatrix} - \begin{pmatrix} -e_k \\ q_k \end{pmatrix} \left(-e_k^T w^{(k)} + q_k^T y^{(k)} \right),$$

which shows that only the kth element of $w^{(k)}$ is changed from $\zeta_k = e_k^T z$ to $\omega_k = q_k^T x^{(k)}$. This gives the computation $y^{(n)} = h = h^{(n+1)}$,

$$\textbf{for } k = n, \ldots, 1 \ \text{do}\{\omega_k := q_k^T y^{(k)}; \quad y^{(k-1)} := y^{(k)} - q_k(\omega_k - \zeta_k)\},$$

so $y = y^{(0)}$, $w = (\omega_1, \ldots, \omega_n)^T$. This costs $2mn$ flops compared with mn flops for $y = b - Q_1(d - z)$ in (51). Since in theory $y = b - Q_1 Q_1^T b + Q_1 z = h + Q_1 z$, the computation of y here can be seen to reorthogonalize $y^{(k)}$ against q_k before adding on $q_k \zeta_k$ to give $y^{(k-1)}$, and a weakness in some other MGS algorithms is that this reorthogonalization is not done.

The complete algorithm is then

1. Carry out MGS on A to give $Q_1 = [q_1, \ldots, q_n]$ and R,

2. Solve $R^T z = c$ for $z = (\zeta_1, \ldots, \zeta_n)^T$,

3. **for** $k = 1, \ldots, n$ do$\{\delta_k := q_k^T b; \quad b := b - q_k \delta_k\}$,

4. **for** $k = n, \ldots, 1$ do$\{w_k := q_k^T b; \quad b := b - q_k(\omega_k - \zeta_k)\}; \quad y := b$,

5. Solve $Rx = d - z$ for y, where $d = (\delta_1, \ldots, \delta_n)^T$.

Note that $w = (\omega_1, \ldots, \omega_n)^T$ is zero in exact computation, but can be significant when x is large. This algorithm is different from the MGS algorithms and the Huang algorithm studied in Arioli and Laratta [3].

The following implementation issues and specializations of the algorithm are fairly obvious. For (47), step 5 can often be omitted, while for (48), step 4 can often be omitted. If $c = 0$, corresponding to LLS, then $z = 0$ and step 2 and 4 can be omitted. This algorithm is the recommended approach for LLS, see Björck [9]. If $b = 0$, corresponding to LUS, then $d = 0$ and step 3 can be omitted, and step 5 if the Lagrange parameters are not needed. This constitutes a new algorithm which is superior to the usual approach that omits the ω_k in step 4. Nearly every ill-conditioned problem with a significant $\|\bar{y} - b\|_2$ will have a residual $\|A^T \bar{y} - c\|_2$ which is κ times larger than necessary if we use the algorithm with $\omega_k = 0$.

If A is square and nonsingular, (47) becomes the solution of $A^T y = c$, and y is independent of b, so b can be taken as zero in the algorithm, and steps 3 and 5 dropped. For square nonsingular A, (48) becomes $Ax = b$, and x is independent of c, so c can be taken as zero in the algorithm, and steps 2 and 4 dropped. This shows how to solve nonsingular systems using MGS in a backward stable way. Since steps 1, 3 and 5 with $z = 0$ already constitute the recommended algorithm for solving $Ax = b$, the result here is another way of proving the backward stability of this approach, and adds insight into why it works. For $A^T y = c$ however, the numerically stable algorithm made of steps 1,2 and 4 with b initially zero is a new superior algorithm.

3.6 Solution of the ASF using Seminormal Equations

Using the QR factorization, (50) can be rewritten as the *seminormal equations*

$$R^T Rx = A^T b - c, \qquad y = b - Ax. \tag{56}$$

If A is large and sparse, then it is often uneconomical to store and access the matrix Q, or even Q_1 the above seminormal equations may be used.

Björck [13] showed that this algorithm is *not* acceptable-erro stable for LLS, that is $c = 0$, in (56). However, he also showed one step of iterative refinement will usually give an acceptable-error stable result.

The corrected seminormal equations (CSNE) for the ASF take the form

$$\begin{aligned} \bar{R}^T \bar{R}\bar{x} &= A^T b - c, \\ \bar{y} &= b - A\bar{x}, \\ \bar{R}^T \bar{R}\delta\bar{x} &= A^T \bar{y} - c, \\ \bar{\delta y} &= -A\bar{\delta x}. \end{aligned} \tag{57}$$

and the corrected solution is taken to be $\bar{y} := \bar{y} + \delta\bar{y}$, $\bar{x} := \bar{x} + \delta\bar{x}$. In most situations one correction is enough, but if needed the correction step can be iterated.

Taking $c = 0$ in (58) we get the corrected seminormal equations for LLS used in [13]. There it was shown that one correction step usually yields an acceptable-error stable method. Paige [52] showed that using R from the Householder factorization in (56) with $b = 0$ gives an acceptable-error stable algorithm for LUS. Because of (45) this result holds also for the MGS factorization. In many applications it is important that also the residual $c - A^T\bar{y}$ for LUS corresponding to the computed solution $\bar{y}$ is small. For a backward stable method we have the bound

$$\|c - A^T\bar{y}\|_2 \leq c_3 u \|A\|_2 \|y\|_2, \tag{58}$$

where c_3 is a constant depending on m and n and the details of the arithmetic. This bound holds for method (1.4) if Householder, Givens or MGS is used for the QR factorization. For the method of seminormal equations (56) the norm of the residual vector can be expected to be of size $\kappa u\|A\|_2\|y\|_2$.

To get a small residual for LUS we use CSNE (58) taking $b = 0$. An error analysis for the method of corrected seminormal equations (58) can be carried through as in Björck [13], but we will not describe all details here. Instead we give a simplified analysis, that gives some insight into the performance of the method. Neglecting all roundoff errors, except those in the QR factorization, we get

$$\bar{r} = c - A^T A\bar{R}^{-1}\bar{R}^{-T}c = Mc, \qquad M = \bar{R}^T(I - \hat{Q}_1^T\hat{Q}_1)\bar{R}^{-T}, \tag{59}$$

where $\hat{Q}_1 = A\bar{R}^{-1}$. Using (41) we obtain,

$$\|I - \hat{Q}_1^T\hat{Q}_1\|_2 \leq 2\bar{c}_2\kappa u + O(\kappa^2u^2), \tag{60}$$

and neglecting terms of order κ^2u^2, $\|\bar{r}\|_2 \leq 2c_1\kappa u\|A\|_2\|y\|_2 \leq 2c_1\kappa^2 u\|c\|_2$. If we now perform the correction step, and again neglect roundoff errors except in the QR factorization, the residual $\bar{r}_c$ corresponding to the corrected solution $\bar{y}_c = \bar{y} + \delta\bar{y}$ becomes

$$\bar{r}_c = M^2 c = \bar{R}^T (I - \hat{Q}_1^T \hat{Q}_1)^2 \bar{z}.$$

By (60) we have

$$\|\bar{r}_c\|_2 \leq (\bar{c}_2 \kappa u)^2 \|A\|_2 \|y\|_2.$$

It follows that provided an inequality of the form $\sigma = \bar{c}_2 \kappa^2 u < c_3$ holds, then the size of the residuals will be as small as for a backward stable method. If roundoff errors in the other parts of the computations are taken into account this condition only has to be slightly modified.

3.7 Numerical Results

The following tests were run using MATLAB on a SUN 4/60 workstation, with machine precision equal to $2.220 \cdot 10^{-16}$. We took A to be the first six columns of the inverse Hilbert matrix of order nine, scaled to have $\|A\|_2 = 1$,

$$A = H/\|H\|_2, \qquad H = H_9^{-1} \begin{pmatrix} I_6 \\ 0 \end{pmatrix}, \qquad \kappa = 1.44 \cdot 10^{10}.$$

The solution y was taken to be a random vector rand(9, 1), and $c = A^T y$. Using MGS and (51), and the new algorithm respectively, we obtained

$$\|c - A^T \bar{y}_{mgs}\|_2 = 2.02 \cdot 10^{-10} \|y\|_2, \qquad \|c - A^T \bar{y}_{mgs}\|_2 = 3.23 \cdot 10^{-18} \|y\|_2,$$

illustrating the backward stability of the new algorithm.

To test method CSNE we computed the QR factorization with Householder transformations. The norm of the difference between the computed solutions $\bar{y}_{qr}$ and $\bar{y}_{sne}$ using methods (52) and (56), respectively, was

$$\|\bar{y}_{sne} - \bar{y}_{qr}\|_2 = 7.25 \cdot 10^{-10} \|y\|_2.$$

The residual corresponding to the solution $\bar{y}_{qr}$ has a small norm, but as predicted this is not true for $\bar{y}_{sne}$

$$\|c - A^T \bar{y}_{qr}\|_2 = 4.21 \cdot 10^{-17} \|y\|_2, \qquad \|c - A^T \bar{y}_{sne}\|_2 = 1.81 \cdot 10^{-10} \|y\|_2.$$

The relative residual norm is here *not* of the order of the machine precision, and the norm $\|\bar{R}^{-T} c\|_2 = 5.0775 \cdot 10^9$ is very large. We now perform one step of iterative refinement of $\bar{y}_{sne}$ using (58) with $b = 0$. The error in the corrected solution does not change much, but the corresponding residual is now of the order of machine precision

$$\|\bar{y}_{csne} - \bar{y}_{qr}\|_2 = 6.92 \cdot 10^{-10} \|y\|_2, \qquad \|c - A^T \bar{y}_{csne}\|_2 = 3.62 \cdot 10^{-17} \|y\|_2.$$

4 ITERATIVE METHODS

4.1 Basic Iterative Methods

For some large sparse least squares problems iterative methods are a useful alternative to direct methods. In iterative methods an initial approximate solution is successively improved until an acceptable solution is obtained. Iterative methods are especially attractive for problems in which the elements of the matrix A can be easily generated on demand. In such cases the matrix A need not be stored at all, but instead can be defined by its action on vectors.

In principle any iterative method for symmetric positive definite (or if rank $(A) < n$, semidefinite) linear systems can be applied to the system of normal equations $A^TAx = A^Tb$. As we will see, explicit formation of the matrix A^TA can be avoided by using the factored form of the normal equations

$$A^T(b - Ax) = 0. \tag{61}$$

For a treatment of iterative methods for symmetric positive definite linear systems see Varga [61] and Young [64]. Surveys of iterative methods for least squares problems are given by Björck [11] and Heath [35].

We now consider some basic iterative methods for solving (61). **Richardson's first order method** for the normal equations can be written

$$x^{(k+1)} = x^{(k)} + \alpha A^T(b - Ax^{(k)}). \tag{62}$$

It can be shown to converge to the least squares solution $x = A^\dagger b$ if $x^{(0)} \in \text{Range}(A^T)$, and $0 < \alpha < 2\sigma_1^{-2}(A)$.

We note that there is no need to form A^TA in the iteration (62). This has two advantages.

(i) A small perturbation in A^TA, e.g. by roundoff, may change the solution much more than perturbations of similar size in A itself. (ii) We avoid the fill which can result in the formation of A^TA. For some problems A^TA may be much less sparse than A. For example if A has approximately $n^{1/2}$ nonzeros randomly distributed in each row, then A^TA will be almost full.

In **Jacobi's** method a sequence of approximations $x^{(k)}$, $k = 1, 2, \ldots$ is computed from

$$x_j^{(k+1)} = x_j^{(k)} + a_j^T(b - Ax^{(k)})/d_j, \quad j = 1, 2, \ldots, n, \tag{63}$$

where

$$A = (a_1, \ldots, a_n) \in R^{m,n}, \quad d_j = a_j^T a_j = \|a_j\|_2^2. \tag{64}$$

Introducing the diagonal matrix $D = \text{diag}(d_1, \ldots, d_n)$ we can write (63) in matrix form

$$x^{(k+1)} = x^{(k)} + D^{-1}A^T(b - Ax^{(k)}). \tag{65}$$

Often the **Gauss-Seidel method** has a better rate of convergence than Jacobi's method. This method is usually defined in terms of the splitting $A^TA = L + D +$

L^T, where L is a strictly lower triangular matrix and D diagonal. The Gauss-Seidel iterations can then be written

$$x^{(k+1)} = x^{(k)} + (A^T b - Lx^{(k+1)} + (D + L^T)x^{(k)}).$$

As remarked above, it is preferable to implement the Gauss-Seidel method working only with the matrix A. In Björck and Elfving [15] a major iteration step is divided into n minor steps. We put $z^{(1)} = x^{(k)}$ and compute

$$z^{(j+1)} = z^{(j)} + e_j a_j^T (b - Az^{(j)})/d_j, \quad j = 1, 2, \ldots, n, \tag{66}$$

where e_j is the j-th unit vector and d_j defined by (64). We then have that $x^{(k+1)} = z^{(n+1)}$. Note that in the j-th minor step only the j-th component of $z^{(j)}$ is changed. Therefore the residual

$$r^{(j)} = b - Az^{(j)},$$

can be cheaply updated. With $r^{(1)} = b - Ax^{(k)}$ we can write (66)

$$z^{(j+1)} = z^{(j)} + \delta_j e_j, \quad r^{(j+1)} = r^{(j)} - \delta_j a_j, \quad \delta_j = a_j^T r^{(j)}/d_j. \tag{67}$$

Hence, in the j-th minor step only the j-th column of A is accessed. In contrast to the Jacobi method the ordering of the columns of A will influence the convergence. For a square matrix A the method (66) was developed by de la Garza [28]. It is also a special case of a class of residual reducing projection methods studied by Householder and Bauer [41].

We obtain the **Successive Over-Relaxation (SOR) method** for the normal equations from (67) simply by taking

$$\delta_j = \omega a_j^T r^{(j)}/d_j, \quad 0 < \omega < 2. \tag{68}$$

where ω is an acceleration parameter. The SOR method always converges provided that the relaxation parameter satisfies $0 < \omega < 2$. It has the advantage of simplicity and small storage requirements. Ashkenazi [5] has applied it to very large systems of geodetic normal equations.

The Jacobi method has the advantage over Gauss-Seidel and SOR that it is easily adopted to parallel computation. Further it does not require A to be stored (or generated) columnwise, since products of the form Ax and $A^T r$ can conveniently be computed also if A is accessed by rows. If $A = (a_1, \ldots, a_m)^T$ then we have

$$(Ax)_i = a_i^T x, \quad i = 1, \ldots, n, \quad A^T r = \sum_{i=1}^{m} a_i r_i,$$

i.e. for Ax we use an inner product and for $A^T r$ an outer product formulation.

The methods discussed above are special cases of the general stationary iterative method

$$x^{(k+1)} = Gx^{(k)} + c, \tag{69}$$

which can be obtained by the splitting $A^T A = M - N$, where M is nonsingular, and taking $G = M^{-1} N$. For the case when rank$(A) < n$ convergence of the iteration (69) has been investigated by Keller [43] and Young [64].

The concept of splitting has been extended to rectangular matrices by Plemmons [54]. Berman and Plemmons [7] define $A = M - N$ to be a proper splitting if the ranges and nullspaces of A and M are equal. They show that for a proper splitting the iteration

$$x^{(k+1)} = M^{\dagger}(Nx^{(k)} + b) \tag{70}$$

converges to the pseudoinverse solution $x = A^{\dagger}b$ for every $x^{(0)}$ if and only if the spectral radius $\rho(M^{\dagger}N) < 1$. The iterative method (70) avoids the explicit recourse to the normal system.

Tanabe [60] considers iterative methods of the form (70) for computing more general solutions $x = A^{-}b$, where A^{-} is any generalized inverse of A, $(AA^{-}A = I)$. He shows that the iteration can always be written

$$x^{(k+1)} = x^{(k)} + B(b - Ax^{(k)})$$

for some matrix B, and characterizes the solution in terms of the Range(AB) and Null(BA).

Splittings of rectangular matrices has also been investigated by Chen [19]. Chen shows that if rank$(A) = n$, then for any consistent iterative method (69) there exist a splitting such that the method can be written in the form (70), and for every such method $G = I - C^{-T}A^{T}A$ for some nonsingular matrix C. Hence the most general iterative method of the form (69) for the least squares problem is equivalent to Richardson's first order method applied to the preconditioned normal system $C^{-T}A^{T}(Ax - b) = 0$.

Young [64] notes that when $A^{T}A$ does not have property 'A' or is an L-matrix, then the SOR method may not be effective for any choice of ω. A powerful class of methods can be described by the recursion

$$x^{(k+1)} = x^{(k-1)} + \omega_{k+1}\left(\alpha A^{T}r^{(k)} + x^{(k)} - x^{(k-1)}\right), \tag{71}$$

where ω_{k+1} and α are parameters, see Golub and Varga [32]. If the singular values of A satisfy

$$a \le \sigma_i^2 < b,$$

we get the Chebyshev semi-iterative method by taking $\alpha = 2/(a+b)$, $\mu = (b-a)/(b+a)$, and recursively computing the parameters ω_k from

$$\omega_1 = 1, \quad \omega_{k+1} = (1 - \frac{\mu^2}{4}\omega_k)^{-1}.$$

The second order Richardson method is obtained by taking α as above and

$$\omega_k = 2/(1 + (1 - \mu^2)^{1/2}).$$

4.2 Conjugate Gradient Methods

For the methods (71) it is necessary to know lower and upper bounds $[a, b]$ on the singular values of A, and the rate of convergence is sensitive to the quality of these bounds. We now describe the **conjugate gradient method** of Hestenes and Stiefel [38], which does not have this drawback. For a general discussion of this method see Golub and Van Loan [31, Sec. 10.2-10.3]. A slight algebraic rearrangement given below is needed to make the method perform well for the least squares problem.

Let $x^{(0)}$ be an initial approximation. Take

$$r^{(0)} = b - Ax^{(0)}, \quad p^{(0)} = s^{(0)} = A^T r^{(0)}, \quad \gamma_0 = \|s^{(0)}\|_2^2, \tag{72}$$

and for $k = 0, 1, 2, \ldots$ compute

$$\begin{aligned} q^{(k)} &= Ap^{(k)}, \quad \alpha_k = \gamma_k / \|q^{(k)}\|_2^2, \\ x^{(k+1)} &= x^{(k)} + \alpha_k p^{(k)}, \quad r^{(k+1)} = r^{(k)} - \alpha_k q^{(k)}, \\ s^{(k+1)} &= A^T r^{(k+1)}, \quad \gamma_{k+1} = \|s^{(k+1)}\|_2^2, \\ \beta_k &= \gamma_{k+1}/\gamma_k, \quad p^{(k+1)} = s^{(k+1)} + \beta_k p^{(k)}. \end{aligned} \tag{73}$$

We call this algorithm CGLS. It can be used also in the rank deficient case. Provided that $x^{(0)} \in \text{Range}(A^T)$, which holds if e.g. $x^{(0)} = 0$, $x^{(k)}$ will converge to the pseudoinverse solution $A^\dagger b$. In absence of rounding errors it will compute the exact solution in at most t iterations, where $t \leq n$ equals the number of distinct nonzero singular values of A. This finite termination property does not hold under roundoff and therefore the conjugate gradient method is best regarded as an iterative method. The error measures $\|x^{(k)} - A^\dagger b\|_2$ and $\|r^{(k)}\|_2$ will both decrease monotonically in exact computation and in general also under roundoff.

When A is ill conditioned and the least squares solution x has substantial components along singular vectors corresponding to small singular values, then many more than n iterations may be needed. However, more important is that in some practical applications one is satisfied with approximations that are obtained in far less than n iterations. This is true when A is well conditioned, or when the right hand side is such that the effective condition number of the system $Ax = b$ is small, see below.

The algorithm (74) requires the storage of two n-vectors x and p and two m vectors r and q. (Note that s can share storage with q.) Each iteration requires about $2nz(A) + 3n + 2m$ flops, where $nz(A)$ are the number of nonzero elements in A.

Elfving [22] has compared the implementation (74) with several other variants of the conjugate gradient algorithm, and found this to be the most accurate. Paige and Saunders [53] have developed an algorithm LSQR based on the Lanczos bidiagonalization algorithm of Golub and Kahan [29], which is described at the end of this section. LSQR is mathematically equivalent to CGLS but is more accurate when A is ill-conditioned and many iterations are carried out.

The basic drawback with iterative methods is that for general right hand sides the rate of convergence will depend on the spectrum of A and can therefore be very slow. For the methods of Jacobi and Gauss-Seidel the number of iterations needed for a fixed reduction in the accuracy is proportional to $\kappa^2(A)$, where $\kappa(A)$ is the condition number of A. For the Chebyshev semi-iteration and second order Richardson method this is reduced to order $\kappa(A)$ iterations, provided that accurate bounds for the singular values of A are known. For the conjugate gradient method no such bounds are needed and for this method an upper bound for the number of iterations k to reduce the relative error by a factor of ϵ is given by

$$k < 1/2\kappa(A)\log\frac{2}{\epsilon}.$$

In some least squares problems the underlying model is such that the solution will have only small projections along the singular vector directions corresponding to small singular values. Let $A = U\Sigma V^T$ be the singular value decomposition of A. From (4.2) the least squares solution can be written

$$x = Vz = \sum_{i=1}^{n}(c_i/\sigma_i)v_i, \qquad c_i = u_i^T b.$$

Let δ be an error tolerance for x and assume that

$$\|c_i/\sigma_i\| < \delta, \qquad i = p+1,\ldots,n.$$

Then, if we take $x^{(0)} = 0$ as the initial approximation in an iterative method, the errors along v_i, $i > p$, are already sufficiently small. Therefore the effective condition number for the convergence of the error along the other singular vectors will be

$$\kappa_e(A,b) = \sigma_1/\sigma_p \leq \kappa(A).$$

Problems, where A is very ill-conditioned, but the effective condition number is small arise in certain reconstruction problems e.g. in tomography. Perturbations to the right hand side often make it undesirable to continue the iteration process in order to find its limiting solution. Such problems are often ill-posed in a mathematical sense.

4.3 Preconditioners for Least Squares Problems

Because of potentially slow convergence one main emphasis in the development of iterative methods is on convergence acceleration. A general technique to improve convergence is by preconditioning, which for the linear least squares problem is equivalent to a transformation of variables. Let $M \in R^{n,n}$ be a nonsingular matrix and consider the problem

$$\min_y \|(AM^{-1})y - b\|_2, \quad Mx = y. \tag{74}$$

If M is chosen so that AM^{-1} has a more favorable spectrum than A this will improve convergence of an iterative method applied to (74).

It is important to note that the product AM^{-1} should not be explicitly formed but treated as a product of two operators. In an iterative method preconditioned by the matrix M matrix times vector products of the form $AM^{-1}y$ and $M^{-T}A^Tr$ will occur. Thus the extra cost of preconditioning will be in solving linear systems of the form $Mx = y$ and $M^Tq = s$. Hence M has to be chosen so that such systems can be easily solved.

Several different preconditioners have been suggested for the least squares problem. We first note that if A has full rank and we take $M = R$, where R is the Cholesky factor of A^TA, then $\kappa(AM^{-1}) = \kappa(Q_1) = 1$. This is the ideal choice for M and all the above iterative methods will converge in only one step. Thus, a preconditioner should in some sense approximate R.

The simplest preconditioner corresponds to a diagonal scaling of the columns of A,

$$M = D^{1/2} = \mathrm{diag}(d_1^{1/2}, \dots, d_n^{1/2}), \quad d_j = \|a_j\|_2^2. \tag{75}$$

Since AM^{-1} has columns of unit length this approximately minimizes $\kappa(AM^{-1})$ over all diagonal scalings.

A more general preconditioner is the SSOR-preconditioner. Here we take

$$M = D^{-1/2}(D + \omega L^T), \quad 0 \leq \omega < 2. \tag{76}$$

where $A^T A$ has been split so that

$$A^T A = L + D + L^T,$$

with L strictly lower triangular. In Björck and Elfving [15] it is shown how to implement this preconditioner without actually forming $A^T A$ or L. Note that taking $\omega = 0$ in (76) gives (75).

Another approach is to take $M = \bar{R}$, where $\bar{R}$ is an incomplete Cholesky factor of $A^T A$, i.e.

$$A^T A = \bar{R}^T \bar{R} + E,$$

with E small and $\bar{R}$ sparse. One way to compute such a matrix $\bar{R}$ is to use a direct method for sparse Cholesky factorization, but only keep those elements in R which lie within a predetermined sparsity structure. Meijerink and van der Vorst [48] proved the existence and stability of the incomplete Cholesky factorization for the case when $A^T A$ is an M-matrix and the structure of $R^T + R$ equals that of $A^T A$. Positive definiteness of $A^T A$ alone is not sufficient to guarantee the existence of R. Modifications to prevent the breakdown of the incomplete factorization has been suggested by Manteuffel [46] amongst others.

Another approach is taken by Zlatev and Nielsen [65]. They compute a sparse incomplete orthogonal factorization of A, neglecting all elements which in the course of the computations become smaller than a certain *drop-tolerance* $T \geq 0$. This can sometimes substantially reduce the number of nonzero elements during the factorization process and. The drop-tolerance $T = 2^{-3}$ is recommended, and T is successively reduced if the iterations fail to converge. This approach can be very efficient for some classes of problems, in particular when storage is a limiting factor.

Other choices of preconditioners result from using elimination methods on A. This seems to have been first suggested by Läuchli [44]. If A has full column rank, we may assume that the rows of A have been permuted so that A has the form

$$A = \begin{pmatrix} A_1 \\ A_2 \end{pmatrix}, \qquad A_1 \in R^{n,n},$$

where A_1 is nonsingular. Läuchli suggests using A_1 as preconditioner, applying the semi-iterative method (71) to the system a

$$AA_1^{-1} y = b, \qquad y = A_1 x. \tag{77}$$

If the matrix

$$AA_1^{-1} = \begin{pmatrix} I \\ C \end{pmatrix}, \qquad C = A_2 A_1^{-1},$$

is computed by Gauss-Jordan elimination. The parameters can then be estimated from the elements of C using

$$a = 1 \le \lambda(I + CC^T) \le 1 + \text{trace}(C^T C) = b..$$

Alternatively the conjugate gradient method can be applied to (77). This method requires two matrix-vector products ($A_2 v$ and $A_2^T w$) and two solves of a n, n system ($A_1^{-1} v$ and $A_1^{-T} v$) per iteration. Hence we only need a sparse factorization of the matrix A_1. The convergence of this method has been studied by Freund [25]. Evans and Li [24] point out that one only needs to solve the system

$$\begin{pmatrix} I & C^T \\ -C & I \end{pmatrix} \begin{pmatrix} r_1 \\ r_2 \end{pmatrix} = \begin{pmatrix} 0 \\ b_2 - Cb_1 \end{pmatrix}, \tag{78}$$

where b and the residual vector r have been partitioned conformally with A. When the residual vector $r^T = (r_1^T, r_2^T)$ has been computed the least squares solution x is obtained from the system

$$A_1 x = b - r_1.$$

Because of the special form of (78) a variant of the generalized conjugate gradient method can be devised which requires approximately half the work per iteration.

Saunders [55] suggests using Gaussian elimination with row interchanges to compute a stable, sparse factorization

$$P_r A = LU,$$

where P_r is a permutation matrix, $L \in R^{m,n}$ unit lower trapezoidal, and U upper triangular. As preconditioner one takes $M = U$, the rationale for this choice being that any ill-conditioning in A is usually reflected in U and L tends to be well conditioned. A preliminary pass through the rows of A are made to select a triangular subset with maximal diagonal elements. The matrix L is not saved, and subsequent use of the operator AU^{-1} involves backsubstitution with U. This approach has the advantage that often there is very little fill-in in U, and so U is likely to contain less nonzeros than A.

4.4 Block Preconditioners

In many large sparse least squares problems arising from multidimensional models the matrix A has a natural column block structure

$$A = (A_1, A_2, \ldots, A_N), \tag{79}$$

where

$$A_j \in R^{m,n_j}, \quad n_1 + \ldots + n_N = n.$$

A special example of such a block structure is the dual block angular form.

For problems of the structure (79) block versions of the preconditioners (75) and (76) are particularly suitable. Let the QR decompositions of the blocks be

$$A_j = Q_j R_j, \quad Q_j \in R^{m,n_j}, \quad j = 1, \ldots, N. \tag{80}$$

Then to (75) corresponds the block diagonal preconditioner

$$M = R_B = \operatorname{diag}(R_1, R_2, \ldots, R_N). \tag{81}$$

For this choice we have $AM^{-1} = (Q_1, Q_2, \ldots, Q_N)$, i.e. the columns of each block are mutually orthogonal.

If we split $A^T A$ according to

$$A^T A = L_B + D_B + L_B^T,$$

where L_B is strictly lower block triangular then the block SSOR preconditioner becomes

$$M = R_B^{-1}(D_B + \omega L_B^T). \tag{82}$$

This preconditioner was introduced for least squares problems by Björck [12]. As with the corresponding point preconditioner it can be implemented without actually forming $A^T A$.

We now consider the use of the block diagonal preconditioner (81). We will partition x and $y = Mx$ conformingly with (79)

$$x = (x_1, x_2, \ldots, x_N)^T, \quad y = (y_1, y_2, \ldots, y_N)^T.$$

Jacobi's method (65) applied to the preconditioned problem (74) can then be written

$$y_j^{(k+1)} = y_j^{(k)} + Q_j^T(b - AM^{-1}y^{(k)}), \quad j = 1, \ldots, N,$$

or in terms of the original variables

$$x_j^{(k+1)} = x_j^{(k)} + R_j^{-1}Q_j^T(b - Ax^{(k)}), \quad j = 1, \ldots, N. \tag{83}$$

This is the block Jacobi method for the normal equations. Note that the correction $z_j = x_j^{(k+1)} - x^{(k)}$ equals the solution to the problem

$$\min_{z_j} \|A_j z_j - r^{(k)}\|_2, \quad r^{(k)} = b - Ax^{(k)}, \tag{84}$$

and that these corrections can be computed in parallel. Often Q_j is not available and we have to use $Q_j = A_j R_j^{-1}$. This is equivalent to using the method of seminormal equations (56) for solving (84). This can lead to some loss of accuracy and a correction step is recommended unless all the blocks A_j are well conditioned.

Similarly we can derive a block SOR method for the normal equations. Let $r_1^{(k)} = b - Ax^{(k)}$ and for $j = 1, \ldots, N$ compute

$$x_j^{(k+1)} = x_j^{(k)} + \omega z_j, \quad r_{j+1}^{(k)} = r_j^{(k)} - \omega A_j z_j, \tag{85}$$

where z_j is the solution to

$$\min_{z_j} \|A_j z_j - r_j^{(k)}\|_2. \tag{86}$$

Taking $\omega = 1$ in (85) gives the block Gauss-Seidel method.

The choice of partitioning of A into blocks is important for the storage and computational efficiency of the methods. An important criterion is that it should be possible to compute the factorizations $A_j = Q_j R_j$ (or at least the factors R_j) without too much fill-in. To be able to achieve efficiency through parallelism. Note that if A_j is the direct sum of blocks, then in the SOR method the computation of z_j splits up in independent subproblems.

For grid problems a high degree of parallelism can be achieved. For example, Kamath and Sameh [42] give a scheme for a three dimensional $n, \times n \times n$ mesh and a seven point difference star, for which $N = 9$ and each block consists of $n^2/9$ separate subblocks of columns. Hence each subproblem can be solved with a parallelism of $n^2/9$. For some ordering algorithms based on graph-theoretic algorithms see Dennis and Steihaug [20].

The case $N = 2$, $A = (A_1, A_2)$, is of special interest. For the block diagonal preconditioner (81) we have $AM^{-1} = (Q_1, Q_2)$ and the matrix of normal equations for the preconditioned system becomes

$$(AM^{-1})^T AM^{-1} = \begin{pmatrix} I & K \\ K^T & I \end{pmatrix}, \quad K = Q_1^T Q_2. \tag{87}$$

This matrix has "Property A", Young [64]. This means that it is possible to reduce the work required per iteration to approximately half for many iterative methods. This preconditioner is also called the Cyclic Jacobi Preconditioner.

For matrices with "Property A" the SOR theory holds. As shown by Elfving [23] it follows that the optimal ω in the block SOR method (85) is given by (for $N = 2$)

$$\omega_{\text{opt}} = 2/(1 + \sin\theta_{\min}), \qquad \cos\theta_{\min} = \sigma_{\max}(Q_1^T Q_2).$$

Here $\theta_{\min}$ is the smallest principal angle between Range(A_1) and Range(A_2), see Golub and Van Loan [31, 12.4.3]. Using ω_{opt} in the block SOR method reduces the number of iterations by a factor of $2/\sin\theta_{\min}$ compared to using $\omega = 1$.

For $N = 2$, the preconditioner (82) with $\omega = 1$ also has special properties, see Golub, Manneback and Toint [30]. We have

$$M = \begin{pmatrix} R_1 & \omega Q_1^T A_2 \\ 0 & R_2 \end{pmatrix},$$

and it follows that for $\omega = 1$

$$(A_1, A_2)M^{-1} = (Q_1, (I - P_1)Q_2), \tag{88}$$

where $P_1 = Q_1 Q_1^T$ is the orthogonal projector onto Range(A_1). It follows that the two blocks in (88) are mutually orthogonal and thus the preconditioned problem (74) can be split into two problems which are

$$\min_{y_2} \|(I - P_1)Q_2 y_2 - b\|_2, \quad y_1 = Q_1^T b. \tag{89}$$

This effectively reduces the original system to a system of size n_2. Hence this preconditioner is also called the Reduced System Preconditioner. The matrix of normal equations becomes

$$(AM^{-1})^T AM^{-1} = \begin{pmatrix} I & 0 \\ 0 & Q_2^T(I-P_1)Q_2 \end{pmatrix} = I - \begin{pmatrix} 0 & 0 \\ 0 & K^T K \end{pmatrix},$$

where $K = Q_1^T Q_2$. This reduction of the variables corresponding to the first block columns can also be performed in the case when $N > 2$.

We now consider preconditioning the conjugate gradient method (74). It is convenient to formulate the preconditioned method in terms of the original variables x. Let $x^{(0)}$ be an initial approximation and $r^{(0)} = b - Ax^{(0)}$. Take

$$p^{(0)} = s^{(0)} = M^{-T} A^T r^{(0)}, \quad \gamma_0 = \|s^{(0)}\|_2, \tag{90}$$

and for $k = 0, 1, 2, \ldots$ compute

$$\begin{aligned} q^{(k)} &= AM^{-1}p^{(k)}, \quad \alpha_k = \gamma_k / \|q^{(k)}\|_2^2, \\ x^{(k+1)} &= x^{(k)} + \alpha_k M^{-1} p^{(k)}, \quad r^{(k+1)} = r^{(k)} - \alpha_k q^{(k)}, \\ s^{(k+1)} &= M^{-T} A^T r^{(k+1)}, \quad \gamma_{k+1} = \|s^{(k+1)}\|_2^2, \\ \beta_k &= \gamma_{k+1}/\gamma_k, \quad p^{(k+1)} = s^{(k+1)} + \beta_k p^{(k)}. \end{aligned} \tag{91}$$

In order to use the block SSOR preconditioner (82) for the conjugate gradient method we have to be able to compute vectors $q = AM^{-1}p$ and $s = M^{-T}A^T r$ efficiently, given $p = (p_1, \ldots, p_N)^T$ and r. The following algorithms for this are derived in Björck [12]:

Put $q^{(N)} = 0$ and compute $q = q^{(0)}$, and $M^{-1}p = (z_1, \ldots, z_N)^T$

$$z_j = R_j^{-1}(p_j - R_j^{-T} A_j^T q^{(j)}), \quad q^{(j-1)} = q^{(j)} - A_j z_j, \qquad j = N, N-1, \ldots, 1.$$

Put $r^{(1)} = r$, and compute $s = (s_1, \ldots, s_N)^T$

$$s_j = R_j^{-T} A_j^T r^{(j)}, \quad r^{(j+1)} = r^{(j)} - A_j R_j^{-1} s_j, \qquad j = 1, 2, \ldots, N.$$

For the case $N = 2$ the SSOR preconditioned conjugate gradient method has been carefully studied by Manneback [45]. It is shown that the choice $\omega = 1$, i.e. using the Reduced System Preconditioning is optimal with respect to the number of iterations. Further, Hageman, Luk and Young [33] have shown the equivalence of Reduced System Preconditioning and Cyclic Jacobi preconditioning ($\omega = 0$) for Chebyshev semi-iteration and the conjugate gradient method. The reduced system preconditioning will essentially generate the same approximations in half the number of iterations. Since the work per iteration is about doubled for $\omega \neq 0$, this means that cyclic Jacobi preconditioning is optimal for the conjugate gradient method in the class of SSOR preconditioners.

Golub, Manneback and Toint [30] have applied block SSOR preconditioned conjugate gradient method to the Doppler positioning problem. Here the matrix is of dual block angular form but is partitioned into two blocks (A, B) where A consists of all the diagonal blocks and B of the last block column.

Some experimental results of block SSOR preconditionings for the case $N > 2$ are given by Björck [12]. In this case $\omega = 1$ is not necessarily optimal. However, in the tests the optimal value of ω was close to one and the number of iterations required was not sensitive to the choice of ω around $\omega = 1$.

References

[1] Arioli, M.; Demmel, J.W.; Duff, I.S.: Solving sparse linear systems with sparse backward error, SIAM J. Matrix Anal. Appl., **10**, 165–190 (1989)

[2] Arioli, M.; Duff, I.S.; de Rijk, P.P.M.: On the augmented system approach to sparse least-squares problems, Numer. Math., **55**, 667–684 (1989)

[3] Arioli, M.,; Laratta, A.: Error analysis of an algorithm for solving an underdetermined linear system. Numer. Math. **46**, 255–268 (1985)

[4] Arioli, M.; Laratta, A.: Error analysis of algorithms for computing the projection of a point onto a linear manifold. Linear Algebra Appl. **82**, 1–26 (1986)

[5] Ashkenazi, V.: Geodetic normal equations. In **Large Sparse Sets of Linear Equations**, J.K Reid, ed., Academic Press, 57–74 (1971)

[6] Bauer, F.L.: Genauigkeitsfragen bei der Lösung linearer Gleichungssysteme, ZAMM, **46**, 409–421 (1966)

[7] Berman; Plemmons, R.J.: Cones and iterative methods for best least squares solutions of linear systems, SIAM J. Numer. Anal. **11**, 145–154 (1974)

[8] Bischof, C.H.: Incremental condition estimation. To appear in SIAM J. Matrix Analysis Appl.

[9] Björck, Å.: Solving linear least squares problems by Gram-Schmidt orthogonalization, BIT **7**, 1–21 (1967)

[10] Björck, Å.: Iterative Refinement of linear least squares solutions I, BIT **7**, 257–278 (1967)

[11] Björck, Å.: Methods for sparse least squares problems, in **Sparse Matrix Computations**, J.R. Bunch; D. J. Rose, eds., Academic Press, New York, 1976, 177–199 (1976)

[12] Björck, Å.: SSOR preconditioning methods for sparse least squares problems. In **Proceedings of the Computer Science and Statistics 12th Annual Symposium on the Interface**, 21–25. University of Waterloo, Canada, 1979

[13] Björck, Å.: Stability analysis of the method of seminormal equations for linear least squares problems, Linear Algebra Appl., **88/89**, 31–48 (1987)

[14] Björck, Å.: Least squares methods. in **Handbook of Numerical Analysis, Vol.I: Finite Difference Methods-Solution of Equations in R^n.** Eds. P.G. Ciarlet and J.L. Lions, Elsevier/North Holland, 1990

[15] Björck, Å., Elfving, T.: Accelerated projection methods for computing pseudoinverse solutions of systems of linear equations. BIT **19**, 145–163 (1979)

[16] Björck, Å.; Paige, C.C.: Stable uses of the Modified Gram-Schmidt algorithm and solution of augmented systems. In preparation.

[17] Bunch, J.R.: The weak and strong stability of algorithms in numerical linear algebra. Linear Algebra Appl., **88/89**, 49–66 (1987)

[18] Chan, T.F.: Rank revealing QR-factorizations, Linear Algebra Appl. **88/89**, 67–82 (1987)

[19] Chen, Y.T.: Iterative methods for linear least squares problems, Report CS-75-04, University of Waterloo, Canada, 1975

[20] Dennis, J.E.; Steihaug, T.: On the successive projections approach to least-squares problems. SIAM J. Numer. Anal. **23**, 717–733 (1986)

[21] Eckhard, C.; Young, G.: The approximation of one matrix by another of lower rank, Psychometrica **1**, 211–218 (1936)

[22] Elfving, T.: Some numerical results obtained with two gradient methods for solving the linear least squares problem, Report LiTH-MAT-R-75-5, Department of Math. Linköping University, Sweden, 1978

[23] Elfving, T.: Block-iterative methods for consistent and inconsistent linear equations. Numer. Math. **35**, 1–12 (1980)

[24] Evans, D.J.; Li, C.: Numerical aspects of the generalized cg-method applied to least squares problems. Computing **41**, 171–178 (1989)

[25] Freund, R.: A note on two block SOR methods for sparse least squares problems. Linear Algebra Appl. **88/89**, 211–221 (1987)

[26] Foster, L.V.: Rank and null space calculations using matrix decompositions without column interchanges. Linear Algebra Appl. **74**, 47–71 (1986)

[27] Foster, L.V.: The probability of large diagonal elements in the QR factorization. SIAM J. Sci. Stat. Comput. **11**, 531–544 (1990)

[28] de la Garza, A.: An iterative method for solving systems of linear equations, Union Carbide, Oak Ridge, Report K-731, Tennessee, 1951

[29] Golub, G.H.; Kahan, W.: Calculating the singular values and pseudo-inverse of a matrix, SIAM J. Num. Anal. Ser. B **2**, 205–224 (1965)

[30] Golub, G.H.; Manneback, P.; Toint, P.: A comparison between some direct and iterative methods for large scale geodetic least squares problems, SIAM J. Sci. Stat. Comput. **7**, 799–816 (1986)

[31] Golub, G.H.; Van Loan, C.F.: **Matrix Computations. 2nd ed.**, Johns Hopkins Press, Baltimore, MD, 1989

[32] Golub, G.H.; Varga, R.S.: Chebyshev semi-iterative methods, successive overrelaxation iterative methods and second order Richardson iterative methods, Parts I and II. Numer. Math. **3**, 147–168 (1961)

[33] Hageman, L.A.; Luk, F.T.; Young, D.M.: On the equivalence of certain iterative acceleration methods, SIAM J. Numer. Anal. **17**, 852–873 (1980)

[34] Hager, W.W.: Condition Estimates. SIAM J. Sci. Stat. Comput. **5**, 311–316 (1984)

[35] Heath, M.T.: Numerical methods for large sparse linear least squares problems, SIAM J. Sci. Stat. Comput., **5**, 497–513 (1984)

[36] Hestenes, M.R.: Inversion of matrices by biorthogonalization and related results. J. SIAM **6**, 51–90 (1958)

[37] Hestenes, M.R.: Pseudoinverses and conjugate gradients. Comm. ACM **18**, 40–43 (1975)

[38] Hestenes, M.R.; Stiefel E.: Methods of conjugate gradients for solving linear system, J. Res. Nat. Bur. Stds., **B49**, 409–436 (1952)

[39] Higham, N.J.: FORTRAN codes for estimating the one-norm of a real or complex matrix, with applications to condition estimation. *ACM Trans. Math. Soft.*, **14**, 381–396 (1988)

[40] Higham, N.J.: Iterative refinement enhances the stability of QR-factorization methods for solving linear equations. Numerical Analysis Report No. 182, University of Manchester/UMIST, April 1990.

[41] Householder, A.S. and Bauer, F.L.: On certain iterative method for solving linear systems. Numer. Math. **2**, 55–59 (1960)

[42] Kamath, C.; Sameh, A.: A projection method for solving nonsymmetric linear systems on multiprocessors, Parallel Computing **9**, 291–312 (1988/89)

[43] Keller, J.: On the solution of singular and semidefinite linear systems by iterations, J. SIAM Ser. B **2**, 281–290 (1965)

[44] Läuchli, P.: Jordan-Elimination und Ausgleichung nach kleinsten Quadraten, Numer. Math., **3**, 226–240 (1961)

[45] Manneback, P.: On some numerical methods for solving large sparse linear least squares problems, Ph. D. thesis, Facultés Universitaires Notre-Dame de la Paix, Namur, Belgium, 1985

[46] Manteuffel, T.: An incomplete factorization technique for positive definite linear system, Math. Comp. **34**, 473–497 (1980)

[47] Markham, T.L.; Neumann, M.; Plemmons, R.J.: Convergence of a direct-iterative method for large-scale least-squares problems. Linear Algebra Appl. **69**, 155–167 (1985)

[48] Meijerink, J.A.; van der Vorst, H.A.: An iterative method for linear systems of which the coefficient matrix is a symmetric M-matrix. Math. Comp. **31**, 148–162 (1977)

[49] Mirsky, L.: Symmetric gauge functions and unitarily invariant norms, Quart. J. Math. Oxford **11**, 50–59 (1960)

[50] Niethammer, W.; de Pillis, J.; Varga, R.S.: Convergence of block iterative methods applied to sparse least-squares problems. Linear Algebra Appl. **58**, 327–341 (1984)

[51] Oettli, W.; Prager, W.: Compatibility of approximate solutions of linear equations with given error bounds for coefficients and right-hand sides, Numer. Math. **6**, 405–409 (1964)

[52] Paige, C.C.,: An error analysis of a method for solving matrix equations, Math. Comp., **27**, 355–359 (1973)

[53] Paige, C.C.; Saunders, M.A.: LSQR. An algorithm for sparse linear equations and sparse least squares, ACM Trans. Math. Software. **8**, 43–71 (1982)

[54] Plemmons, R.J.: Monotonicity and iterative approximations involving rectangular matrices, Math. Comp., **26**, 853–858 (1972)

[55] Saunders, M.A.: Sparse least squares by conjugate gradients: a comparison of preconditioning methods, **Proceedings Computer Science and Statistics: Twelfth Annual Symposium on the Interface**, 15–20, 1979

[56] Skeel, R.D.: Scaling for numerical stability in Gaussian elimination. J. of the ACM, **26**, 494–526 (1979)

[57] Skeel, R.D.: Iterative refinement implies stability for Gaussian elimination. Math. of Comput., **35**, 817–832 (1980)

[58] van der Sluis, A.: Condition numbers and equilibration of matrices. Numer. Math. **14**, 14–23 (1969)

[59] Stewart, G.W.; Sun, Ji-guang.: **Matrix Perturbation Theory**, New York-London: Academic Press 1990

[60] Tanabe, K.: Projection method for solving a singular system of linear equations and its applications, Numer. Math. **17**, 203–214 (1971)

[61] Varga, R.S.: **Matrix Iterative Analysis**, New Jersey: Prentice-Hall 1962

[62] Wedin, P-Å .: Perturbation theory for pseudo-inverses. BIT **13**, 217–232 (1973)

[63] Wilkinson, J.H.: **The Algebraic Eigenvalue Problem.** Oxford: Clarendon Press 1965

[64] Young, D.M.: **Iterative Solution of Large Linear Systems**, New York-London: Academic Press 1971

[65] Zlatev, Z.; Nielsen, H.B.: Solving large and sparse linear least-squares problems by conjugate gradient algorithms. Comput. Math. Applic. **15**, 185–202 (1988)

ABS ALGORITHMS FOR GENERAL LINEAR SYSTEMS

E. Spedicato
Department of Mathematics
University of Bergamo
Piazza Rosate 2, 24100 Bergamo

ABSTRACT

In this paper we present the main properties of the class of direct methods for solving linear equations proposed by Abaffy, Broyden and Spedicato in 1984 and called the ABS class.

1. INTRODUCTION

The ABS algorithms are a very general class of direct methods of finite iterative form, named after Abaffy, Broyden and Spedicato, who presented the algorithms, in the unscaled form, in a paper in Numerische Mathematik in 1984. While the first algorithms of ABS form had appeared at least already in the seventies, see Huang (1975), Abaffy (1979) and also Adachi (1971), the development of ABS methods has taken place almost entirely in the eighties, in the framework of a collaboration between researchers of the University of Bergamo (Spedicato, Bertocchi, Bodon, Vespucci) and researchers in Hungary (Abaffy, Galantai), in UK (Broyden, Dixon) and in China (particularly Xia, Deng, Zhu). In 1989 a monograph by Abaffy and Spedicato, later referred to as the AS monograph, has appeared presenting the state of knowledge until that year, documented in more than sixty papers. At the end of 1990 the AS monograph had been translated into Chinese and Russian. The amount of research done recently in this field is documented, in the Russian version of the AS monograph, by an update section summarizing more than thirty papers which have appeared after the publication of the original AS monograph.

NATO ASI Series, Vol. F 77
Computer Algorithms for Solving Linear Algebraic Equations
Edited by E. Spedicato

In this paper we present synthetically the main properties of the ABS algorithms for general linear systems, avoiding proofs, which can be found, if not otherwise stated, in the AS monograph. We assume exact arithmetic and sequential formulation, since work on roundoff analysis and vector/parallel formulation is still at a preliminary stage. We do not consider the formulation for structured large problems, since it is considered, for some important cases, in the paper of Abaffy in this volume. We do not consider the extension to nonlinear equations (see, in addition to the results in the AS monograph, the papers of Galantai (1990), Deng and Chen (1990), Abaffy (1990), Huang (1990)), or to nonlinear optimization (see for instance the papers of Spedicato and Xia (1990, 1991), Zhu (1990), Xia (1990, 1991a, 1991b)).

The problem to be solved is $Ax = b$, $A \in R^{m,n}$, $x \in R^n$, $b \in R^m$, $m \leq n$, and rank $(A) = q \leq m$. For $m < n$ we shall consider the solution in the least squares sense. The system $Ax = b$ is also represented as $a_i^T x = b_i$, $i = 1, \dots m$, $a_i \in R^n$, $A = (a_1, \dots a_m)^T$. The vector x^+ indicates a solution of the system.

Unless otherwise stated, vector norms are Euclidean norms and matrix norms are the norms induced by Euclidean vector norms. By Diag (v_i) we indicate the diagonal matrix whose i-th diagonal element is v_i. A nonsingular matrix will be called strongly nonsingular if all its principal leading submatrices are nonsingular.

2. THE BASIC (UNSCALED) ABS ALGORITHM: FORMULATION AND MAIN PROPERTIES

The basic (unscaled) ABS algorithm is defined by the following procedure, see Abaffy, Broyden and Spedicato (1984).

(A) Initialization

Let x_1 be an arbitrary vector in R^n, H_1 be an arbitrary nonsingular matrix in $R^{n,n}$ and set $i = 1$.

(B) Test of dependency and compatibility

Compute s_i and μ^*_i by

(1) $s_i = H_i a_i$

(2) $\mu^*_i = a_i^T x_i - b_i$

If $s_i \neq 0$, proceed to (C). If $s_i = 0$ and $\mu^*_i = 0$, set $x_{i+1} = x_i$, $H_{i+1} = H_i$ and proceed to (F), the i-th equation is a linear combination of the previous equations. If $s_i = 0$ and $\mu^*_i \neq 0$ stop, the i-th equation is not compatible.

(C) Computation of the search vector.

Define the search vector $p_i \in R^n$ by

(3) $p_i = H_i^T z_i$

where $z_i \in R^n$ is arbitrary subject to

(4) $z_i^T H_i a_i \neq 0$

(D) Update of the approximation x_i to the solution.

Compute the stepsize μ_i and the new approximation x_{i+1} by

(5) $\mu_i = \mu^*_i / p_i^T a_i$

(6) $x_{i+1} = x_i - \mu_i p_i$

Stop if i=m, x_{m+1} solves the system.

(E) Update of H_i.

Compute the new matrix H_{i+1} by

(7) $H_{i+1} = H_i a_i w_i^T H_i$

where $w_i \in R^n$ is arbitrary subject to

(8) $w_i^T H_i a_i = 1$

(F) Increment i by one and return to (B).

The ABS algorithm deals with one equation at a time, in the originally given order. Clearly this order might be modified on grounds of numerical stability or of computational cost, a problem which has yet to be developed. In presence of roundoff the dependency and compatibility test should also be formulated in terms of a suitable tolerance μ. The correct determination of μ is also a problem to be analyzed.

Some basic properties of the ABS algorithm are the following:

- the algorithm is well defined, i.e. at every iteration vectors z_i and w_i exist such that conditions (4) and (8) are satisfied. A particular algorithm corresponds to specific admissible choices of H_1, z_i, w_i
- $H_i a_i = 0$ iff a_i is a linear combination of $a_1, \ldots a_{i-1}$
- $\mathrm{Null}(H_i) = \mathrm{Span}(a_1, \ldots a_{i-1})$, $\mathrm{Null}(H_i^T) = \mathrm{Span}(w_1, \ldots w_{i-1})$

In the following we assume, unless otherwise stated and for simplicity of formulation, that A has full row rank. Then we have:

- $\mathrm{rank}(H_{i+1}) = \mathrm{rank}(H_i) - 1 = n-i+1$

- $H_i H_1^{-1} H_j = H_{\max(i,j)}$

- if $P_i = (p_1, \ldots p_i)$, then P_i has full column rank

- $a_j^T x_{i+1} - b_j = 0, j \le i$, i.e. x_{i+1} is a special solution of the first i equations; moreover any particular solution of the first i equations can be obtained by a suitable choice of z_i

- the linear variety S_i containing all solutions of the first i equations is the set of vectors x of the following form, with $s \in R^n$ arbitrary

$$(9) \quad x = x_{i+1} + H_{i+1}^T s$$

- define A_i by $A_i = (a_1, \dots a_i)$ and L_i by

$$(10) \quad L_i = A_i^T P_i$$

then L_i is nonsingular lower triangular. Hence, if m=n and $L = L_n$, $P = P_n$, we get the factorization

$$(11) \quad A = LP^{-1}$$

There exist alternative formulations of the linear algebra associated with the ABS algorithm, which are useful in theoretical analysis and may also be preferable in terms of storage or computational cost. Here we give four such formulations, see Bodon (1989, 1990) for other formulations:

- the matrix H_{i+1} can be written in the form

$$(12) \quad H_{i+1} = H_1 - H_1 A_i Q_i W_i^T H_1$$

where $W_i = (w_1, \dots, w_i)$ and Q_i is the strongly nonsingular matrix defined by

$$(13) \quad Q_i = (W_i^T H_1 A_i)^{-1}$$

The above formulation has been used by Bodon (1988), Spedicato (1988) and Spedicato and Yang (1990) in the study of optimally conditioned ABS algorithms.

- the search vectors p_i can be determined by the relation

$$(14) \quad p_i = H_1^T z_i - \sum_{j<i} u_j s_j^T z_i$$

where $s_1, \dots s_{i-1}, u_1, \dots u_{i-1}$ are stored and updated by

$$(15) \quad s_i = H_1 a_i - \sum_{j<i} s_j u_j^T a_i$$

$$(16) \quad u_i = H_1^T w_i - \sum_{j<i} u_j s_j^T w_i$$

for $i>1$ and $s_1 = H_1 a_1$, $u_1 = H_1^T w_1$. With the above formulas there is saving in storage for $m<n/2$ and the number of multiplications is in general $3nm^2$ instead of $4n^2m$. They are therefore convenient for underdetermined systems with $m<<n$.

- let $z_1, \dots z_m$ be a sequence of vectors satisfying (4). Then $w_i = z_i / z_i^T H_i z_i$ is admissible and the search vector p_i satisfies the relation $p_i = u_i^i$, where one defines $u_j^1 = H_1^T z_j$ for $j = 1, \dots m$ and for $i = 1, \dots m-1$

$$(17) \quad u_j^{i+1} = u_j^i - (a_i^T u_j^i / a_i^T u_i^i) u_i^i \qquad (j = i+1, \dots m)$$

If $H_1 = I$, the storage requirement is mn and the number of multiplications is $3m^2n/2$ instead of $3mn^2$.

3. SPECIAL ALGORITHMS: THE HUANG, THE MODIFIED HUANG AND THE IMPLICIT LU/LL^T (IMPLICIT GAUSS/CHOLESKY) ALGORITHMS

The Huang algorithm, originally considered by Huang (1975), corresponds to the choices $H_1 = I$, $z_i = a_i$, $w_i = a_i / a_i^T H_i a_i$, giving $p_i = H_i a_i$ and $H_{i+1} = H_i - H_i a_i a_i^T H_i / a_i^T H_i a_i$. Now as $H_i^2 = H_i$, we can also write $H_{i+1} = H_i - p_i p_i^T / p_i^T p_i$, a form shown by Broyden (1989) to be preferable from the point of view of error propagation. The Huang algorithm has also the following properties:

- the matrix H_i is symmetric and is the orthogonal projector onto the orthogonal complement of Span $(a_1, \dots, a_{i-1})$

- the search vectors are orthogonal and coincide with the vectors obtained by the unnormalized Gram-Schmidt procedure applied on the rows of A
- if $x_1 = \mu a_1$, with μ an arbitrary scalar, then for i = 1, ... m, x_{i+1} is the vector of minimal Euclidean norm satisfying the first i equations and moreover it is true that $||x_i|| \le ||x_{i+1}||$ and $||x_{i+1} - x^+|| \le ||x_i - x^+||$.

The modified Huang algorithm, later referred to as MHA, differs from the Huang algorithm in the following. The search vector is obtained by explicitly premultiplying the Huang search vector by H_i, $p_i = H_i(H_i a_i)$. The stepsize is given by $\mu_i = \mu^*_i/s_i^T s_i$, $s_i = H_i a_i$. The update of H_i is $H_{i+1} = H_i - s_i p_i^T/s_i^T s_i$ or also $H_{i+1} = H_i - p_i p^T_i/p^T_i p_i$, see Spedicato and Vespucci (1989) for a numerical study of several variations in the definition of MHA. The computational experience and some theoretical analysis, for instance in Broyden (1989), indicate that MHA is numerically more stable. If MHA is formulated in terms of the recursions (14) - (16), the formulas defining the search vector become

$$(18)\quad p_i = s_i - \sum_{j<i} (s_j^T s_i/s_j^T s_j)p_j$$

$$(19)\quad s_i = a_i - \sum_{j<i} (p_j^T a_i/s_j^T s_j)s_j$$

Using the identity $s_i = p_i$, valid in exact arithmetic, the above formulas correspond to the doubly iterated unmodified Gram-Schmidt iteration, proposed by Daniel et al. (1976).

A natural question is if there exists an algorithm in the basic ABS class such that in the factorization (11) one has $P^{-1} = U$, with U upper triangular. It can be shown that the algorithm exists iff A is strongly nonsingular. It can be implemented by different parameter choices, the simplest being given by

$$(20)\quad H_1 = I,\ z_i = e_i,\ w_i = e_i/e_i^T H_i a_i$$

Under strong nonsingularity of A the scalar product $e_i^T H_i a_i = p_i^T a_i$ is nonzero and is equal to the i-th diagonal element in the matrix obtained at the i-th step of Gaussian elimination applied on A. If A is symmetric and positive definite, then $e_i^T H_i a_i$ is positive and $P^{-1} = L^T$ if z_i is given by $z_i = (e_i^T H_i a_i)^{\frac{1}{2}}$. If A is nonsingular, but not strongly nonsingular, the algorithm is well defined after suitable row or column pivoting.

The algorithm defined by the choices (20) is called the implicit LU/LL^T or the implicit Gauss/Cholesky algorithm. Some of its properties are the following:

- the search vectors are A-semiconjugate, i.e. $p_j^T A p_i = 0$ for $j<i$; if A is symmetric, then the search vectors are A-conjugate. They coincide with the vectors obtained by applying the Fox-Huskey-Wilkinson conjugation procedure to the unit vectors $e_1, \dots e_n$
- if A is symmetric and positive definite and $x_1 = 0$, then x_{i+1} minimizes the quadratic function $F(x) = (x - x^+)^T A(x - x+)$ over the linear variety $x_1 + \mathrm{Span}(p_1, \dots p_i)$, hence μ_i minimizes $f(\mu) = F(x_i - \mu p_i)$
- if A is symmetric and positive definite and A^*_i is the principal leading submatrix of order i, then the following relation is true

$$(21)\quad (A^*_i)^{-1} = \sum_{j\le i} p_j p_j^T / a_j^T p_j$$

- only the first i components of p_i can be nonzero and the matrix H_{i+1} has the following structure, with $S_i \varepsilon R^{n-i,i}$

$$(22)\quad H_{i+1} = \begin{bmatrix} 0_{i,i} & 0_{i,n-i} \\ S_i & I_{n-i} \end{bmatrix}$$

- if A is strongly nonsingular, the algorithm requires $n^3/3 + O(n^2)$ multiplications and the storage requirement is $n^2/4 + O(n)$, without overwriting.

In contrast with the classical Cholesky algorithm, symmetry does not allow reduction in the number of multiplications. However, if A is symmetric and positive definite, use of (21) allows the computation of the inverse with only $n^3/6 + 0(n^2)$ multiplications.

4. THE SCALED ABS ALGORITHM

We consider now an extension of the basic ABS algorithm which has an extra parameter at each iteration. The extension results in a class which contains all possible algorithms of a rather general finite termination type for solving linear equations. The class is derived as follows. Let $V = (v_1, \ldots v_m) \varepsilon R^{m,m}$ be an arbitrary nonsingular matrix and apply the basic ABS algorithm to the scaled system

(23) $V^T Ax = V^T b$

whose set of solutions is the same as that of the original system. Let $r = r(x) \varepsilon R^m$ be the residual vector, defined by

(24) $r(x) = Ax - b$

Then we obtain the following recursions: (A) and (F) unchanged and

(B) Test of dependency and compatibility.

Compute $r_i = r(x_i)$. If $r_i = 0$ stop, x_i solves the system. Otherwise compute $s_i = H_i A^T v_i$ and $\mu^*_i = v_i^T r_i$. If $s_i \neq 0$ proceed to (C). If $s_i = 0$ and $\mu^*_i = 0$, set $x_{i+1} = x_i$, $H_{i+1} = H_i$ and proceed to (F). If $s_i = 0$ but $\mu^*_i \neq 0$ stop, the system is not compatible.

(C) Computation of the search vector.

Compute the search vector by (3) with z_i arbitrary save that

(25) $z_i^T H_i A^T v_i \neq 0$

(D) Update of the approximation x_i to the solution.

Compute the new approximation x_{i+1} to the solution by (6) with μ_i given by

(26) $\mu_i = \mu^*_i / v_i^T A p_i$

Stop if $i = m$, x_{m+1} solves the system.

(E) Update of H_i

Compute the new matrix H_{i+1} by

(27) $H_{i+1} = H_i - H_i A^T v_i w_i^T H_i$

where w_i is arbitrary save that

(28) $w_i^T H_i A^T v_i = 1$

The above recursions are formally obtained by substituting in the basic ABS algorithm a_i with $A^T v_i$ and $b^T e_i$ with $b^T r_i$. It must be noted that at the i-th iteration only the i-th column of V is used. Hence it is not necessary to define V in advance and v_i can be considered as an extra parameter available at the i-th iteration, arbitrary save that $V_i = (v_1, \dots v_i)$ must have full column rank. The vector v_i is called the scaling parameter.

We give some properties of the scaled ABS algorithm, many of them being an immediate reformulation of the properties valid for the basic ABS algorithm. It is assumed as usual that A has full row rank.

- the scaled ABS algorithm is a realization of the finitely terminating algorithm of Broyden (1985), implying that any algorithm of the form $x_{i+1} = x_i + \mu_i p_i$, which solves a linear system from an arbitrary starting point in a number of steps no greater than the number of equations, can be obtained by some parameter choices in the scaled ABS algorithm

- if $L_i = V_i^T A P_i$, then L_i is nonsingular lower triangular, hence, for m = n, one has the factorization

(29) $A = V^{-T} L P^{-1}$

- the residual $r_{i+1} = r(x_{i+1})$ is orthogonal to the first i scaling vectors, i.e.

(30) $V_i^T r_{i+1} = 0$

- the set of vectors x such that r(x) is orthogonal to the columns of V_i is the set of the vectors having the form (9)

In the given formulation the basic and the scaled ABS algorithms deal with one equation at each iteration. It is possible to modify these algorithms to deal with p blocks of m_i equations, $m_1 + \ldots + m_p = m$. The relating recursions are the following, with $V = (V_1, \ldots V_p)$ arbitrary nonsingular:

(31) $P_i = H_i^T Z_i$

where $Z_i \in R^{n,m_i}$ is such that $V_i^T A H_i^T Z_i$ is nonsingular;

(32) $x_{i+1} = x_i - P_i d_i$

where $d_i \in R^{m_i}$ is the unique solution of the system

(33) $V_i^T A P_i d_i = V_i^T r_i$

and finally

(34) $H_{i+1} = H_i - H_i A^T V_i W_i^T H_i$

where $W_i^T \in R^{n,m_i}$ is any left inverse of $H_i A^T V_i$.

It can be shown that the pairs $(V_1, \dots V_i)$, $(P_1, \dots P_i)$ in the above recursions are A-semiconjugate, i.e. that $V_j^T AP_i = 0$ for j<i and that $V_i^T AP_i$ is nonsingular. Moreover all possible such block A-semiconjugate pairs can be generated by the scaled block ABS algorithm.

The scaled block ABS algorithm is not more general than the standard scaled ABS algorithm in the sense that all possible sequences $x_1, \dots x_{p+1}$ that it can generate are contained in the set of sequences generated by the standard scaled ABS algorithm. It may however be more suitable in terms of storage or complexity on certain problems or on vector/parallel computers, see for instance Bertocchi and Spedicato (1990), Spedicato, Xia and Gao (1990) or Xia and Gao (1990). In the nonlinear ABS algorithm the block formulation provides however an essential generalization (the method of Newton is then the algorithm corresponding to a single block of n equations).

5. SUBCLASSES OF THE SCALED ABS ALGORITHM

We consider now some subclasses of the scaled ABS algorithm defined by some choices of the scaling parameter. In the first three subclasses the matrix L_i is diagonal and the scaling vector has the form

(35) $v_i = A^{-T} Y p_i$

where $Y \in R^{n,n}$ is symmetric and positive definite. For the above choice of v_i the following properties are true:

- the search vectors are Y-conjugate
- if x_1 is an arbitrary multiple of p_1, then x_{i+1} minimizes the quadratic function $F(x) = (x - x^+)^T Y(x - x^+)$ over the linear variety $x_1 + \text{Span}(p_1, \dots p_i)$

For the first subclass, named the conjugate directions subclass, we assume m = n and A symmetric and positive definite. The scaling vector is defined as $v_i = p_i$. From (29) it follows that $L = D$ is diagonal and that the search vectors are A-conjugate. Three particular algorithms in this subclass are the following:

- the Hestenes-Stiefel algorithm, corresponding to $H_1 = I$, $z_i = r_i$, $w_i = z_i/p_i^T A p_i$
- the Lanczos algorithm, corresponding to $H_1 = I$, $z_1 = r_1$ and, for $i>1$, $z_i = Ap_{i-1}$, $w_i = z_i/p_i^T A p_i$
- an algorithm equivalent to the implicit LU/LL^T algorithm, in the sense of generating the same iterates p_i and x_i, corresponding to $H_1 = I$, $z_i = e_i$ and $w_i = e_i/p_i^T A p_i$.

In the second subclass, named the orthogonally scaled subclass, we assume $m = n$ and A nonsingular. The scaling vector is defined by $v_i = Ap_i$. From (29) we obtain $V^T AP = P^T A^T AP = V^T V = D$, hence the scaling vectors are orthogonal and the search vectors are $A^T A$-conjugate. The minimization property associated with the choice (35) implies that x_{i+1} minimizes the squared Euclidean norm of the residual. Two particular algorithms in this subclass are the following:

- an algorithm, called the implicit QR algorithm, corresponding to the choices $H_1 = I$, $z_i = e_i$, $w_i = e_i/v_i^T v_i$. This algorithm is well defined since it corresponds to applying the implicit LU/LL^T algorithm to the scaled system $P^T A^T Ax = P^T A^T b$, where P is upper triangular and $P^T A^T A$ is strongly nonsingular. The associated factorization has the form $A = V^{-T} DP^{-1}$, which is of the QR type after suitable normalization of v_i or p_i
- an algorithm of the conjugate gradient type (i.e. where p_i can be written as a linear combination of the residual and of another vector), called the implicit minimum residual conjugate gradient method, corresponding to the choices $H_1 = I$, $z_i = A^T r_i$, $w_i = z_i/v_i^T v_i$. This algorithm can be obtained

by applying the Hestenes-Stiefel method on the normal system $A^TAx = A^Tb$.

In the third subclass, named the optimally stable subclass, we again assume that $m = n$ and that A is nonsingular in the initial definition of the algorithm. These restrictions can however be dropped once the algorithm is defined. The scaling vector is defined by $v_i = A^{-T}p_i$. From (29) we obtain $V^TAP = P^TP = D$, hence the search vectors are orthogonal. Moreover, since $p_i = A^Tv_i$, we obtain

(36) $V^TAA^TV = D$

which is the condition found by Broyden (1985) to characterize the subclass with optimal stability with respect to a single error in the sequence x_i. More precisely, Broyden's result is the following: if x_i is perturbed by an error dx_i and no additional errors are made, then the error $d_n = x_{n+1} - x^+$ in the computed solution satisfies the inequality

(37) $||d_n|| \le \text{Cond}(V^TA)\ ||dx_i||$

The above bound is minimized iff V satisfies (36), in which case it can be shown that $||d_n|| \le ||dx_i||$, hence no error amplification occurs. A similar analysis applied on the residual norm gives a bound which is minimized iff

(38) $V^TV = D$

in which case one can show that $||r_{n+1}|| \le ||dx_i||$, hence again there is no error amplification. Condition (38) is satisfied by all algorithms in the unscaled ABS class and in the orthogonally scaled subclass.

The inverse A^{-T} appears in the definition of v_i in the optimally stable subclass. This fact is not computationally acceptable, but it can be shown that in the actual formulation of the algorithm the explicit presence of the inverse can be eliminated. Indeed it is seen immediately that the update of H_i

has the form $H_{i+1} = H_i - H_i p_i w_i^T H_i$, where the inverse does not appear. We can also set without loss of generality $H_1 = I$, $w_i = z_i/z_i^T H_i p_i = z_i/p_i^T p_i$, since these choices keep symmetry in H_i. The update takes then the form $H_{i+1} = H_i - p_i p_i^T/p_i^T p_i$, which is the same as in the Huang algorithm, but now represents a class of symmetric updates, with z_i the parameter. The definition (26) of the stepsize gives $\mu_i = r_i^T A^{-T} p_i/p_i^T p_i$. It can be shown that the presence of the inverse in the formula for μ_i can be removed if, without loss of generality, we set $z_i = A^T q_i$, $q_i \in R^n$ an arbitrary nonzero vector, in which case the stepsize formula becomes $\mu_i = r_i^T q_i/p_i^T p_i$.

It can be shown that the optimally stable subclass also coincides with the subclass generating symmetric matrices, with $H_1 = I$, where p_i is given by $p_i = H_i A^T q_i$. This subclass intersects the basic ABS class, having in common with it the Huang algorithm, corresponding to $q_i = e_i$. By setting $q_i = r_i$ an algorithm is obtained of the conjugate gradient type, which is equivalent to the method of Craig (1955) for nonsymmetric systems.

The reprojection technique (premultiplication of p_i by H_i) used in the definition of MHA can be applied to stabilize all algorithms in the optimally stable subclass. Experiments of Bodon (1991) show for instance that the reprojected ABS formulation of the Craig method can accurately solve severely ill-conditioned problems. Notice finally that all algorithms in this subclass monotonically reduce at each iteration the Euclidean distance from the solution.

Other subclasses of the scaled ABS algorithms, which are considered in the monograph of Abaffy and Spedicato (1989), are not discussed here.

6. APPLICATION TO LINEAR LEAST SQUARES

If m>n there are several ABS approaches for solving the overdetermined system Ax = b in the least squares sense without explicitly forming the normal equations. One approach is based upon the equivalence of the normal equations with the following extended system in the unknowns x and $y \in R^m$

(39) $Ax = y$

(40) $A^T y = A^T b$

In order that (39) be solvable, y must belong to the range of A, which implies that it must be the unique minimal Euclidean norm solution of the underdetermined system (40). This solution can be found by using the Huang algorithm with $y_1 = 0$. If A has full column rank, any basic ABS algorithm can be used to solve (39), the m-n dependent equations being recognized and removed at the step (B). If A is rank deficient, the minimal Euclidean norm solution can be obtained by solving (39) again with the Huang or MHA with $x_1 = 0$.

A second approach is based upon a subclass of ABS algorithms which can be directly applied on the overdetermined system. Suppose that A has full column rank and define the scaling vector by $v_i = Au_i$, where $U = (u_1, \ldots u_n) \in R^{n,n}$ is nonsingular. Then, after n iterations, it follows from (30) that $UA^T r_{n+1} = 0$ or $A^T Ax_{n+1} = A^T b$, implying that x_{n+1} is the unique least squares solution. If we take $u_i = p_i$, we get the orthogon8

ally scaled subclass, implying that the implicit QR algorithm and the implicit minimum residual conjugate gradient method compute the least squares solution in the overdetermined case. It can be shown moreover that the choice $v_i = Au_i$ is well defined even if A is rank deficient. In such a case the minimal Euclidean norm solution is obtained by the algorithm corresponding to $H_1 = I$, $z_i = A^T Au_i$, $H_{i+1} = H_i - p_i p_i^T / p_i^T p_i$, with stepsize $\mu_i = r_i^T A_i / p_i^T p_i$. Since u_i is a parameter, this is actually a class of algorithms which coincides with the optimally stable subclass if in that subclass q_i is restricted to be an arbitrary vector in the range of A.

A third approach consists in explicitly computing the LQ or QR factorization associated with the Huang, the MHA or the implicit QR algorithm and then using the factorization in the usual way.

Finally one can compute the Moore-Penrose pseudoinverse by two approaches using properties of the Huang algorithm. In the first approach, which is equivalent to a classic method of Greville, see Spedicato and Xia (1990b), the pseudoinverse of $A^+ = A^+_i$, $i \le n$, is computed by updating recursively A^+_j, $j = 1, \dots i-1$. In the second approach the pseudoinverse is computed by solving for the minimal Euclidean norm solution a certain underdetermined linear system, see Spedicato and Bodon (1989). Extensive numerical experiments have shown that several of the considered methods are competitive with LINPACK or NAG codes in terms of accuracy on ill conditioned or rank deficient problems.

6. REFERENCES

ABAFFY J. (1979) A Lineàris Egyenletrendszerek Altalànos Megoldàsànak Egy Mòdszerosztàlya, Alkalmaz. Mat. Lapok 5, 223-240.

ABAFFY J. (1990) Superlinear Convergence Theorems for Newton Type Methods for Nonlinear Systems of Equations, Report DMSIA 90/10, University of Bergamo

ABAFFY J., BROYDEN C.G. and SPEDICATO E. (1984) A Class of Direct Methods for Linear Equations, Numer. Math. 45, 361-376

ABAFFY J. and SPEDICATO E. (1989) ABS Projection Algorithms, Mathematical Techniques for Linear and Nonlinear Equations, Ellis Horwood, Chichester

ADACHI N. (1971) On Variable Metric Algorithms, J.Optim.Theory Appl. 7, 391-410

BERTOCCHI M. and SPEDICATO E. (1990) Vectorizing the Implicit Gauss/Cholesky Algorithm of the ABS Class on the IBM 3090 VF, Technical Report 1/37, PF Sistemi Informatici e Calcolo Parallelo, University of Bergamo

BROYDEN C.G. (1985) On the Numerical Stability of Huang's Update and Related Methods, J. Optim. Theory Appl. 47, 401-412

BROYDEN C.G. (1989) On the Numerical Stability of Huang's Update, Report DMSIA 89/19, University of Bergamo

BODON E. (1988) Globally Optimally Conditioned Updates in the ABS Class for Linear Systems, Preprint, University of Bergamo

BODON E. (1989) Biconjugate Algorithms in the ABS Class I: Alternative Formulation and Theoretical Properties, Report DMSIA 89/3, University of Bergamo

BODON E. (1990) Derivation and Numerical Performance of Factorized ABS Algorithms for Solving Linear Equations, Preprint, University of Bergamo

BODON E. (1991) On some Conjugate Gradient Type ABS Algorithms for Nonsymmetric Linear Systems of Equations, Preprint, University of Bergamo

CRAIG E.J. (1955) The N-Step Iterations Procedures, J. Math. Phys. 34, 64-73
DANIEL J., GRAGG W.B., KAUFMAN L. and STEWART G.W. (1976) Reorthogonalization and Stable Algorithms for Updating the Gram-Schmidt QR Factorization, Math. Comput. 30, 772-795
DENG N. and CHEN Z. (1990) Truncated Nonlinear ABS Algorithm and its Convergence Property, Computing 45, 169-173
GALANTAI A. (1990) A Global Convergence Theorem for Nonlinear ABS Methods, Report DMSIA 90/9, University of Bergamo
HUANG H.Y. (1975) A Direct Method for the General Solution of a System of Linear Equations, J. Optim. Theory Appl. 16, 429-445
HUANG Z. (1990) On the Convergence Analysis of the Nonlinear Block Scaled ABS Methods, Report DMSIA 90/14, University of Bergamo
SPEDICATO E. (1988) Optimal Conditioning Parameter Selection in the ABS Class for Linear Algebraic Systems, Mat. Applic. Computac. 3, 187-199
SPEDICATO E. and BODON E. (1989) Solving Linear Least Squares by Orthogonal Factorization and Pseudoinverse Computation via the MHA in the ABS Class, Computing 42, 195-205
SPEDICATO E. and VESPUCCI M.T. (1989) Computational Performance of the Implicit Gram-Schmidt (Huang) Algorithm for Linear Algebraic Systems, Report DMSIA 89/6, University of Bergamo
SPEDICATO E. and XIA Z. (1990a) Application of ABS Algorithms to Constrained Optimization I: Linear Equality Constraints, Report DMSIA 90/17, University of Bergamo
SPEDICATO E. and XIA Z. (1990b) On some ABS Methods for the Computation of the Pseudoinverse, Report DMSIA 90/24, University of Bergamo
SPEDICATO E. and YANG Z. (1990) Optimally Conditioned Scaled ABS Algorithms for Linear Systems, J. Optim. Theory Appl. 67, 141-150
SPEDICATO E., XIA Z. and GAO Z. (1990) The Basic Block ABS Method I: Formulation and Theoretical Properties, Report DMSIA 90/15, University of Bergamo
XIA Z. (1990) Application of ABS Algorithms to Constrained Optimization II: Linear Inequality Constraints, Report DMSIA 90/29, University of Bergamo
XIA Z. (1991a) The ABS Class and Quadratic Programming I: Equality Constraints, Report DMSIA 91/1, University of Bergamo
XIA Z. (1991b) The ABS Class and Quadratic Programming II: General Constraints, Report DMSIA 91/5, University of Bergamo
XIA Z. and GAO Z. (1990) The Basic Block ABS Method II: Application to the Rutheford Matrix Equation, Report DMSIA 90/3, University of Bergamo
ZHU M. (1990) Finding General Solution to Perturbed Linear Systems and its Application in Optimization, Report DMSIA 90/5, University of Bergamo

ABS Algorithms for Sparse Linear Systems

Jozsef Abaffy
University of Economics
Dimitrov ter 8
1093 Budapest IX
Hungary

Abstract

We consider the application of ABS algorithms to linear systems with sparse coefficient matrices. The implicit LL^T algorithm is applied to matrices having the Nested Dissection form. The implicit LU and QR algorithms and the Huang method are applied to various types of band type matrices. In several cases the ABS algorithms are advantageous in terms of storage with respect to methods using standard factorizations.

1. Introduction

In this paper the application of the implicit LU, QR and the symmetric (Huang) algorithms to some types of sparse and possibly very large linear system is considered. The problem to solve is the following:

$$(1.1) \qquad Ax = b, \qquad \text{where } A \in R^{n,n}, \quad x,b \in R^n$$

(the (i,j)-th element of A in the sequel will be denoted by $a_{i,j}$).

The sparse matrices that are considered are the following

a) Positive definite matrices having m-block nested dissection form (later referred to as ND matrices), which arise in particular when the ND ordering technique of George (1973) is applied to matrices obtained from regular finite element meshes.
b) Standard q-banded matrices, upper q-banded matrices and modified q/p banded matrices or upper q/p-banded matrices. We consider the regular q/p-cyclic matrices too.

With regard to ND matrices we show that the implicit LU algorithm generates matrices where the nonzero elements are found only in positions corresponding to the nonzero blocks of

NATO ASI Series, Vol. F 77
Computer Algorithms for Solving Linear Algebraic Equations
Edited by E. Spedicato

the ND matrix A. From this fact it follows that the memory requirement needed by the implicit LU algorithm is less than that required by the Cholesky factorization.

With regard the q-banded matrices we show that only O(q) vectors are needed to represent tha matrix H_i when the implicit LU, LQ or QR algorithms are applied. For this cases we give some preliminary numerical results too.

The paper summarizes results in Abaffy [1,2], Abaffy and Dixon [3], Deng [5] and Zhu [8] without proofs of the theorems and lemmas.

2. Application of the implicit LL^T algorithm to ND matrices

In this section first we consider the Nested Dissection form matrices. Then we give the implicit LL^T algorithm of the ABS class. Finally we examine the structure of the projection matrix H_i when the coefficient matrix A is sparse and generated by the ND method, see George [6] and George and Liu [7].

2.1 The implicit LL^T algorithm

We begin the definitions of the Nested Dissection matrices.

Definition 2.1
Let A be an n by n symmetric positive definite matrix of the form

$$(2.1) \qquad A = \begin{bmatrix} A_1 & G & D \\ G^T & B_1 & E \\ D^T & E^T & F \end{bmatrix}$$

Then A is called a 1-block ND matrix if G=0 and both A_1 and B_1 are square submatrices. A_1 and B_1 are called square blocks and C_1 is called a rectangular block, where C_1 is the submatrix given by

$$C_1 = (D^T, E^T, F).$$

Definition 2.2
Let A be an n by n symmetric positive definite matrix of the form (2.3). Then A is called an m-block ND matrix if G=0 and both A_1 and B_1 are (m-1)-block ND matrices.

Now we give the implicit LL^T algorithm in the ABS class. For the whole class see for example Abaffy and Spedicato [4].

Algorithm 2.1

Step 1. Select $x_1 \varepsilon R^n$ arbitrary, set $H_1=I$, where I is the n by n identity matrix and let i=1, iflag=0.

Step 2. Compute the vector $s_i=H_ia_i$.
If $s_i\neq 0$ proceed to Step 3.
If $s_i=0$ and $a_i^Tx_i-b^Te_i=0$, where a_i^T is the i-th row of A and e_i is the i-th unit vector set $x_{i+1}=x_i$, $H_{i+1}=H_i$, iflag=iflag+1 and proceed to Step 6 (a_i is linearly dependent on the previous rows of A).
If $s_i=0$ and $a_i^Tx_i-b^Te_i\neq 0$, then set iflag=-1 and stop, the system is incompatible.

Step 3. Compute the search vector p_i by

$$p_i = H_i^Te_i/|e_i^TH_ia_i|^{\frac{1}{2}} \tag{2.2}$$

Step 4. Update the approximation of the solution by

$$x_{i+1} = x_i - \alpha_ip_i, \tag{2.3}$$

where

$$\alpha_i = (a_i^Tx_i - b^Te_i)/a_i^Tp_i.$$

Step 5. If i=n, stop, x_{n+1} solves the system (1.1), otherwise update the matrix H_i by

$$H_{i+1} = H_i - H_ia_ie_i^TH_i/e_i^TH_ia_i. \tag{2.4}$$

Step 6. Increment the index i by one and go to Step 2.

2.2 Structure of the projection matrix H_i

For the matrix H_{i+1} the following theorem can be stated which does not depend on the structure of the coefficient matrix A.

Theorem 2.1

If the matrix A is strongly nonsingular (that is all principal leading submatrices are nonsingular) then the matrix H_{i+1} has the following properties.

a) The first i rows of H_{i+1} are identically zero.
b) The last n-i columns of H_{i+1} are equal to the last n-i columns of $H_1=I$.
Hence H_{i+1} has the following form

$$(2.5) \qquad H_{i+1} = \begin{bmatrix} 0 & 0 \\ S_i & I_i \end{bmatrix}$$

Suppose now that the sparse linear system (1.1) is generated by the ND method. Let us suppose that the coefficient matrix A has the form as follows:

$$(2.6) \qquad A = \left[\begin{array}{cc|c} A_1 & 0 & \\ \hline 0 & A_2 & A_4^T \\ \hline \multicolumn{2}{c|}{A_4} & A_3 \end{array}\right]$$

where A_1, A_2 and A_3 are square submatrices with order p, q and r respectively, and p+q+r=n. Since the structure of the projection matrix H_i depends on the iteration i, the nonzero structure of H_i is shown in the following three lemmas for $i \varepsilon [2,\dots,p+1]$, $i \varepsilon [p+2,\dots,p+q+1]$ and $i \varepsilon [p+q+2,\dots,n+1]$ separately.

<u>Lemma 2.1</u>
Suppose that the coefficient matrix of the problem (1.1) has the form (2.6). Then for i=2,...,p+1 the matrices H_i have the following structure:

$$(2.7) \qquad H_i = \left[\begin{array}{cc|c|c} 0 & 0 & 0 & 0 \\ \hline D_i & I_{p-i+1} & & \\ \hline \multicolumn{2}{c|}{0} & I_q & 0 \\ \hline C_i & & 0 & I_r \end{array}\right] \qquad i=2,\dots,p+1$$

where the first i-1 rows are zero and I_k is the k by k unit matrix when k=p-i+1, q and r respectively. D_i and C_i are p-i+1 by i-1 and r by i-1 matricés respectively. Moreover D_i and C_i satisfy the recurrence relations:

$$D_2 = -1/a_{1,1}\begin{bmatrix} a_{1,2} \\ \cdot \\ \cdot \\ a_{1,p} \end{bmatrix} = \begin{bmatrix} d_2 \\ \bar{D}_2 \end{bmatrix}$$

$$C_2 = -1/a_{1,1}\begin{bmatrix} a_{1,p+q+1} \\ \cdot \\ \cdot \\ a_{1,n} \end{bmatrix}$$

and for $i=2,\ldots,p$

$$D_{i+1} = [\bar{D}_i - u_i d_i^T,\ -u_i],$$

$$C_{i+1} = [C_i - v_i d_i^T,\ -v_i],$$

where d_i^T and $\bar{D}_i$ respectively contains the first row and the last $p-i$ rows of D_i and

$$u_i = -1/\mu_i\{\bar{D}_i\begin{bmatrix} a_{i,1} \\ \cdot \\ \cdot \\ a_{i,i-1} \end{bmatrix} + \begin{bmatrix} a_{i,i+1} \\ \cdot \\ \cdot \\ a_{i,p} \end{bmatrix}\}$$

$$v_i = -1/\mu_i\{C_i\begin{bmatrix} a_{i,1} \\ \cdot \\ \cdot \\ a_{i,i-1} \end{bmatrix} + \begin{bmatrix} a_{i,p+q+1} \\ \cdot \\ \cdot \\ a_{i,n} \end{bmatrix}\}$$

where

$$\mu_i = [d_i^T, 1]\begin{bmatrix} a_{i,1} \\ \cdot \\ \cdot \\ a_{i,i} \end{bmatrix}$$

<u>Lemma 2.2</u>
Suppose that the coefficient matrix of the problem (1.1) has the form (2.6). Then for $i=p+2,\ldots,p+q+1$ the matrices H_i have the following structure:

$$H_i = \left[\begin{array}{c|c|c|c} 0 & \multicolumn{2}{c|}{0} & 0 \\ \hline & 0 & 0 & \\ \cline{2-3} 0 & D_i & I_{p+q-i+1} & 0 \\ \hline C_i & F_i & 0 & I_r \end{array}\right]$$

where the first i-1 rows are zero, and the I_k is the k by k unit matrix when k=p+q-i+1 and r respectively. D_i is p+q-i+1 by i-1-p, $C_i=C_{p+1}$ for i=p+2,...,p+q+1, and F_i is an r by i-1-p matrix. Moreover D_i and C_i satisfy the recurrence relations:

$$D_{p+2} = -1/a_{p+1,p+1}\begin{bmatrix} a_{p+1,p+2} \\ \vdots \\ a_{p+1,p+q} \end{bmatrix} = \begin{bmatrix} d_{p+2} \\ \bar{D}_{p+2} \end{bmatrix}$$

$$F_{p+2} = -1/a_{p+1,p+1}\begin{bmatrix} a_{p+1,p+q+1} \\ \vdots \\ a_{p+1,n} \end{bmatrix}$$

and for i=p+2,...,p+q

$$D_{i+1} = [\bar{D}_i-u_i d_i^T, \ -u_i],$$

$$F_{i+1} = [F_i \ -v_i d_i^T, \ -v_i],$$

where d_i^T and $\bar{D}_i$ respectively contain the first row and the last p+q-i rows of D_i and

$$u_i = -1/\mu_i\{\bar{D}_i\begin{bmatrix} a_{i,p+1} \\ \vdots \\ a_{i,i-1} \end{bmatrix} + \begin{bmatrix} a_{i,i+1} \\ \vdots \\ a_{i,p+q} \end{bmatrix}\}$$

$$v_i = -1/\mu_i\{F_i\begin{bmatrix} a_{i,p+1} \\ \vdots \\ a_{i,i-1} \end{bmatrix} + \begin{bmatrix} a_{i,p+q+1} \\ \vdots \\ a_{i,n} \end{bmatrix}\}$$

where

$$\mu_i = [d_i^T, 1] \begin{bmatrix} a_{i,p+1} \\ \vdots \\ a_{i,i} \end{bmatrix}$$

Lemma 2.3
Suppose that the coefficient matrix of the problem (1.1) has the form (2.6). Then for $i=p+q+2,\ldots,n+1$ the matrices H_i have the following structure:

$$H_i = \left[\begin{array}{c|c} 0 & 0 \\ \hline G_i & I_{n-i+1} \end{array}\right]$$

where the first $i-1$ rows are zero and I_k is the k by k unit matrix when $k=n-i+1$. G_i is an $n-i+1$ by $i-1$ matrix that satisfies the recurrence relations:

$$G_{p+q+1} = [C_{p+q+1}, F_{p+q+1}] = \begin{bmatrix} g_{p+q+1} \\ \bar{G}_{p+q+1} \end{bmatrix}$$

where G_{p+q+1} is an r by $p+q$ matrix and for $i=p+q+1,\ldots,n$

$$G_{i+1} = [\bar{G}_i - v_i g_i^T, \ -v_i],$$

is an $n-i$ by i matrix, and g_i and $\bar{G}_i$ respectively contain the first row and the last rows of G_i where

$$v_i = -1/\mu_i \left\{ \bar{G}_i \begin{bmatrix} a_{i,1} \\ \vdots \\ a_{i,i-1} \end{bmatrix} + \begin{bmatrix} a_{i,i+1} \\ \vdots \\ a_{i,n} \end{bmatrix} \right\}$$

and

$$\mu_i = [g_i^T, 1] \begin{bmatrix} a_{i,1} \\ \vdots \\ a_{i,i} \end{bmatrix}.$$

The following theorem is a direct consequence of the three Lemmas.

Theorem 2.5
Suppose that Algorithm 2.1 is used to solve the sparse linear system

$$Ax = b$$

where A is an m-block ND matrix. Then the projection matrices H_i have the following structure

$$(2.8) \qquad H_i = \left[\begin{array}{c|c} 0 & 0 \\ \hline R_i & I_{n-i+1} \end{array}\right]$$

where I_{n-i+1} is an identity matrix, R_i is an n-i+1 by i-1 matrix and moreover the nonzero elements of R_i only appear in the parts corresponding to nonzero square and rectangular blocks of A.

The following figure shows the matrix A obtained by the ND method and the part R_i of H_i together.

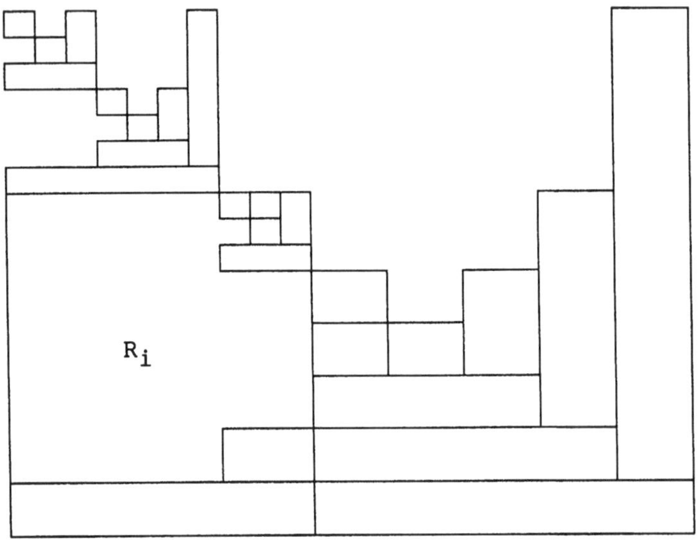

i-1 columns

Finally we note that the algorithm for the ND method can be found in the book of George and Liu [7], while a flowchart for the application of the ND method to the implicit LL^T method can be found in Zhu [8].

Note that R_i can be stored in the first rows of A because we do not use more the first i-1 rows of A and the structure of R_i is the same as the suitable structure of A.

3. The implicit LL^T, LU algorithm for q band matrices

In this section we use the implicit LL^T and LU algorithms for band matrices and for some generalization.

Further on we introduce the following notation. Let h_i denote the vector $H_i^T e_i$.

We begin with some definitions.

Definition 3.1

An n by n matrix A is called a q-band matrix if $a_{i,j}=0$ for $|i-j| > q$ while $a_{i,j} \neq 0$ for at least one pair i,j such that $|i-j| = q$, $1 \leq i,j \leq n$.

Definition 3.2

An n by n matrix is called an upper q-band matrix if $a_{i,j}=0$ for $j>i+q$, while $a_{i,i+q} \neq 0$ for at least one i, $1 \leq i \leq n$.

Definition 3.3

An n by n matrix A is called an upper q/p-band matrix if $a_{i,j}=0$ for $n-p>j>i+q$, while $a_{i,i+q} \neq 0$ for at least one i, $1 \leq i \leq n-p$ and $a_{i,j} \neq 0$ for at least one i, $1 \leq i \leq n-q$ and $j=n-p$.

The following figure shows an upper q/p-band matrix when q=1 and p=2.

$$A = \begin{bmatrix} * & * & & & & & * & * \\ * & * & * & & & & * & * \\ & * & * & * & & & * & * \\ & & * & * & * & & * & * \\ & & & * & * & * & * & * \\ & & & & * & * & * & * \\ & & & & & * & * & * \\ & & & & & & * & * \end{bmatrix}$$

Figure 3.1

Definition 3.4

A q band matrix is said to be t-cyclic if there are indexes j_k, $0 \leq k \leq t \geq 1$, $0 \leq j_1 \leq \ldots \leq j_t = q$, such that $a_{i,i+j}$ and $a_{i,i-j}$ are zero for $0 \leq j \leq q$, unless j equals one of the j_k.

The following figure shows a t cyclic q-band matrix when $j_1=1$, $j_2=4$ that is q=4.

$$A = \begin{bmatrix} * & * & & * & & & \\ * & * & * & & * & & \\ & * & * & * & & * & \\ * & & * & * & * & & * \\ & * & & * & * & * & \\ & & * & & * & * & * \\ & & & * & & * & * \end{bmatrix}$$

Figure 3.2

Using the notations of (2.8) we can prove the following theorem.

<u>Theorem 3.1</u>
Let A be an upper q-band matrix. Then the nonzero elements of R_i in (2.8) can only lie in the first q rows of R_i.

Observe that, when the matrix A is a lower triangular matrix then $q=0$, $p_i=e_i$ and the projection matrix is eliminated from the algorithm.

We may extend the result of Theorem 3.1 still further by noting that if in the i-th step the k-th row of C_i, $i+1 \le k \le i+q-1$ is still zero and $a_{i,k}=0$, then $(H_i a_i)_k=0$ and so the k-th row of C_{i+1} is also zero. It follows that the k-th row of C_{i+1} is only nonzero if $a_{j,k} \ne 0$ for some $j \le i$ and $j<k$.

Reversing the above argument, if $a_{j,k} \ne 0$ for some $j \le i$ and $j<k$, then row k of C_{i+1} will be nonzero.

Combining these results we may state

<u>Theorem 3.2</u>
When using the implicit LL^T method every element $a_{k,j} \ne 0$ $j>k$ implies that that corresponding part of C_{i-1} in row j of H_i is nonzero if $j>i>k$.

In particular, if A is an upper q/p band matrix

q p

$$A = \begin{bmatrix} * & * & & & & & * & * \\ * & * & * & & & & * & * \\ * & * & * & * & & & * & * \\ * & * & * & * & * & & * & * \\ * & * & * & * & * & * & * & * \\ * & * & * & * & * & * & * & * \\ * & * & * & * & * & * & * & * \\ * & * & * & * & * & * & * & * \end{bmatrix}$$

Figure 3.3

then the number of vectors needed to be stored is q+p (the additional p rows being the last rows of H_i).

As the lower triangular part of the coefficient matrix A has no influence in the structure of the projection matrix H_i, observe that the Theorem 3.2 includes the case of the ND type of matrices. Theorem 3.2 is slightly more general in the sense that it does not require the symmetry of the coefficient matrix, but the nonsingularity of the principal leading submatrices.

Let us now consider the arithmetic operations involved at each iteration. We have two cases to consider:

- upper q band matrix
- q band matrix.

Upper q band case

In the core operation the formation of $H_i a_i$ only involves qi operations. Forming the dyad requires also qi multiplications and then we require q divisions, assuming $H_i a_i$ is divided before the dyad is formed. The number of operations is 2iq+q if $n-q \geq i \geq q$.
Similarly for the outer rows we obtain $2i^2+i$ if $i<q$ and $2(n-i)i+(n-i)$ if $i>n-q$.
The total number of operations is therefore of the order

$$(q-1)q(2q-1)/3 + (q(n-q(n-q+1)-q^2(q+1)) + 2n^2q-2n^2(n+1)+2n(n-q)(n-q+1)+(n(n+1)(2n+1)/3 - (n-q)(n-q+1)(2n-2q+1)/3$$

which is dominated by $n^2q-2nq^2+4q^3/3 \approx n^2q$.

Q-band case

In the q-band matrix case, fewer multiplications are required since we have $iq+q^2+q$ if $n-q \geq i \geq q$, $2i^2+i$ if $i<q$ and $(n-i)q+(n-i)i+(n-i)$ if $i>n-q$. These operations are dominated by the term $\frac{1}{2}n^2q$.

The analysis in this section also applies to a number of related members of the ABS family, e.g. the implicit LU algorithm.

It was observed that the reprojection of H_i in some cases gives better numerical results. In this respect we note that if in the numerator we write

$$H_i H_i a_i h_i^T$$

then no additional multiplications are required in general for any matrix. If in the numerator we write

$$H_i a_i h_i^T H_i$$

this involves i*q extra multiplications for the upper q/p band and (i-q)q for the q-band case.

Similarly in the denominator if we write

$$(h_i^T) H_i H_i a_i$$

this involves no additional arithmetic, whilst

$$(h_i^T H_i)(H_i a_i)$$

involves the same additional arithmetic as above.

Let us turn now to the t cyclic band matrix case. The simplest case is the tridiagonal band matrix, say a q-band tridiagonal matrix, with $a_{i,j}=0$ unless
(1) j=i
(2) j=i-q
(3) j=i+q.

<u>Theorem 3.3</u>
If A is a q band tridiagonal matrix, then C_{i-1} in (3.1) contains at most q rows and row m can only contain nonzeros in column m-kq, where k is integer k≤m/q.

If we take the case q=3, the matrices C_{i-1} are build as follows

C_2	C_3	C_4	C_5	C_6
0	00	*00	0*00	00*00
0	*0	0*0	00*0	*00*0
*	0*	00*	*00*	0*00*

which illustrates clearly the way additional waves enter C_i.

To generalise the above result, we note that it follows from Theorem 3.2 that if $a_{j,k}$ is the first nonzero in column k of A, then the k-th row of H_i in C_{i-1} is zero for i≤j. Now for C_{j+1} this row will possess the structure of h_j which we will write as $str(h_j)$.

<u>Theorem 3.4</u>

In subsequent iterations $k \ge m \ge j+1$, the k-th row of H_{m+1} in C_m has the structure

$$str(C_{m,k}^T) = \sum_{i=j+1}^{m} \delta_i str(H_{i,i}^T)$$

where $H_{i,i}$ is the i-th row of H_i and

$$\delta_i = \begin{cases} 0 \text{ if } C_{m,k}^T a_i = 0 \\ 1 \text{ if } C_{m,k}^T a_i \neq 0. \end{cases}$$

As a special case, if we have $a_{i,j} = 0$ unless $j=i \pm kq$, $k=1,2,\dots,k_{max}$, where k_{max} and q are integers then the structure of C_m reflects the structure of A.

The above theorem enables us to predict the structure of each row in advance of any arithmetic. This is of course a similar result to that used in many sparse matrix codes. For instance if q=3 and $k_{max}=2$, then the C_i become

C_2	C_3	C_4	C_5	C_6
0	00	*00	0*00	00*00
0	*0	0*0	00*0	*00*0
*	0*	00*	*00*	0*00*
0	00	*00	0*00	00*00
0	*0	0*0	00*0	*00*0
*	0*	00*	*00*	0*00*

As if A is symmetric then rows of the inverse Cholesky matrix L^{-1} are simply the rows denoted $H_{i,i}$ above, these are readily predictable. In particular, we have shown that for a cyclic A matrix, L^{-1} is cyclic. We now consider the arithmetic count for the tridiagonal band matrix $H_i a_i$ only involves one operation, forming the dyad is int(i/q) multiplications, where int(*) is the integer below i/q, hence

$$\text{total operation count} < n(n+1)/q + 2n.$$

If we consider the cyclic case with k_{max}, q as the band size, $H_i a_i$ involves k_{max} operations. Forming the dyad is k_{max}int(i/q) multiplications, so we now only have

$k_{max}+k_{max}int(i/q)+k_{max}$ operations. Hence

$$\text{total operation count} < k_{max}(n(n+1)/q + 2n).$$

4. The implicit QR algorithm

One of the principal disadvantages of the implicit LL^T method discussed in sections 3 and 4 is the weakness that it only is defined without pivoting on matrices with nonsingular principal leading submatrices. Its great advantage is that the matrix H_i has the structure

$$(4.1) \qquad H_i = \left[\begin{array}{c|c} 0 & 0 \\ \hline R_i & I_{n-i+1} \end{array}\right]$$

It is therefore natural to consider what other algorithms have this property. We will now consider the scaled ABS class in which an additional set of vectors v_i are introduced and Step 4 in Algorithm 2.1 becomes

$$(4.2) \qquad x_{i+1} = x_i - (v_i^T(Ax_i-b)/p_i^Ta_i)p_i$$

Step 5 becomes

$$(4.3) \qquad H_{i+1} = H_i - H_iA^Tv_iu_i^TH_i/u_i^TH_iA^Tv_i$$

Theorem 4.1
If $u_i=e_i$ then H_i retains the required structure (4.1) for arbitrary v_i.

In this family, the implicit Cholesky algorithm considered in the previous section corresponds to $v_i=e_i$. In this section we will considered the implicit QR algorithm which is defined by

$$v_i=Ah_i.$$

We no longer need to worry about pivots as the denominator is now $v_i^Tv_i>0$ unless $h_i=0$ or AA^T is singular.

Theorem 4.2
If the A matrix is a q-band matrix then for the implicit QR algorithm, the C_{i-1} matrix has at most 2q nonzero rows.

We can also relate this result to those obtained in the previous sections by noting that we are now solving

$$V^TAx = V^Tb$$

and that if we form the implicit LL^T (LU) matrices for VA we obtain precisely the implicit QR matrices for A^TA. As we have just shown that VA now has a band width 2q, all the results of the previous section for q-band matrices apply to the implicit QR with q replaced by 2q. The results for upper q-band matrices do not apply as then VA is full. If we turn now to the results of the t-cycle matrices, we need the following general Lemma.

Lemma 4.1
If a matrix has a sparse cyclic structure i.e.

$$a_{i,j} = 0 \text{ if } j \neq i \pm kq, \; k=0,1,\ldots,k_{max}$$

and a vector v_j has the same cyclic structure

$$v_j = 0 \text{ if } j \neq m+kq, \; k=0,1,\ldots$$

then the resulting vector $y_j = Av_j$ has the same cyclic structure, i.e.

$$y_j = 0 \text{ if } j \neq m+kq, \; k=0,1,\ldots$$

Now let us apply this to prove

Theorem 4.3
If A^T and A have the same cyclic structure, namely $a_{i,j} \neq 0$ if $j=i \pm kq$ where $k=0,1,\ldots,k_{max}$ and q is integer, then the matrix C_{i-1} is also cyclic, for all i, in the QR algorithm.

Now let us generalise Theorem 3.2.

Theorem 4.4
The portion of the k-th row of H_{i+1} in C_i becomes nonzero, when the iteration index i equals the column index m of the first nonzero element in row k of A^TA, if m>k.

<u>Note</u>
The structure of this row is the same as the structure of h_i, hence a parallel theorem to Theorem 3.4 also holds.

An interesting special case for the above result follows if we consider the matrix

$$
\begin{array}{c}
\quad\quad q \quad\quad\quad\quad\quad k \\
A = \left[\begin{array}{cccccccc}
* & * & & & & & * & * \\
* & * & * & & & & * & * \\
 & * & * & * & & & * & * \\
 & & * & * & * & & * & * \\
 & & & * & * & * & * & * \\
 & & & & * & * & * & * \\
 & & & & & * & * & * \\
 & & & & & & * & *
\end{array}\right]
\end{array}
$$

Figure 4.1

then

$$
\begin{array}{c}
\quad\quad 2q \quad\quad\quad\quad\quad k \\
A^TA = \left[\begin{array}{cccccccc}
* & * & * & & & & * & * \\
* & * & * & * & & & * & * \\
* & * & * & * & * & & * & * \\
 & * & * & * & * & * & * & * \\
 & & * & * & * & * & * & * \\
 & & & * & * & * & * & * \\
* & * & * & * & * & * & * & * \\
* & * & * & * & * & * & * & *
\end{array}\right] \; k \text{ rows}
\end{array}
$$

Figure 4.2

So R_i's will contain nonzeros in at most the first 2q rows and the last k rows.

<u>5. The Symmetric (Huang) Algorithm</u>

If the parameter choices $z_i=a_i$ (that is $p_i=H_i^Ta_i$), $u_i=ka_i$ are made, then the symmetric Huang [4] algorithm is obtained. The advantage of this choice is that at Step 5 of Algorithm 2.2

$$H_{i+1} = H_i - H_ia_ia_i^TH_i/a_i^TH_ia_i$$

so we have a symmetric projection matrix with its associate well-known properties. H_{i+1} has i zero eigenvalues associated with the space spanned by a_j, $j=1,\ldots,i$.

For this method we need to modify our definition of a q-band matrix.

<u>Definition 5.1</u>
An n by n matrix A is called a q_1/q_2-band matrix if $a_{i,j}=0$ for $i>j+q_1$ while $a_{j,j+q1}\neq 0$ for at least one j, $1\leq j\leq n$ and $a_{i,j}=0$ for $j>i+q_2$, while $a_{i,i+q2}\neq 0$ for at least one i, $1\leq i\leq n$.

<u>Theorem 5.1</u>
Let A be a q_1/q_2-band matrix and consider the matrices H_i generated by the Huang algorithm. Then for the computation of the search vectors it is sufficient to store the first $i+q_2-1$ elements of the columns from $(i-q_1)$th to the $(i+q_2-1)$th.

Let us now again consider the special case

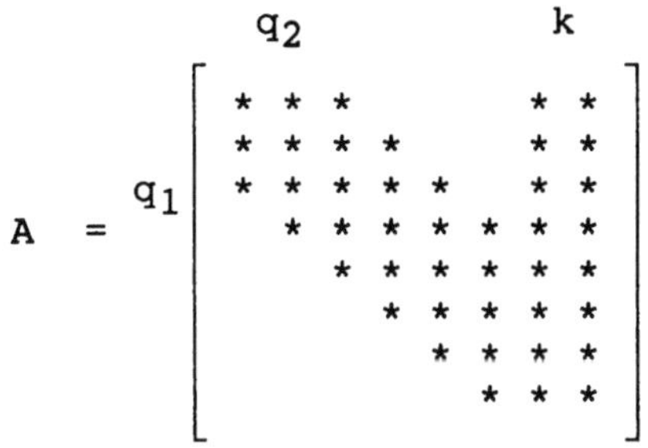

Figure 5.1

Now it is easy to prove the following theorem.

<u>Theorem 5.2</u>
If A has the special structure shown above, then the structure of H_i is similar to that given in the previous theorem, but contains $(i+q_2-1)$ nonzeros in each of the last k rows and columns and the last k diagonal values are no longer ones.

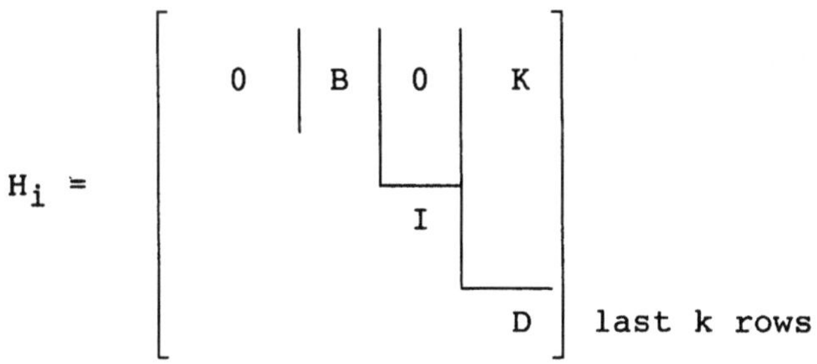

We note that this algorithm again needs more storage than that required by the implicit LU algorithm.

Theorem 5.3
If A is an n by n tridiagonal ($q_1=q_2$) and the Huang algorithm is used, then the matrix H_i only differs from the identity in the $i+q_2-1$ square principal submatrix and this submatrix has a cyclic diagonal structure of period q_2. i.e.

$$(H_i)_{p,m} \neq 0 \text{ only if } p=1+k_1q_2 \text{ and } m=1+k_2q_2$$

where k_1 and k_2 are integers.

Corollary
This result can be extended to cyclic band matrices where $a_{i,j}=0$ if $j\neq i+kq_2$, $k=0,\ldots,k_{max}$. The band is therefore $k_{max}(q_2+1)$. The matrix H_i has the same structure as that given in Theorem 5.3, but as now there are $k_{max}(2q_2)+1$ columns, each row can now have $2k_{max}+1$ nonzeros.

6. Numerical experiments

Five different versions of the above subclasses were implemented in FORTRAN 77 in simple precision on the computer DEC 1091 using the operation system TOPS 10. Three special cases from the implicit LU and related subclasses one from the QR and one from the Huang (LQ) subclasses were implemented, which were as follows.

a) Implicit LU decomposition:

$$v_i=e_i, \quad u_i=e_i, \quad z_i=e_i/h_i^Ta_i$$

b) Implicit Cholesky (LL^T) method:

$$v_i=e_i, \quad u_i=e_i \quad z_i=e_i/(h_i^Ta_i)^{\frac{1}{2}}$$

c) The so called E-method:

$$v_i=e_i, \quad u_i=e_i, \quad z_i=e_i$$

d) Implicit QR method:

$$v_i=A^Tp_i, \quad u_i=e_i, \quad z_i=e_i$$

e) Huang method:

$$v_i=e_i, \quad u_i=a_i, \quad z_i=a_i$$

The so called Harwell problems were chosen in dimensions 115, 185, 216 and 343. The right hand sides were calculated by setting $x^*=(1,2,\ldots,n)$ to obtain the vector b in simple precision.

The coefficient matrix A was stored by the well-known pointer technique.

Instead of the matrix H_i a vector was implemented in which the suitable rows of H_i were stored.

The q-band width of the problems was determined by the program.

We mention that the reduction of the band width was tried by a certain trivial strategy but this type of width reduction was very time consuming. Here we mention that the Harwell problems contain test problems in dimension 512 and 1107 too. For these cases the q value could not be reduced enough by the program therefore we could not try them because of insufficient central memory.

The results with the LU and related methods were much worse than the results of the Huang and QR methods so they are not reported.

Table 1 and 2 demonstrate the results obtained by the symmetric (Huang) and the implicit QR algorithms. In Table 1 the nonsymmetric sparse problems can be seen where the number after G means the dimension of the problem. Table 2 contains the results obtained on the symmetric versions of the Harwell problems, introduced by Maany and Dixon. We mention further that the relative residuals and relative errors were also calculated in simple precision.

PROBL.		REL. RES.	REL. ERR.	EPS
	Hu	.17220E-7	.59394E-7	1.E-6
G115	QR	.38494E-7	.81247E-7	1.E-6
	Hu	.41539E-7	.11674E-2	1.E-6
G185	QR	.11179E-4	.17433E-2	1.E-6
	Hu	.26733E-7	.63936E-7	1.E-6
G216A	QR	.96064E-7	.15282E-6	1.E-6
	Hu	.24037E-6	.10041E-1	1.E-9
G216B	QR	.78279E-2	.50873E+6	1.E-16
	Hu	.31207E-7	.61967E-7	1.E-6
G343	QR	.13554E-6	.15949E-7	1.E-6

Table 1

PROBL.		REL. RES.	REL. ERR.	EPS
	Hu	.19556E-7	.23948E-7	1.E-6
S115	QR	.31380E-7	.40728E-7	1.E-6
	Hu	.18255E-7	.31185E-7	1.E-6
S185	QR	.34858E-7	.56840E-7	1.E-6
	Hu	.16329E-7	.29081E-7	1.E-6
S216A	QR	.41554E-7	.59718E-7	1.E-6
	Hu	.18601E-7	.55358E-7	1.E-6
S216B	QR	.40155E-7	.12457E-6	1.E-6
	Hu	.21391E-7	.36854E-7	1.E-6
S343	QR	.54472E-7	.73829E-7	1.E-6

Table 2

From the Tables it follows that the symmetric update algorithm seems to be better than the QR algorithm. As the problems are in general well-conditioned except the non-symmetric G216B problem, all were solved successfully. In that case the first ε at which results were obtained was 1.E-9 and 1.E-16 for the symmetric and the QR methods respectively. For larger ε values an error message was obtained indicating that the denominator at a certain step became too small (ill conditioning). Finally we mention that the symmetric update has the property that it gives the minimum norm solution, a fact which can explain the good solution of the problem G216B.

References

[1] Abaffy, J., Some Special Cases of the ABSg Class for Sparse Matrices , Proceedings of Second Conference on Automata Languages and Mathematical Systems, University of Economics, Budapest (1986) 7-11

[2] Abaffy, J., The Description of a Package for Sparse Linear Systems Using 6 Algorithms in the ABSg Class, Report No.TR 192, NOC, Hatfield Polytechnic, (1987)

[3] Abaffy, J. and Dixon, L. C. W., On Solving Sparse Band Systems with Three Algorithms of the ABS Family, Report No. TR 191, NOC, Hatfield Polytechnic, (1987)

[4] Abaffy, J. and Spedicato, E., ABS Projection Algorithms: Mathematical Techniques for Linear and Nonlinear Equations, Ellis Horwood Ltd, Chichester, (1989)

[5] Deng, N. Y., Nested Dissection Methods for Sparse Positive Definite Linear Systems of Equations, Report No. TR 197, NOC, Hatfield Polytechnic, (1987)

[6] George, A., Nested Dissection of a Regular Finite Element Mesh, SIAM J. Numer. Anal., 10 (1973) 345-363

[7] George, A. and Liu, J. H., Computer Solution of Large Sparse Positive Definite Systems, Prentice-Hall: Englewood Cliffs, NJ, (1981)
[8] Zhu, M., The Implicit LL^T Algorithm for Sparse Nested Dissection Linear Systems, Report No. TR 196, NOC, Hatfield Polytechnic, (1987)

Multi-Grid Algorithms
Introduction, Survey and Convergence Analysis

W. Hackbusch
Christian-Albrechts-Universität zu Kiel
Institut für Informatik und Praktische Mathematik
Olshausenstr. 40
D-2300 Kiel 1, Germany

Preliminaries

Multi-grid methods belong to the class of the fastest iterations, since their convergence rate is independent of the step size h. Their applicability does not require symmetry or positive definiteness. Further, it is easy to generalise multi-grid methods to nonlinear problems.

We introduce the method in §1 for a one-dimension problem and give a precise convergence analysis in §2. §3 describes the multi-grid algorithm in the general case. It is completed by the nested iteration in §4. The convergence analysis in the general case is in §5. Special results (e.g. V-cycle convergence) are mentioned in §6. §7 is devoted to the nonlinear multi-grid methods. Finally, §8 introduces into the multi-grid methods of the second kind.

The first two-grid method is described by Brakhage [1] in 1960. More precisely, it is a two-grid method of the second kind applying to integral equations (cf. §8). In 1961 Fedorenko [1] published a two- and in 1964 a multi-grid method (Fedorenko [2]) for the Poisson problem in the square. In 1966 Bachvalov [1] proved the typical convergence properties for a more complicated situation. Further early publications are due to Astrachancev [1] (1971), Hackbusch [1] (1976), Bank - Dupont (in a report from 1977, which later has been split into [1,2]), Brandt [1] (1977), Nicolaides [1] (1977). An extensive list of multi-grid literature until 1985 is contained in the book of Hackbusch [14] and in the volume edited by McCormick [1]. Here we mention Bank [1,2], Bank/Dupont/Yserentant [1], Braess [1], Braess/Hackbusch [1], Bramble/Pasciak [1], Brandt [2], D'Jakonov [1], Hackbusch [2-12], Hackbusch/Reusken [1], Hemker [1], Hemker/Schippers [1], Maitre/Musy [1], Mandel [1], Nicolaides [2], Sonneveld et al. [1], Wittum [1-6], Yserentant [1].We also refer to the proceedings volumes Hackbusch / Trottenberg [1,2], Braess / Hackbusch/Trottenberg [1], Hackbusch [17], [18].

1. Introductory Model Problem

We introduce the two-grid method for a one-dimensional boundary value problem. The general case is considered in §5.

1.1 One-dimensional Model Problem

The system

(1.1a) $$Ax=b \quad \text{with} \quad A=h^{-2}\,\mathrm{tridiag}\{-1,2,-1\}=h^{-2}\begin{bmatrix}2 & -1 & & \\ -1 & 2 & \ddots & \\ & \ddots & \ddots & -1 \\ & & -1 & 2\end{bmatrix}$$

of $n=N-1$ equations is the discretisation of the one-dimensional Poisson boundary-value problem

(1.1b) $$-u''(x) = f(x) \text{ for } 0<x<1, \qquad u(0)=u_0, \quad u(1)=u_1,$$

with the step size $h=1/N$. It should be emphasized that tridiagonal matrices are easy to solve directly. The analysis of iterative methods for these tridiagonal equations is of interest only by the fact that the convergence properties carry over also to the general case of two or more space dimensions.

NATO ASI Series, Vol. F 77
Computer Algorithms for Solving Linear Algebraic Equations
Edited by E. Spedicato

1.2 Smoothing

The simplest iteration we can think of is *Richardson's iteration*

(1.2a) $x^{m+1} = \Phi_\Theta^{\mathrm{Rich}}(x^m, b) := x^m - \Theta(Ax^m - b)$ with

(1.2b) $\Theta = \frac{1}{4}h^2$.

This method is identical with the Jacobi iteration damped by the factor $\frac{1}{2}$, since the diagonal of the matrix A is $D = \frac{1}{2}h^2 I$. The orthonormal system of eigenfunctions is given by the grid functions e^α with the components

(1.3a) $e_k^\alpha = \sqrt{\frac{h}{2}}\ \sin(\alpha k \pi h)$ $(0 \leqslant k \leqslant N)$.

α is called the *frequency* of e^α. The corresponding eigenvalues are λ_α:

(1.3b) $A_\ell e^\alpha = \lambda_\alpha e^\alpha$ with $\lambda_\alpha := 4h^{-2}\sin^2(\alpha\pi h/2)$ for $1 \leqslant \alpha \leqslant N-1$.

Hence, the eigenvalues of the iteration matrix $M_\Theta^{\mathrm{Rich}} = I - \Theta A$ with Θ from (1.2b) are

(1.3c) $M_\Theta^{\mathrm{Rich}} e^\alpha = (1 - \sin^2(\alpha\pi h/2))\, e^\alpha = \cos^2(\alpha\pi h/2)\, e^\alpha$

The maximum eigenvalue $\cos^2(\pi h/2) \approx 1 - \pi h^2/4$ corresponds to the lowest frequency $\alpha = 1$. By definition the *spectral radius* of the iteration matrix is its maximal eigenvalue: $\rho(M_\Theta^{\mathrm{Rich}}) = \cos^2(\pi h/2)$ and describes the *convergence rate*. We conclude that the very slow convergence of Richardson's iteration is due to the lowest frequency $\alpha = 1$.

Now we change the point of view and ask: How does the iteration behave applied to the subspace of *high* frequencies α? Hitherto, all vectors x, b etc. belonged to the vector space $X = \mathbb{R}^n$. For all $x \in X$ there is a representation (1.2b) by means of the eigenvector basis (1.2a):

(1.4a) $x = \sum_{\alpha=1}^{N-1} \xi_\alpha e^\alpha$ with $\xi_\alpha := \langle x, e^\alpha \rangle$.

Since high frequencies α correspond to strong oscillations of the sine functions (1.2a), we define

(1.4b) $X_{\mathrm{osc}} := \mathrm{span}\{e^\alpha:\ \frac{N}{2} \leqslant \alpha \leqslant N-1\}$.

If we were able to generate an approximation x^0, whose error were in the subspace X_{osc}:

(1.4c) $e^0 := x^0 - x \in X_{\mathrm{osc}}$,

the simple Richardson iteration would yield fast convergence:

Lemma 1.1 Assume the model case with initial error (1.2d). Then all further errors e^m also belong to X_{osc} and satisfy the error estimate

$$\|e^m\|_2 \leqslant \tfrac{1}{2}\|e^{m-1}\|_2,$$

i.e. restricted to X_{osc} the convergence rate is h-independent.

Proof. Let $M := M_\Theta^{\mathrm{Rich}}$. Since the vectors (1.2a) are orthonormal, we have

(1.4d) $\|x\|_2^2 = \sum_{\alpha=1}^{N-1} |\xi_\alpha|^2$ for x from (2b).

Because of $Me^\alpha = (1 - \Theta\lambda_\alpha)e^\alpha$ the application of the iteration matrix $M = I - \Theta A$ to the error e^m with coefficients ξ_α yields:

$$\|e^{m+1}\|_2^2 \leqslant \sum |1-\Theta\lambda_\alpha|^2 |\xi_\alpha|^2 \leqslant \max|1-\Theta\lambda_\alpha|^2 \sum|\xi_\alpha|^2 = \\ = \max|1-\Theta\lambda_\alpha|^2 \, \|e^m\|_2^2$$

with $\lambda_\alpha = 4h^{-2}\sin^2(\alpha\pi h/2)$ (cf. (1.3b)). The maximum is to be taken over all $\frac{N}{2} \leqslant \alpha \leqslant N-1$. For these α

$$2h^{-2} = 4h^{-2}\sin^2(\pi/4) = \lambda_{N/2} \leqslant \lambda_\alpha \leqslant \lambda_{N-1} < 4h^{-2}$$

holds, hence $|1-\Theta\lambda_\alpha| \leqslant \frac{1}{2}$ proves the assertion of the Lemma. QED

The statement of the Lemma is not of direct practical use, since the assumption (4c) cannot be established in practice (at least not with less work than for the exact solution of $Ax=b$). However, we can conclude as follows. Let $X_{smooth} := X_{osc}^{\perp}$ be the orthogonal complement of X_{osc}. In the next remark e does not stand for an eigenvector but denotes the *error* of the m-th iterate x^m: $e^m := x^m - x$. Note that $e^{m+1} = M_\Theta^{Rich} e^m$.

Remark 1.2 Split the starting error e^0 into

(1.5a) $\quad e^0 = e^0_{osc} + e^0_{smooth}, \quad e^0_{osc} \in X_{osc}, \quad e^0_{smooth} \in X_{smooth} := X_{osc}^{\perp}.$

Then after m iterations we have

(1.5b) $\quad e^m = e^m_{osz} + e^m_{smooth} \quad$ with

(1.5c) $\quad e^m_{osc} = M^m e^0_{osc} \in X_{osc}, \quad e^m_{smooth} = M^m e^0_{smooth} \in X_{smooth},$

(1.5d) $\quad \|e^m_{osc}\|_2 \leqslant (\frac{1}{2})^m \|e^0_{osc}\|_2,$

while e^m_{smooth} converges only very slowly to 0. Since e^m_{osc} decreases faster than e^m_{smooth}, the smooth part of e^m has increased and one may call e^m «smoother» than e^0.

Figure 1 shows the (piecewise linearly connected) values e^0_i $(0 \leqslant i \leqslant N=8)$ as first solid line. The further Richardson iterates have errors e^m $(m=1, 2, 3)$, which are insignificantly smaller but clearly smoother.

Because of this observation we will call iterative methods like the presented Richardson iteration (2a,b) *smoothing iterations* and use the symbol $\mathcal{S}$ for this iteration.

In the following an iteration Ψ with a complementary property is desired: Ψ should effectively reduce the smooth components in

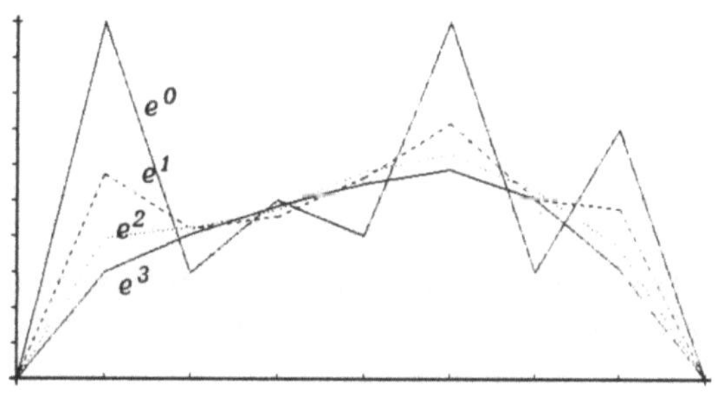

Figure 1.1 Errors e^m for example (1a)

(1.6) $\quad X_{smooth} := X_{osc}^{\perp} = \mathrm{span}\{e^\alpha : 1 \leqslant \alpha < N/2\}.$

Ψ is not required to have good convergence with respect of the subspace X_{osc}. Then the product method $\Psi \circ \mathcal{S}$ would have the property that for both subspaces X_{osc} and X_{smooth} one of the factors in $\Psi \circ \mathcal{S}$ would yield fast convergence.

Unfortunately, none of the classical methods have this property. To construct such an iteration Ψ, we follow the idea that smooth grid functions can be well approximated by means of a coarser grid. After the introduction of coarser grids in §§1.3-5 we will return to the construction of the so-called «coarse-grid correction» Ψ in §1.6.

1.3 Hierarchy of Systems of Equations

For the following considerations we have to embed the problem $Ax=b$ into a family of systems of equations. In the model case, we obtain for all step sizes $h=1/N$ a system $Ax=b$ depending on N or h, respectively. Let

(1.7a) $\quad h_0 > h_1 > \ldots > h_{\ell-1} > h_\ell > \ldots \qquad$ with $\lim_{\ell\to\infty} h_\ell = 0$

be a sequence of step sizes, which may be generated e.g. by

(1.7b) $\quad h_\ell := h_0/2^\ell \qquad (\ell \geq 0).$

The index ℓ is the <u>level number</u>. $\ell=0$ corresponds to the coarsest grid. In the model case, where the grid $\Omega_\ell := \Omega_{h_\ell} = \{h_\ell, 2h_\ell, \ldots, 1-h_\ell\}$ is contained in $(0,1)$, the step size

(1.7c) $\quad h_0 = \frac{1}{2}$

is the coarsest one. Then $\Omega_0 = \Omega_{h_0}$ contains only *one* interior grid point.

Each step size h_ℓ (i.e. each level ℓ) corresponds to a system

(1.8a) $\quad A_\ell x_\ell = b_\ell \qquad (\ell = 0,1,2,\ldots)$

of equations of the dimension n_ℓ, which in the model case amounts to

(1.8b) $\quad n_\ell = N_\ell - 1 = 1/h_\ell - 1.$

The family of systems (1.8a) for $\ell=0,1,2,\ldots$ represents the <u>hierarchy of systems of equations</u>. The actual problem $Ax=b$ to be solved corresponds to one special level $\ell=\ell_{\max}$. During the solution of $A_\ell x_\ell = b_\ell$ at $\ell=\ell_{\max}$ the lower levels $\ell<\ell_{\max}$ will be involved.

1.4 Prolongation

The vectors x_ℓ and b_ℓ from (1.8a) are elements of the vector space

(1.9) $\quad X_\ell \cong \mathbb{R}^{n_\ell}.$

A connection between the different levels $\ell=0,1,2,\ldots$ is needed. The <u>prolongation</u>

(1.10) $\quad p: X_{\ell-1} \to X_\ell \qquad (\ell=1,2,\ldots)$

is assumed to be a linear and injective mapping (more precisely: a family of mappings for all $\ell \geq 1$) from the coarse grid into the fine one.

In the one-dimensional case (1.1a), the vector x_ℓ can be considered as grid function defined on $\Omega_\ell = \{\mu h_\ell : 0 \leq \mu \leq N_\ell = 1/h_\ell\}$. We write x_ℓ as function u_ℓ connected with x_ℓ via

(1.11a) $\quad u_\ell(\mu h_\ell) = x_{\ell,\mu} \qquad (1 \leq \mu \leq N_\ell - 1),$

i.e. for all step sizes the arguments of u_ℓ belong to the interval $\Omega=(0,1)$ and are restricted to Ω_ℓ. For ease of notation we add the boundary values

(1.11b) $\quad u_\ell(0) = u_\ell(1) = 0.$

An obvious proposal for the prolongation p is the piecewise linear interpolation between the grid points of $\Omega_{\ell-1}$:

(1.12a) $\quad (pu_{\ell-1})(\xi) := u_{\ell-1}(\xi) \qquad$ for $\xi \in \Omega_{\ell-1} \subset \Omega_\ell$,

(1.12b) $\quad (pu_{\ell-1})(\xi) := \frac{1}{2}[u_{\ell-1}(\xi+h_\ell) + u_{\ell-1}(\xi-h_\ell)] \quad$ for $\xi \in \Omega_\ell \backslash \Omega_{\ell-1}$,

where the definition (1.11b) is used at $\xi=h_\ell$ und $\xi=1-h_\ell$. A shorter characterisation of the prolongation p is the symbol (1.12c):

(1.12c) $\quad p = [\frac{1}{2}\ 1\ \frac{1}{2}].$

(1.12c) indicates that the unit vector $x_{\ell-1}=(\ldots,0,1,0,\ldots)^T$ is mapped into $x_\ell = p x_{\ell-1} = (\ldots,0,\frac{1}{2},1,\frac{1}{2},0,\ldots)^T$.

1.5 Restriction

The restriction r is a linear and surjective mapping

(1.13) $\quad r:\ X_\ell \to X_{\ell-1} \qquad (\ell \geq 1),$

which maps fine-grid functions into coarse-grid functions. If $\Omega_{\ell-1} \subset \Omega_\ell$ holds as in the model case, the simplest choice is the *trivial restriction* $(r_{\text{triv}} u_\ell)(\xi) = u_\ell(\xi)$ for $\xi \in \Omega_{\ell-1}$. However, because of certain disadvantages, we advise against its use (cf. Hackbusch [14,§3.5]). Instead we define $(r u_\ell)(\xi)$ as weighted mean of the neighbouring values. The stencil

(1.14a) $\quad r = [\rho_{-1}\ \rho_0\ \rho_1]$

characterises the restriction

(1.14b) $\quad (r u_\ell)(\xi) = \sum_{i=-1}^{1} \rho_i u_\ell(\xi + i h_\ell) \qquad$ for $\xi \in \Omega_{\ell-1}$.

The restriction

(1.15a) $\quad r = [\frac{1}{4}\ \frac{1}{2}\ \frac{1}{4}]$

is the adjoint of p from (1.12c) with respect to the scalar products

(1.15b) $\quad \langle\cdot,\cdot\rangle = \langle\cdot,\cdot\rangle_\ell \quad$ with $\langle u_\ell, v_\ell\rangle_\ell = h_\ell \sum_{\alpha=1}^{N_\ell-1} u_{\ell,\alpha} \overline{v_{\ell,\alpha}}$ for $u_\ell, v_\ell \in X_\ell$.

1.6 Coarse-Grid Correction

Let $\bar{x}_\ell$ be the result of few steps of the smoothing iteration (1.2a,b). The corresponding error

(1.16a) $\quad \bar{e}_\ell := \bar{x}_\ell - x_\ell$

is the exact correction by which the solution can be computed:

(1.16a') $\quad x_\ell = \bar{x}_\ell - \bar{e}_\ell$

Since $A_\ell \bar{e}_\ell = A_\ell(\bar{x}_\ell - x_\ell) = A_\ell \bar{x}_\ell - A_\ell x_\ell = A_\ell \bar{x}_\ell - b_\ell$, x_ℓ satisfies the equation

(1.16b) $\quad A_\ell \bar{e}_\ell = d_\ell \quad$ with the *defect* $d_\ell := A_\ell \bar{x}_\ell - b_\ell$.

According to the considerations of §1.1 $\bar{e}_\ell$ is smooth. Therefore, it should be possible to approximate $\bar{e}_\ell$ by means of the coarse grid: $\bar{e}_\ell \approx p e_{\ell-1}$. As ansatz for $e_{\ell-1}$ we take the coarse-grid equation corresponding to (1.16b):

(1.16c) $\quad A_{\ell-1} e_{\ell-1} = d_{\ell-1} \qquad$ with $d_{\ell-1} := r d_\ell$.

Assume that we are able to solve equation (1.16c) exactly in the coarse grid:

(1.16d) $\quad e_{\ell-1} = A_{\ell-1}^{-1} d_{\ell-1}$

Its image $p e_{\ell-1}$ under the prolongation p should approximate the solution $\bar{e}_\ell$ of (1.16b), so that the coarse-grid correction is completed by

(1.16e) $\quad x_\ell^{\text{new}} := \bar{x}_\ell - p e_{\ell-1}$.

In compact form, the *coarse-grid correction* (1.16b-e) reads as follows:

(1.17) $\bar{x}_\ell \mapsto x_\ell^{\text{new}} := \bar{x}_\ell - pA_{\ell-1}^{-1}\, r\,(A_\ell \bar{x}_\ell - b_\ell).$

Renaming $\bar{x}_\ell$ and x_ℓ^{new} by x_ℓ^m and x_ℓ^{m+1}, (1.17) defines an iterative method, the so-called coarse-grid correction:

(1.17') $\Phi_\ell^{\text{CGC}}(x_\ell, b_\ell) := x_\ell - pA_{\ell-1}^{-1}\, r\,(A_\ell x_\ell - b_\ell).$

M is called the *iteration matrix* of a consistent iteration Φ and N is denoted as matrix of the second normal form of Φ if

$$\Phi(x,b) = Mx + Nb = x - N(Ax-b)$$

(cf. Hackbusch [20,§3]).

Remark 1.3 The iteration matrix M and the matrix N of the second normal form of the coarse-grid correction are

(1.18) $M_\ell^{\text{CGC}} = I - pA_{\ell-1}^{-1}\, r\, A_\ell, \quad N_\ell^{\text{CGC}} = pA_{\ell-1}^{-1}\, r.$

Φ_ℓ^{CGC} is not an interesting iteration as such, since it is divergent: $\rho(M_\ell^{\text{CGC}}) \geq 1$.

For the model problem, the so-called *Galerkin product* representation is valid:

(1.19) $A_{\ell-1} = r A_\ell p.$

The next remark shows that the coarse-grid equation (1.16c) is a reasonable ansatz for $e_{\ell-1}$.

Remark 1.4 Let A_ℓ be positive definite. The best approximation of $\bar{e}_\ell \in X_\ell$ with respect to the A_ℓ-norm $\|x_\ell\|_A := \langle A_\ell x_\ell, x_\ell \rangle_\ell^{1/2}$ is $pe_{\ell-1}$, where p and r are adjoint according to (1.15b) and where $e_{\ell-1}$ is the solution of (1.16c) with the Galerkin matrix (1.19).

1.7 The Two-Grid Algorithm

The so-called smoothing iteration $\mathcal{S}_\ell$ is defined in §1.2 and the coarse-grid correction Φ_ℓ^{CGC} is constructed in §1.6. The two-grid iteration is the product method

(1.20) $\Phi_\ell^{\text{TGM}} := \Phi_\ell^{\text{CGC}} \circ \mathcal{S}_\ell^{\nu} \qquad (\ell \geq 1,\ \nu \geq 1).$

ν is the number of smoothing steps. In algorithmical notation (1.20) takes the form

(1.21) procedure $\Phi_\ell^{\text{TGM}}(x_\ell, b_\ell)$;
(1.21a) begin for $i := 1$ to ν do $x_\ell := \mathcal{S}_\ell(x_\ell, b_\ell)$;
(1.21b) $d_{\ell-1} := r\,(A_\ell x_\ell - b_\ell)$;
(1.21c) $e_{\ell-1} := A_{\ell-1}^{-1}\, d_{\ell-1}$;
(1.21d) $x_\ell := x_\ell - p\, e_{\ell-1}$;
(1.21e) $\Phi_\ell^{\text{TGM}} := x_\ell$
end;

Instead of (1.21a) a *semiiterative* smoothing method can also be applied (cf. Hackbusch [20,§10.8.1]).

It is easy to see that $\Phi_\ell^{\text{CGC}} \circ \mathcal{S}_\ell^{\nu}$ has the same convergence behaviour as

(1.22a) $\Phi_\ell^{\text{TGM}(\nu_1,\nu_2)} := \mathcal{S}_\ell^{\nu_2} \circ \Phi_\ell^{\text{CGC}} \circ \mathcal{S}_\ell^{\nu_1}$ with $\nu = \nu_1 + \nu_2$.

In this case, ν_1 *pre-* and ν_2 *post-smoothing* steps are applied. Iteration (1.22a) contains algorithm (1.20) for $\nu_1 = \nu$ and $\nu_2 = 0$ as special case. Therefore, in the sequel we will use the version (1.22a).

One may also use different iterations $\mathcal{S}_\ell$ and $\hat{\mathcal{S}}_\ell$ as pre- and post-smoothers:

(1.22b) $\hat{\mathcal{S}}_\ell^{\nu_2} \circ \Phi_\ell^{\mathrm{CGC}} \circ \mathcal{S}_\ell^{\nu_1}$ with $\nu = \nu_1 + \nu_2$.

Lemma 1.5 Let $\mathcal{S}_\ell$ be a consistent iteration with iteration matrix S_ℓ. Then $\Phi_\ell^{\mathrm{TGM}(\nu_1,\nu_2)}$ is a consistent iteration with the iteration matrix

(1.23) $M_\ell^{\mathrm{TGM}}(\nu_1,\nu_2) = S_\ell^{\nu_2}(I - pA_{\ell-1}^{-1}\, r A_\ell) S_\ell^{\nu_1}.$

Proof. M_ℓ^{TGM} is the product of the iteration matrices of $\mathcal{S}_\ell^{\nu_2}$, Φ_ℓ^{CGC}, $\mathcal{S}_\ell^{\nu_1}$. ∎

2. Analysis for the One-Dimensional Example

2.1 Fourier Analysis

We abbreviate the quantities at the levels ℓ and $\ell-1$ by

$$N = N_\ell, \quad N' = N_{\ell-1}, \quad h = h_\ell = 1/N, \quad h' = h_{\ell-1} = 2h.$$

The vector $x = (x_k)_{1 \le k \le N-1}$ is formally extended by the components

(2.1) $x_0 = x_N = 0.$

The eigenvectors (grid functions) e^α introduced in (1.3a) satisfy the condition (2.1) for all $\alpha \in \mathbf{Z}$. Since the vectors $\{e^\alpha : 1 \le \alpha \le N-1\}$ form an orthonormal basis, the matrix Q built by e^α as columns is unitary: $Q^H Q = I$:

(2.2) $Q := [e^1, e^{N-1}, e^2, e^{N-2}, \dots, e^\alpha, e^{N-\alpha}, \dots, e^{\frac{N}{2}-1}, e^{\frac{N}{2}+1}, e^{\frac{N}{2}}]$

Since multiplication by Q or $Q^H = Q^{-1}$ does not change the spectral norm and spectral radius, we have

(2.3) $\|\hat{M}\|_2 = \|M\|_2$ and $\rho(\hat{M}) = \rho(M)$ for $\hat{M} := Q^{-1} M Q$.

It turns out that the Fourier transformed iteration matrix $\hat{M}$ of the two-grid method is a block-diagonal matrix of the form

(2.4) $\hat{M} = \mathrm{blockdiag}\{M_1, M_2, \dots, M_{N'-1}, M_{N'}\}$ with

(2.5) M_α 2×2-matrices for $1 \le \alpha \le N'-1$, $M_{N'}$ 1×1-matrix for $\alpha = N'$.

A well-known property of block-diagonal matrices yields

Lemma 2.1 For a matrix of the form (2.4), (2.5) it holds that

$$\|\hat{M}\|_2 = \max\{\|M_\alpha\|_2 : 1 \le \alpha \le N'\}, \quad \rho(\hat{M}) = \max\{\rho(M_\alpha) : 1 \le \alpha \le N'\}.$$

Choose Richardson's iteration with $\Theta = h^2/4 \approx 1/\rho(A_\ell)$ as smoothing iteration. For the proof of the block structure (2.4), (2.5) we transform the iteration matrix

$$M = (I - pA_{\ell-1}^{-1} r A_\ell) S_\ell^\nu \quad \text{with} \quad S_\ell = I - \Theta A_\ell,\ \Theta = h^2/4$$

into

(2.6a) $\hat{M} = (I - \hat{p}\hat{A}_{\ell-1}^{-1}\hat{r}\hat{A}_\ell)\hat{S}_\ell^\nu$ with $\hat{A}_\ell := Q^{-1} A_\ell Q$, $\hat{S}_\ell := Q^{-1} S_\ell Q$.

To define the quantities

(2.6b) $\hat{p} := Q^{-1}pQ', \quad \hat{A}_{\ell-1} := Q'^{-1}A_{\ell-1}Q', \quad \hat{r} := Q'^{-1}rQ,$

we have to introduce the Fourier transformation Q' at level $\ell-1$. Replacing $h=h_\ell$ in (1.3a) by $h'=h_{\ell-1}$, one obtains the vectors e'^{α} with

(2.7) $$e'^{\alpha}_k = \sqrt{h}\ \sin(2\alpha k\pi h) \qquad (0\leq k\leq N').$$

Analogously to (1.3a), the vectors e'^{α} form the matrix

$$Q' = [e'^{1}, e'^{2}, \dots, e'^{N'-1}].$$

2.2 Transformed Quantities

We recall $A_\ell e^\alpha = \lambda_\alpha e^\alpha$ with $\lambda_\alpha := 4h^{-2}\sin^2(\alpha\pi h/2)$ and introduce

$$s_\alpha^2 = \sin^2(\alpha\pi h/2), \qquad c_\alpha^2 = \cos^2(\alpha\pi h/2).$$

Noting that $\lambda_{N-\alpha} = s_{N-\alpha}^2 = c_\alpha^2$, we obtain

(2.8a) $$\hat{A}_\ell := Q^{-1}A_\ell Q = \operatorname{blockdiag}\{A_1,\dots,A_{N'}\} \quad \text{with the blocks}$$

(2.8b) $$A_\alpha = 4h^{-2}\begin{bmatrix} s_\alpha^2 & 0 \\ 0 & c_\alpha^2 \end{bmatrix} \quad \text{for } 1\leq\alpha\leq N'-1, \quad A_{N'} = 2h^{-2}.$$

Since $S_\ell = I-\frac{1}{4}h^2A_\ell$ and $1-s_\alpha^2 = c_\alpha^2$, $1-c_\alpha^2 = s_\alpha^2$, (2.8a,b) yields the result

(2.9a) $$\hat{S}_\ell := Q^{-1}S_\ell Q = \operatorname{blockdiag}\{S_1,\dots,S_{N'}\} \quad \text{with the blocks}$$

(2.9b) $$S_\alpha = \begin{bmatrix} c_\alpha^2 & 0 \\ 0 & s_\alpha^2 \end{bmatrix} \quad \text{for } 1\leq\alpha\leq N'-1, \qquad S_{N'} = \tfrac{1}{2}.$$

Because of $A_{\ell-1}e'^{\alpha} = \lambda'_\alpha e'^{\alpha}$ with $\lambda'_\alpha = 4h'^{-2}\sin^2(\alpha\pi h'/2) = h^{-2}\sin^2(\alpha\pi h)$, and using $\sin^2(\alpha\pi h) = 4s_\alpha^2c_\alpha^2$, one obtains the diagonal matrix

(2.10) $$\hat{A}_{\ell-1} := Q'^{-1}A_{\ell-1}Q' = \operatorname{diag}\{A'_1,\dots,A'_{N'}\} \quad \text{with } A'_\alpha = 4h^{-2}s_\alpha^2c_\alpha^2.$$

Next, we transform p and r. Let p be defined by (1.12a-c). For r we choose the adjoint mapping $r=p^*$ from (1.15a):

(2.11a) $$r = \tfrac{1}{2}[\tfrac{1}{2}\ 1\ \tfrac{1}{2}], \quad \text{i.e. } (ru_\ell)(\xi) = \tfrac{1}{4}u_\ell(\xi-h)+\tfrac{1}{2}u_\ell(\xi)+\tfrac{1}{4}u_\ell(\xi+h).$$

r and $\hat{r}$ are matrices of the format $(N'-1)\times(N-1) = (N'-1)\times(2N'-1)$. The representation

(2.11b) $$\hat{r} := Q'^{-1}rQ = [\operatorname{blockdiag}\{r_1,\dots,r_{N'-1}\}, 0] \quad \text{with } r_\alpha = \sqrt{\tfrac{1}{2}}\,[c_\alpha^2, -s_\alpha^2]$$

means, that the last column of $\hat{r}$ vanishes (follows from $re^{N'}=0$) and that the remaining part of the format $(N'-1)\times 2(N'-1)$ consists of $N'-1$ 1×2-blocks r_α. For the proof of (2.11b) it is to be shown that

$$re^\alpha = c_\alpha^2 e'^{\alpha}/\sqrt{2}, \qquad re^{N-\alpha} = -s_\alpha^2 e'^{\alpha}/\sqrt{2} \qquad \text{for } 1\leq\alpha\leq N'-1.$$

p from (1.12a-c) and r from (2.11a) are connected by $r=p^*$. From $p^* = \frac{1}{2}p^H$ one finds the representation

$$\hat{p} := Q^{-1}pQ' = Q^{-1}(2r)^H Q' = Q^H(2r)^H Q' = 2[Q'^H rQ]^H = 2\hat{r}^H.$$

Therefore, the result (2.11b) for $\hat{r}$ proves

(2.12) $$\hat{p} := Q^{-1}pQ' = \begin{bmatrix} \operatorname{diag}\{p_1,\dots,p_{N'-1}\} \\ 0 \end{bmatrix} \quad \text{with } p_\alpha = \sqrt{2}\begin{bmatrix} c_\alpha^2 \\ -s_\alpha^2 \end{bmatrix}.$$

2.3 Convergence Results

Since all factors in (2.6a) have a block-diagonal structure, this carries over to $\hat{M}$ and proves the structure (2.4), (2.5). For the 2x2-blocks M_α $(1 \leq \alpha \leq N'-1)$ and for the 1x1-block $M_{N'}$, (2.8b), (2.9b), (2.10), (2.11b), (2.12) yield

(2.13a) $$M_\alpha = (I - p_\alpha A_\alpha'^{-1} r_\alpha A_\alpha) S_\alpha^\nu \qquad (1 \leq \alpha \leq N'-1), \quad M_{N'} = 2^{-\nu}.$$

Inserting the representations of p_α, A'_α, r_α, A_α, S_α^ν, one obtains

$$M_\alpha = \left(\begin{bmatrix} 1 & 0 \\ 0 & 1 \end{bmatrix} - \begin{bmatrix} c_\alpha^2 \\ -s_\alpha^2 \end{bmatrix} \frac{h^2}{4 s_\alpha^2 c_\alpha^2} [c_\alpha^2, -s_\alpha^2]\, 4h^{-2} \begin{bmatrix} s_\alpha^2 & 0 \\ 0 & c_\alpha^2 \end{bmatrix} \right) \begin{bmatrix} c_\alpha^2 & 0 \\ 0 & s_\alpha^2 \end{bmatrix}^\nu$$

(2.13b) $$= \left(\begin{bmatrix} 1 & 0 \\ 0 & 1 \end{bmatrix} - \begin{bmatrix} c_\alpha^2 & -c_\alpha^2 \\ -s_\alpha^2 & s_\alpha^2 \end{bmatrix} \right) \begin{bmatrix} c_\alpha^2 & 0 \\ 0 & s_\alpha^2 \end{bmatrix}^\nu = \begin{bmatrix} s_\alpha^2 & c_\alpha^2 \\ s_\alpha^2 & c_\alpha^2 \end{bmatrix} \begin{bmatrix} c_\alpha^2 & 0 \\ 0 & s_\alpha^2 \end{bmatrix}^\nu.$$

The block M_α describes the application of M to the two functions e^α, $e^{N-\alpha}$ (respective columns of the matrix Q, cf. (1.3a)). Since $\alpha < N' < N-\alpha$, e^α corresponds to a *smooth* grid function and $e^{N-\alpha}$ to an *oscillatory* one. Obviously, the inequalities $0<\alpha<N'<N-\alpha<N$ lead to

(2.14) $$0 < s_\alpha^2 < \tfrac{1}{2} < c_\alpha^2 < 1.$$

The two 2x2-matrices in (2.13b) characterise the coarse-grid correction and the smoothing iteration, respectively. The entries $c_\alpha^2 > s_\alpha^2$ express, that the smooth e^α-components converge more slowly than the non-smooth $e^{N-\alpha}$-components. The first matrix reflects the complementary behaviour for the coarse-grid correction: The smooth components (s_α^2 in the first column) are better reduced than the oscillatory ones (c_α^2 in the second column).

Remark 2.2 $\rho(M_\alpha) = \rho_\nu(s_\alpha^2)$ and $\|M_\alpha\|_2 = \zeta_\nu(s_\alpha^2)$ hold with

$$\rho_\nu(\xi) := \xi(1-\xi)^\nu + (1-\xi)\xi^\nu,$$
$$\zeta_\nu(\xi) := \sqrt{2[\xi^2(1-\xi)^{2\nu} + (1-\xi)^2\xi^{2\nu}]}.$$

The combination of (2.3), Lemma 1 and Remark 2 yields

$$\rho(M) = \max\{\rho_\nu(s_\alpha^2) : 1 \leq \alpha \leq N'\}, \quad \|M\|_2 = \max\{\zeta_\nu(s_\alpha^2) : 1 \leq \alpha \leq N'\}.$$

Since the values of the arguments s_α^2 for $1 \leq \alpha \leq N'$ lie between 0 and $\frac{1}{2}$ (cf. (2.14)), the following estimates are valid:

(2.15a) $$\rho(M) \leq \rho_\nu := \max\{\rho_\nu(\xi) : 0 \leq \xi \leq \tfrac{1}{2}\},$$

(2.15b) $$\|M\|_2 \leq \zeta_\nu := \max\{\zeta_\nu(\xi) : 0 \leq \xi \leq \tfrac{1}{2}\}.$$

The bounds ρ_ν and ζ_ν of the convergence rate and the contraction numbers depend on the smoothing number ν, however, they do not depend on the step size h. Since ρ_ν and ζ_ν decrease monotonously with increasing ν and since $\rho_1 = \zeta_1 = \frac{1}{2} < 1$, the convergence of the two-grid method for the one-dimensional model problem (1.1a) is proved. A detailed discussion of the functions $\rho_\nu(\xi)$ and $\zeta_\nu(\xi)$ and their maxima in $[0, \frac{1}{2}]$ yields the following

ν	ρ_ν	ζ_ν
1	1/2	1/2
2	1/4	1/4
3	1/8	0.150
4	0.0832	0.1159
5	0.0671	0.0947
10	0.0350	0.0496

Table 2.1 ρ_ν, ζ_ν

Theorem 2.3 Let the two-grid method for solving system (1.1a) be characterised by Richardson's iteration with $\Theta = h^2/4$ (identical to the Jacobi iteration damped by $\frac{1}{2}$) as smoother, by the piecewise linear prolongation p, and the adjoint restriction (1.15a). Then the two-grid method with $\nu \geq 1$ smoothing steps converges with the rate ρ_ν from (2.15a), which is h-independent. The contraction number (with respect to the Euclidean norm) is bounded by ζ_ν from (2.15b). For increasing ν these bounds have the asymptotical behaviour

(2.16) $$\rho_\nu = \frac{1}{e\nu} + O(\nu^{-2}), \quad \zeta_\nu = \frac{\sqrt{2}}{e\nu} + O(\nu^{-2}).$$

Some values of ρ_ν and ζ_ν are listed in Table 2.1. Obviously, the quantities $\rho(M)$ and $\|M\|_2$ converge with decreasing step size parameter to their bounds ρ_ν and ζ_ν, hence the given estimates are sharp. In §5 we will derive a convergence rate for general problems, which also behaves like $O(1/\nu)$. Theorem 3 demonstrates that such results about the asymptotical behaviour for large ν are not too pessimistic.

3. The Multi-Grid Iteration in the General Case

3.1 Two-dimensional Model Problem, Prolongation, Restriction

In the following $Ax=b$ or $A_\ell x_\ell = b_\ell$, respectively, is a discrete boundary value problem in two (or even more) dimensions. In particular, we may consider the five-point scheme for the Poisson equation $-\Delta u = f$ in the unit square. The generalisation of the one-dimensional piecewise linear interpolation (1.12a,b) is the *bilinear interpolation*

(3.1a) $$(pu_{\ell-1})(\xi,\eta) := u_{\ell-1}(\xi,\eta) \qquad \text{for } (\xi,\eta)\in\Omega_{\ell-1}\subset\Omega_\ell,$$

(3.1b) $$(pu_{\ell-1})(\xi,\eta) := \tfrac{1}{2}[u_{\ell-1}(\xi+h_\ell,\eta)+u_{\ell-1}(\xi-h_\ell,\eta)] \quad \text{for } \xi/h_\ell \text{ odd}, \eta/h_\ell \text{ even},$$

(3.1c) $$(pu_{\ell-1})(\xi,\eta) := \tfrac{1}{2}[u_{\ell-1}(\xi,\eta+h_\ell)+u_{\ell-1}(\xi,\eta-h_\ell)] \quad \text{for } \xi/h_\ell \text{ even}, \eta/h_\ell \text{ odd},$$

(3.1d) $$(pu_{\ell-1})(\xi,\eta) := \tfrac{1}{4}[u_{\ell-1}(\xi+h_\ell,\eta)+u_{\ell-1}(\xi-h_\ell,\eta) + u_{\ell-1}(\xi,\eta+h_\ell)+u_{\ell-1}(\xi,\eta-h_\ell)] \quad \text{for } \xi/h_\ell, \eta/h_\ell \text{ odd},$$

using $u_\ell(\xi,\eta) := 0$ for $\xi=0$ or $\xi=1$ or $\eta=0$ or $\eta=1$.

The abbreviating symbol for p from (3.1a-d) is the star

(3.1e) $$p = \begin{bmatrix} 1/4 & 1/2 & 1/4 \\ 1/2 & 1 & 1/2 \\ 1/4 & 1/2 & 1/4 \end{bmatrix}$$ («nine-point prolongation»).

In general, the stencil

(3.2a) $$p = \begin{bmatrix} \pi_{-1,1} & \pi_{0,1} & \pi_{1,1} \\ \pi_{-1,0} & \pi_{0,0} & \pi_{1,0} \\ \pi_{-1,-1} & \pi_{0,-1} & \pi_{1,-1} \end{bmatrix}$$

describes the mapping (3.2b), where the summation is taken over all i,j with $(\xi-ih_\ell,\eta-jh_\ell)\in\Omega_{\ell-1}$:

(3.2b) $$(pu_{\ell-1})(\xi,\eta) := \textstyle\sum_{i,j} \pi_{ij} u_{\ell-1}(\xi-ih_\ell,\eta-jh_\ell) \quad \text{for } (\xi,\eta)\in\Omega_\ell.$$

Further linear interpolations as well as prolongations of higher order are discussed by Hackbusch [14, §3.4]. Similarly, the stencil

(3.3a) $$r = \begin{bmatrix} \rho_{-1,1} & \rho_{0,1} & \rho_{1,1} \\ \rho_{-1,0} & \rho_{0,0} & \rho_{1,0} \\ \rho_{-1,-1} & \rho_{0,-1} & \rho_{1,-1} \end{bmatrix}$$

characterises the restriction

$$(3.3b)\qquad (ru_\ell)(\xi,\eta) = \sum_{i,j=-1}^{1} \rho_{ij}\, u_\ell(\xi+ih_\ell,\eta+jh_\ell) \qquad \text{for } (\xi,\eta)\in\Omega_{\ell-1},$$

The nine-point prolongation (3.1e) corresponds to the _nine-point restriction_

$$(3.3c)\qquad r = \frac{1}{4}\begin{bmatrix} 1/4 & 1/2 & 1/4 \\ 1/2 & 1 & 1/2 \\ 1/4 & 1/2 & 1/4 \end{bmatrix},$$

which can be considered as adjoint to (3.1e) based onto the scalar products (1.15b) with h_ℓ replaced by h_ℓ^d with the dimension $d=2$.

Remark 3.1 Assume the two-dimensional case $d=2$. Then the adjoint of p from (3.2a) is r from (3.3a) with $\rho_{ij}=\pi_{ij}/4$.

Having fixed the prolongation, one can always choose the adjoint

$$(3.4)\qquad r := p^*$$

as restriction.

3.2 Algorithm

The two-grid method is not yet suited for practical applications, since one still has to solve one system of equations per iteration at level $\ell-1$. The problem to be solved in (1.21c) has the form

$$(3.5)\qquad A_{\ell-1}e_{\ell-1} = d_{\ell-1},$$

hence it is of the same structure as the original problem $A_\ell x_\ell = b_\ell$. Instead of solving equation (3.5) exactly, one may approximate the solution iteratively. The iteration of choice is again the two-grid method, now applied to the levels $\ell-1$ and $\ell-2$ instead of ℓ and $\ell-1$. Then new auxiliary problems $A_{\ell-2}e_{\ell-2}=d_{\ell-2}$ arise, for which again the two-grid method (at the level $\ell-2$) is to be applied until equations $A_0e_0=d_0$ arise at the coarsest grid. The arising recursive method is the multi-grid iteration $\Phi_\ell^{\mathrm{MGM}(\nu_1,\nu_2)}$, which has the following algorithmical form.

(3.6) _procedure_ $\Phi_\ell^{\mathrm{MGM}(\nu_1,\nu_2)}(x_\ell,b_\ell)$;

(3.6a) _if_ $\ell=0$ _then_ $x_0 := A_0^{-1}b_0$ _else_

(3.6b) _begin for_ $i:=1$ _to_ ν_1 _do_ $x_\ell := \mathcal{S}_\ell(x_\ell,b_\ell)$;

(3.6c) $\quad d_{\ell-1} := r(A_\ell x_\ell - b_\ell)$;

(3.6d₁) $\quad e_{\ell-1}^{(0)} := 0$;

(3.6d₂) $\quad$ _for_ $i:=1$ _to_ γ _do_ $e_{\ell-1}^{(i)} := \Phi_{\ell-1}^{\mathrm{MGM}(\nu_1,\nu_2)}(e_{\ell-1}^{(i-1)},d_{\ell-1})$;

(3.6e) $\quad x_\ell := x_\ell - p\,e_{\ell-1}^{(\gamma)}$;

(3.6f) $\quad$ _for_ $i:=1$ _to_ ν_2 _do_ $x_\ell := \mathcal{S}_\ell(x_\ell,b_\ell)$;

(3.6g) $\quad \Phi_\ell^{\mathrm{MGM}(\nu_1,\nu_2)} := x_\ell$

end;

Obviously, the recursive procedure calls terminate after ℓ steps, when level $\ell=0$ is reached. Hence, the algorithm is well-defined.

ν_1 and ν_2 denote again the number of pre- and post-smoothing steps. In general, $\nu:=\nu_1+\nu_2>0$ is assumed. For the iterative solution of the coarse-grid equation (3.5), γ steps of the iteration $\Phi_{\ell-1}^{\mathrm{MGM}(\nu_1,\nu_2)}$ are applied to the starting value (3.6d₁). We will see that $\gamma=2$ is sufficient. Therefore, only the cases $\gamma=1$ and $\gamma=2$ are of practical interest. The multi-grid iteration with $\gamma=1$ has got the name «V-cycle», while the iteration with $\gamma=2$ is called «W-cycle».

The exact solution of linear equations is not completely avoided in the multi-grid algorithm (3.6). In (3.6a) the system $A_0 x_0 = b_0$ corresponding to the coarsest grid is to be solved. Since the coarsest grid has the smallest number of grid points, the solution involves no practical difficulties. According to (1.7c) $h_0 = \frac{1}{2}$ would be a possible choice of the coarsest grid size. In this case, $A_0 x_0 = b_0$ represents a single scalar equation.

3.3 Numerical Examples

The model problem with step size $h = h_5 = 1/64$ is taken as example. Table 1 shows the Euclidean norm $\|e^m\|_2$ of the errors and the reduction factors $\rho_{m+1,m} = \|e^m\|_2 / \|e^{m-1}\|_2$. The parameters are $\nu_1 = 2$, $\nu_2 = 0$, $h_0 = \frac{1}{2}$, the nine-point prolongation (3.1e) and nine-point restriction (3.3c). As smoothing iteration the chequer-board Gauß-Seidel method is chosen. Table 3.2 shows that the convergence rate behaves like $O(1/\nu)$.

m	$\gamma = 1$ (V-cycle) $\|e^m\|_2$	$\rho_{m+1,m}$	$\gamma = 2$ (W-cycle) $\|e^m\|_2$	$\rho_{m+1,m}$	$\gamma = \infty$ (two-grid algorithm) $\rho_{m+1,m}$
1	$1.3274_{10}-1$	$1.727_{10}-1$	$2.9984_{10}-02$	$3.902_{10}-2$	$4.124_{10}-2$
2	$2.2223_{10}-2$	$1.674_{10}-1$	$1.3219_{10}-03$	$4.408_{10}-2$	$5.126_{10}-2$
3	$3.7656_{10}-3$	$1.694_{10}-1$	$6.9050_{10}-05$	$5.223_{10}-2$	$5.443_{10}-2$
4	$6.4110_{10}-4$	$1.702_{10}-1$	$3.7824_{10}-06$	$5.477_{10}-2$	$5.741_{10}-2$
5	$1.0941_{10}-4$	$1.706_{10}-1$	$2.1584_{10}-07$	$5.706_{10}-2$	$5.939_{10}-2$
6	$1.8701_{10}-5$	$1.709_{10}-1$	$1.2689_{10}-08$	$5.879_{10}-2$	$6.125_{10}-2$
7	$3.1996_{10}-6$	$1.710_{10}-1$	$7.6788_{10}-10$	$6.051_{10}-2$	$6.280_{10}-2$

Table 3.1 Multi-grid results for the Poisson model problem, $h = 1/64$

ν	1	2	3	4	5	6	10
$\rho_3(\nu)$	0.222	0.062	0.04	0.03	0.023	0.0196	0.0133
$\frac{0.135}{\nu+0.135}$	0.119	0.063	0.043	0.033	0.026	0.022	0.0133

Table 3.2 Convergence factors for different smoothing numbers ν

In order to demonstrate that the multi-grid method works well not only for positive definite problems, the next example is the nonsymmetric differential equation

(3.7a) $$-\Delta u + c\, u_x = f$$

in $\Omega = (0,1) \times (0,1)$ with Dirichlet boundary values $u = 0$, discretised by

(3.7b) $$A_\ell = h_\ell^{-2} \begin{bmatrix} & -1 & \\ -1 & 4 & -1 \\ & -1 & \end{bmatrix} + \tfrac{1}{2} c\, h_\ell^{-1} \begin{bmatrix} & 0 & \\ -1 & 0 & 1 \\ & 0 & \end{bmatrix}.$$

First we choose $c = 4$ (for this value all A_ℓ are M-matrices). $f = u = 0$ are taken as right-hand side and exact solution, while $x(1-x+y)$ serves as starting value. The other parameters are the same as in Table 3.1. The W-cycle (γ=2) shows a convergence rate of ≈ 0.06 and differs hardly from the corresponding rate of the Poisson model case.

As soon as the coefficient c becomes substantially larger, e.g. $c = 100$, a stability problem arises: The discretisation (3.7b) yields an M-matrix for $h_5 = 1/64$, but not for larger h. A remedy is the matrix-dependent prolongation and the corresponding restriction (explained in Hackbusch [20,§10]) together with the Galerkin product (1.19) for $\ell < 5$. Table 3.4 shows the convergence rates $\rho_{m+1,m}$ for $\gamma = 1$ and $\gamma = 2$. Differently from the model case, the results for $\gamma = 2$ are hardly better than those for $\gamma = 1$. Further, the rate ≈ 0.3 is not so favourable. The rate can be improved to ≈ 0.14 by smoothing with the row-wise Gauß-Seidel iteration instead of the (pointwise) Gauß-Seidel iteration (Table 3.4, right column).

For examples of indefinite problems compare Hackbusch [20,§10]).

m	$\rho_{m+1,m}$
1	$3.02452_{10}-2$
2	$4.72247_{10}-2$
3	$5.30817_{10}-2$
4	$5.51039_{10}-2$
5	$5.69383_{10}-2$
6	$5.83535_{10}-2$
7	$5.97048_{10}-2$
8	$6.09182_{10}-2$
9	$6.20609_{10}-2$
10	$6.31206_{10}-2$

Table 3.3 (3.7b) for $c=4$, $h=1/64$

m	pointwise Gauß-Seidel it. $\gamma=1$	$\gamma=2$	line-G.-S. $\gamma=1$
1	$1.58433_{10}-1$	$2.75119_{10}-2$	$4.65472_{10}-2$
2	$2.60220_{10}-1$	$9.54499_{10}-2$	$9.99392_{10}-2$
3	$3.35135_{10}-1$	$2.73365_{10}-1$	$9.51973_{10}-2$
4	$3.47875_{10}-1$	$3.00345_{10}-1$	$1.31862_{10}-1$
5	$3.35962_{10}-1$	$2.94467_{10}-1$	$1.26717_{10}-1$
6	$3.14166_{10}-1$	$3.06239_{10}-1$	$1.47075_{10}-1$
7	$2.92044_{10}-1$	$3.19984_{10}-1$	$1.30377_{10}-1$
8	$2.71982_{10}-1$	$3.34751_{10}-1$	$1.48727_{10}-1$
9	$2.55278_{10}-1$	$3.25717_{10}-1$	$1.32756_{10}-1$

Table 3.4 Equation (3.7b) for $c=100$, $h=1/64$ smoothing by pointwise and line-Gauß-Seidel it.

3.4 Computational Work

To judge the convergence rates from §3.3, one has to take into account the amount of work per iteration. The operations appearing in (3.6) are $\mathcal{S}_\ell$ in (3.6b,f), $r(A_\ell x_\ell - b_\ell)$ in (3.6c) and $x_\ell - p e_{\ell-1}$ in (3.6e). All are proportional to the dimension n_ℓ. It is easy see, that under assumptions which are satisfied in standard situations the total work of one multi-grid iteration is again proportional to n_ℓ (cf. Hackbusch [14] and [20,§10]). There also the optimal choice of the smoothing number ν is discussed.

3.5 Iteration Matrix

Since the iteration is defined recursively, the iteration matrix of the multi-grid method is also determined recursively.

Theorem 3.2 Let S_ℓ be the iteration matrix of the (consistent) pre- and post-smoothing iteration $\mathcal{S}_\ell$. Then the multi-grid iteration $\Phi_\ell^{\mathrm{MGM}(\nu_1,\nu_2)}$ is also consistent. Its iteration matrix $M_\ell^{\mathrm{MGM}} = M_\ell^{\mathrm{MGM}}(\nu_1,\nu_2)$ is defined by

(3.8a) $$M_0^{\mathrm{MGM}} = O, \quad M_1^{\mathrm{MGM}} = M_1^{\mathrm{TGM}}(\nu_1,\nu_2),$$

(3.8b) $$M_\ell^{\mathrm{MGM}} = M_\ell^{\mathrm{TGM}}(\nu_1,\nu_2) - S_\ell^{\nu_2}\, p\, (M_{\ell-1}^{\mathrm{MGM}})^\gamma A_{\ell-1}^{-1}\, r A_\ell S_\ell^{\nu_1} \quad \text{for } \ell \geq 1.$$

4. Nested Iteration

4.1 Algorithm

The presence of a hierarchy of equations (1.8a): $A_\ell x_\ell = b_\ell$ for all levels $0 \leq \ell \leq \ell_{\max}$ is necessary for the applicability of the multi-grid method. This situation can even be exploited for another purpose. Obviously, the result x_ℓ^m of an iteration is the better, the better the starting iterate x_ℓ^0 is. Usually, the solutions x_ℓ approximate a continuous solution x by a certain positive consistency order $O(h_\ell^\varkappa)$. Via the triangle inequality, x_ℓ and $x_{\ell-1}$ should also differ only by $O(h_\ell^\varkappa + h_{\ell-1}^\varkappa)$. This assumption can be expressed as follows:

(4.1) $\|x_\ell - \tilde{p}\, x_{\ell-1}\| \leq C_1 h_\ell^\varkappa$ $\qquad (\varkappa>0,\ x_\ell,\ x_{\ell-1}$ solutions to (1.8a)).

Here, $\tilde{p}: X_{\ell-1} \to X_\ell$ is a suitable prolongation which may not necessarily coincide with p from §1.4 or with p from (3.6e).

Inequality (4.1) suggests using approximations to $x_{\ell-1}$ as starting iterate of the iteration at level ℓ. The algorithm as proposed by Kronsjö/Dahlquist [1] reads

```
         x̃_0 := suitable approximation of the solution of A_0 x_0 = b_0;
         for ℓ := 1 to ℓ_max do
(4.2a)   begin x̃_ℓ := p̃ x̃_{ℓ-1};
               for i := 1 to m_ℓ do  x̃_ℓ := Φ_ℓ(x̃_ℓ, b_ℓ)
         end;
```

Note that (4.2a) is not an iteration in the proper sense but a *finite* process. Here we are only interested in the multi-grid iteration $\Phi_\ell = \Phi_\ell^{\mathrm{MGM}}$, for which the numbers m_ℓ can be chosen independently of ℓ:

(4.2b) $m_\ell = m \qquad (\ell \geq 1).$

In Remark 4.2 we will see that even the smallest number $m=1$ is of practical interest.

4.2 Error Analysis

Multi-grid methods (and other methods with h-independent convergence rate) satisfy the condition

(4.3) $\|M_\ell^\Phi\| \leq \zeta < 1$ for all $\ell \geq 1$, $\quad M_\ell^\Phi$ iteration matrix of Φ_ℓ.

The ratios $n_\ell/n_{\ell-1}$ of the dimensions should be bounded:

(4.4) $n_\ell \leq \bar{C}_h n_{\ell-1}.$

By $n_\ell/n_{\ell-1} \approx (h_{\ell-1}/h_\ell)^d$, inequality (4.4) also gives an estimate of $h_{\ell-1}/h_\ell$ from above. Together with the norm of $\tilde{p}$ one obtains an estimate of the form

(4.5) $\|\tilde{p}\|(h_{\ell-1}/h_\ell)^\varkappa \leq C_2 \qquad (\tilde{p}: X_{\ell-1} \to X_\ell)$

with $\varkappa$ from (4.1).

Theorem 4.1 Assume (4.1), (4.3), (4.5). The iteration number $m_\ell = m$ (cf. (2b)) should be sufficiently large so that

(4.6) $C_2 \zeta^m < 1.$

Then the nested iteration (4.2a,b) produces results $\tilde{x}_\ell$ for all levels $0 \leq \ell \leq \ell_{\max}$ which satisfy the error estimates

(4.7a) $\|\tilde{x}_\ell - x_\ell\| \leq C_3(\zeta^m)\, C_1 h_\ell^\varkappa$ $\qquad$ with

(4.7b) $C_3(\zeta^m) := \zeta^m/(1 - C_2\zeta^m),$

provided that the starting iterate $\tilde{x}_0$ satisfies inequality (4.7a) for $\ell=0$.

The proof is left to the reader (cf. Hackbusch [14],[20]). Estimate (4.7a) has an important practical interpretation. It describes that the iteration error $\tilde{x}_\ell - x_\ell$ coincides up to a factor $C_3(\zeta^m)$ with the bound $C_1 h_\ell^\varkappa$ from (4.1), which is related to the discretisation error. As long as x_ℓ is only considered as an approximation to a continuous solution of a differential equation, it makes no sense to compute x_ℓ more precisely than by the discretisation error. The nested iteration provides a convenient way to compute $\tilde{x}_\ell$ with

(4.8) $\qquad \|\tilde{x}_\ell - x_\ell\| \leqslant C_3(\zeta^m) * \text{discretisation error},$

where the constant $C_3(\zeta^m)$ can be controlled without knowing the discretisation error bound $C_1 h_\ell^\varkappa$ quantitatively.

Remark 4.2 The standard choice $h_\ell = h_{\ell-1}/2$ and the inequality $\|\tilde{p}\| \leqslant 1$, which is valid for the standard interpolations, yield the constant $C_2 = 2^\varkappa$ in (4.5). The consistency order of the model case is $\varkappa = 2$, whence $C_2 = 4$. Therefore, the factor $C_3(\zeta^m)$ has the value

(4.9) $\qquad C_3(\zeta^m) = \zeta^m/(1-4\zeta^m).$

For multi-grid methods with convergence rates $\leqslant \zeta = 0.2$ (cf. results in §3.3) condition (4.6) is satisfied for only *one* iteration step ($m=1$) and produces the value $C_3(0.2) = 1$.

4.3 Amount of Computational Work

Let Cn_ℓ be the work required by the iteration Φ_ℓ at level ℓ and assume (4.7): $n_{\ell-1} \leqslant n_\ell / C_h$. The work for $\tilde{x}_{\ell-1} \mapsto \tilde{p}\,\tilde{x}_{\ell-1}$ is considered as negligible. The total amount of work $C' n_\ell \leqslant m C(n_1 + n_2 + \ldots + n_\ell)$ can be estimated by $C' \leqslant m C C_h/(C_h - 1)$. For the standard case $C_h = 2^d = 4$ (cf. Remark 4.1) we obtain the result

(4.10) $\qquad work_{\text{nested it (4.2a,b)}} \leqslant \frac{4m}{3}\, work(\Phi_{\ell_{\max}}).$

If we tried to achieve the accuracy $\varepsilon = C h^\varkappa$ at the level $\ell = \ell_{\max}$ with the starting iterate $\tilde{x}_\ell := 0$ by iterating with Φ_ℓ, the work would be proportional to $O(\log \varepsilon) = O(|\log h_\ell|)$. According to Remark 2, $m=1$ is a realistic choice. Inequality (4.10) shows that a sufficient accuracy for *all* levels $0 \leqslant \ell \leqslant \ell_{\max}$ can be attained with the $4/3$-fold work of a single Φ_ℓ-step.

From the operation count of the V- and W-cycle (cf. Hackbusch [20, §10.4.4]) and from the rates of Table 3.1 (with $\nu_1 = 2$, $\nu_2 = 0$, $m = 1$) we obtain the following results:

(4.11a) The V-cycle ($\gamma = 1$) requires $34\, n_{\ell_{\max}}$ operations to produce $\|\tilde{x}_\ell - x_\ell\| \leqslant 0.53\; C_1 h_\ell^\varkappa$ for $0 \leqslant \ell \leqslant \ell_{\max}$.

(4.11b) The W-cycle ($\gamma = 2$) requires $51\, n_{\ell_{\max}}$ operations to produce $\|\tilde{x}_\ell - x_\ell\| \leqslant 0.08\; C_1 h_\ell^\varkappa$ for $0 \leqslant \ell \leqslant \ell_{\max}$.

The work given in (11b) corresponds to ≈ 10 Gauß-Seidel iteration steps.

4.4 Numerical Examples

First, the nested iteration is applied to the differential equation

(4.12a) $\qquad -\Delta u = f := -\Delta(e^{x+y^2})$

with boundary values $\varphi = e^{x+y^2}$. The negative Laplacean $-\Delta$ is discretised at all levels by the standard five-point star. The interpolation $\tilde{p}$ is the cubic one. Let x_ℓ^* be the restriction of the exact solution e^{x+y^2} of (4.12a) to the grid Ω_ℓ. Note that x_ℓ^* does not coincide with the discrete solution x_ℓ of the system $A_\ell x_\ell = b_\ell$ corresponding to (4.12a). In Table 4.1 the results $\tilde{x}_\ell$ of the nested iteration (4.2a) are compared with x_ℓ^*, since this is the error interesting in practice. The maximum norm $\|\tilde{x}_\ell - x_\ell^*\|_\infty$ of these errors is given for the cases $m=1$ and $m=2$. For comparison the last column shows the discretisation error $\|x_\ell - x_\ell^*\|_\infty$, which formally corresponds to $m=\infty$. The multi-grid method used in (4.2a) has the same parameters as the W-cycle ($\gamma=2$) in Table 3.1. One

ℓ	h_ℓ	$m=1$	$m=2$	$m=\infty$
0	1/2	$2.6462500_{10}+0$	$2.6462500_{10}+0$	$2.6462500_{10}+0$
1	1/4	$3.0581081_{10}-1$	$2.1228706_{10}-2$	$2.8969487_{10}-2$
2	1/8	$2.8055559_{10}-2$	$8.0669104_{10}-3$	$8.0307789_{10}-3$
3	1/16	$2.3509591_{10}-3$	$2.0768939_{10}-3$	$2.0729855_{10}-3$
4	1/32	$5.6528936_{10}-4$	$5.2253778_{10}-4$	$5.2247389_{10}-4$
5	1/64	$1.3346811_{10}-4$	$1.3093939_{10}-4$	$1.3093954_{10}-4$

Table 4.1 Errors $\|\tilde{x}_\ell - x_\ell\|_\infty$ of the nested iteration for (4.12a)

ℓ	h_ℓ	$m=1$	$m=2$	$m=\infty$
0	1/2	$2.5171762_{10}+0$	$2.5171762_{10}+0$	$2.5171762_{10}+0$
1	1/4	$4.8952876_{10}-1$	$4.6274562_{10}-1$	$4.7880033_{10}-1$
2	1/8	$9.6976746_{10}-2$	$1.0331284_{10}-1$	$1.0308770_{10}-1$
3	1/16	$2.7728844_{10}-2$	$2.6636714_{10}-2$	$2.6689213_{10}-2$
4	1/32	$6.8792352_{10}-3$	$6.6486368_{10}-3$	$6.6506992_{10}-3$
5	1/64	$1.6998756_{10}-3$	$1.6716068_{10}-3$	$1.6714014_{10}-3$

Table 4.2 Errors $\|\tilde{x}_\ell - x_\ell\|_\infty$ of the nested iteration for (4.12b)

understands from Table 4.1, that the choice $m=1$ is sufficient. Doubling the work by using $m=2$, we cannot improve the total error $\|\tilde{x}_\ell - x_\ell^*\|_\infty$ substantially.

Analogous data are given in Table 4.2 for the differential equation

(4.12b) $\quad -\Delta u = f := -\Delta(y\sin(10x))$

with the solution $y\sin(10x)$ being oscillatory in x-direction. By the non-smooth behaviour of the solution, the discretisation error (last column) for problem (4.12b) is by nearly one digit worse than for (4.12a). Therefore, the additional error $O(h_\ell^2)$ of the *linear* interpolation $\tilde{p}$, which is used instead of the cubic one, is of minor weight. For this example, it neither does pay to perform $m=2$ iterations per level.

5. Convergence Analysis

5.1 Summary

The convergence proof of multi-grid methods differs from the convergence considerations used before, since here the relationship between the equations $A_\ell x_\ell = b_\ell$ and $A_{\ell-1}x_{\ell-1} = b_{\ell-1}$ plays an important rôle.

As sufficient criteria we introduce and discuss two conditions in §§5.2-3: the smoothing and the approximation property. The smoothing property is of algebraical nature, whereas the proof of the approximation property involves the continuous problem, whose discretisation is described by $A_\ell x_\ell = b_\ell$. Smoothing and approximation property together yield the convergence statement for the two-grid methods (§5.4). For $\gamma \geqslant 2$, the multi-grid convergence can be concluded directly from the two-grid convergence (§5.5).

For positive definite A_ℓ, the multi-grid method can be designed as symmetric iteration. In §6 we achieve for this case even better convergence results, which include the V-cycle ($\gamma=1$).

5.2 Smoothing Property

In §1.2 we have called a grid function $x_\ell = \sum \xi_\alpha e^\alpha$ (cf. (1.2b)) to be «smooth» if the coefficients ξ_α of high frequencies α corresponding to the large eigenvalues λ_α are small. Quantitatively, one may measure the smoothness by $\|A_\ell x_\ell\|_2 = (\sum |\lambda_\alpha \xi_\alpha|^2)^{1/2}$. If the smoothing step (1.21a) really leads to a smoothing of the errors $e_\ell = x_\ell^0 - x_\ell$, the error $S_\ell^\nu e_\ell$ obtained from the smoothing step must have a better smoothing measure $\|A_\ell S_\ell^\nu e_\ell\|_2$ than e_ℓ. Therefore, the smoothing ability is characterised by the spectral norm $\|A_\ell S_\ell^\nu\|_2$. Before defining the smoothing property, we analyse $\|A_\ell S_\ell^\nu\|_2$ for Richardson's iteration with positive definite A_ℓ:

(5.1a) $\qquad \mathscr{S}_\ell(x_\ell, b_\ell) := x_\ell - \Theta(A_\ell x_\ell - b_\ell) \qquad$ with

(5.1b) $\qquad \Theta = \Theta_\ell = 1/\rho(A_\ell) = 1/\|A_\ell\|_2 .$

We have $\|A_\ell S_\ell^\nu\|_2 = \|A_\ell (I - \Theta A_\ell)^\nu\|_2 = \|X(I-X)^\nu\|_2 / \Theta$ with $X := \Theta A_\ell$.

Lemma 5.1 (a) For all matrices X with $0 \leq X \leq I$ the inequality

(5.2a) $\qquad \|X(I-X)^\nu\|_2 \leq \eta_0(\nu) \qquad (\nu \geq 0)$

holds, where the function $\eta_0(\nu)$ is defined by

(5.2b) $\qquad \eta_0(\nu) := \nu^\nu / (\nu+1)^{\nu+1}.$

(b) The asymptotical behaviour of $\eta_0(\nu)$ for $\nu \to \infty$ is

(5.2c) $\qquad \eta_0(\nu) = \frac{1}{e\nu} + O(\nu^{-2}).$

Remark 5.2 For $A_\ell > 0$, Richardson's method (5.1a,b) leads to

(5.3) $\qquad \|A_\ell S_\ell^\nu\|_2 \leq \eta_0(\nu) \|A_\ell\|_2 \qquad$ for all $\nu \geq 0,\ \ell \geq 0$.

Note that the factor $\eta_0(\nu)$ is independent of h_ℓ and ℓ. The smoothing property, which we are going to define, is an estimate of similar form as (5.3). Instead of $\eta_0(\nu)$ we may take an arbitrary sequence $\eta(\nu) \to 0$. Further, it is neither necessary nor desirable to require an inequality like (5.3) for _all_ $\nu \geq 0$.

Definition 5.3 An iteration $\mathscr{S}_\ell$ $(\ell \geq 0)$ satisfies the _smoothing property_, if there are functions $\eta(\nu)$ and $\bar{\nu}(h)$ independent of ℓ with

(5.4a) $\quad \|A_\ell S_\ell^\nu\|_2 \leq \eta(\nu) \|A_\ell\|_2 \qquad$ for all $0 \leq \nu < \bar{\nu}(h_\ell),\ \ell \geq 1$,

(5.4b) $\quad \lim_{\nu\to\infty} \eta(\nu) = 0,$

(5.4c) $\quad \lim_{h\to 0} \bar{\nu}(h) = \infty \qquad$ or $\qquad \bar{\nu}(h) = \infty.$

$\bar{\nu}(h) = \infty$ in (5.4c) expresses that (5.4a) holds for _all_ ν. This happens only for *convergent* iterations $\mathscr{S}_\ell$ because of

Remark 5.4 The conditions (5.4a,b) with $\bar{\nu}(h) = \infty$ imply the convergence of $\mathscr{S}_\ell$.

From Remark 2 one concludes

Theorem 5.5 For $A_\ell > 0$ Richardson's method (5.1a,b) satisfies the smoothing property (5.4a-c) with $\eta(\nu) := \eta_0(\nu)$ and $\bar{\nu}(h) = \infty$.

The reason of the more general condition (5.4c) instead of $\bar{\nu}(h) = \infty$ is that the smoothing property can also be formulated for non-convergent iterations.

The following theorem can be considered as perturbation lemma. It shows that the smoothing property remains valid under the perturbation of the matrix A'_ℓ into $A_\ell = A'_\ell + A''_\ell$, where A_ℓ may be indefinite and non-symmetric. For the proof compare Hackbusch [14 or 20].

Theorem 5.6 Let $A_\ell = A'_\ell + A''_\ell$. Let $\mathcal{S}_\ell$ and $\mathcal{S}'_\ell$ be the smoothing iterations corresponding to A_ℓ and A'_ℓ, respectively. Their iteration matrices are denoted by S_ℓ and S'_ℓ with $S''_\ell := S_\ell - S'_\ell$. Assume

(5.5a) A'_ℓ and S'_ℓ satisfy the smoothing property with $\eta'(\nu)$, $\bar{\nu}'(h)$,

(5.5b) $\|S'_\ell\|_2 \leqslant C'_S$ for all $\ell \geqslant 1$,

(5.5c) $\lim_{\ell\to\infty} \|S''_\ell\|_2 = 0$,

(5.5d) $\lim_{\ell\to\infty} \|A''_\ell\|_2 / \|A_\ell\|_2 = 0$.

Then the iteration $\mathcal{S}_\ell$ for A_ℓ also satisfies the smoothing property. The corresponding bound $\eta(\nu)$ can be chosen e.g. as $\eta(\nu) := 2\eta'(\nu)$.

The smoothing property can be proved not only for the Richardson method but also for the *damped* (block-)Jacobi iteration, the 2-cyclic Gauß-Seidel iteration (in particular, the chequer-board Gauß-Seidel method for five-point formulae) and the Kaczmarz iteration. Furthermore, for symmetric iterations as the symmetric Gauß-Seidel method, SSOR, and the ILU iteration. The symmetric case will be considered in §10.7.3. The smoothing property does *not* hold e.g. for the non-damped Jacobi method or the SOR method with $\omega \geqslant \omega_{opt}$. For the smoothing analysis of the iterations mentioned above compare Hackbusch [14,§6.2].

5.3 Approximation Property

By the coarse-grid correction the fine-grid solution e_ℓ of $A_\ell e_\ell = d_\ell$ is replaced by $p e_{\ell-1}$ from $A_{\ell-1} e_{\ell-1} = d_{\ell-1} := r d_\ell$. Therefore, $p e_{\ell-1} \approx e_\ell$, i.e. $p A_{\ell-1}^{-1} r d_\ell \approx A_\ell^{-1} d_\ell$ should be valid. We quantify this requirement by

(5.6) $$\|p A_{\ell-1}^{-1} r d_\ell - A_\ell^{-1} d_\ell\|_2 \leqslant C_A \|d_\ell\|_2 / \|A_\ell\|_2 \quad \text{for all } \ell \geqslant 1,\ d_\ell \in X_\ell.$$

(5.6) can be rewritten by means for the matrix norm (spectral norm) as

approximation property		
(5.7)	$\|A_\ell^{-1} - p A_{\ell-1}^{-1} r\|_2 \leqslant C_A / \|A_\ell\|_2$	for all $\ell \geqslant 1$.

Proofs of the approximation property (5.7) are not of algebraical nature but use (at least indirectly) properties of the underlying boundary-value problem. The easiest proof of the approximation property can be given for Galerkin discretisations (cf. Hackbusch [14] or [20, §10.6.3]).

5.4 Convergence of the Two-Grid Iteration

As mentioned in §1.7, $\rho(M_\ell^{TGM}(\nu_1,\nu_2)) = \rho(M_\ell^{TGM}(\nu,0))$ holds for $\nu = \nu_1 + \nu_2$, so that we may restrict our considerations to $\nu_1 > 0$, $\nu_2 = 0$. This choice is optimal for statements concerning the contraction number $\|M_\ell^{TGM}(\nu,0)\|_2$ with respect to the spectral norm. The following Theorems 7 and 8 correspond to the cases $\bar{\nu}(h) = \infty$ and $\bar{\nu}(h) < \infty$.

Theorem 5.7 Assume the smoothing and approximation properties (5.4a-c), (5.7) with $\bar{\nu}(h) = \infty$. For given $0 < \zeta < 1$ there exists a lower bound $\underline{\nu}$, such that

(5.8) $$\|M_\ell^{TGM}(\nu,0)\|_2 \leqslant C_A \eta(\nu) \leqslant \zeta \quad \text{for all } \nu \geqslant \underline{\nu},\ \ell \geqslant 1.$$

Here, C_A and $\eta(\nu)$ are the quantities from (5.7) and (5.4a,b). By $\zeta<1$, (5.8) implies the convergence of the two-grid iteration. The contraction bound $C_A\eta(\nu)$ is h_ℓ-independent!

Theorem 5.8 Assume the smoothing and the approximation properties (5.4a-c), (5.7), possibly with $\bar{\nu}(h)=\infty$ or $\bar{\nu}(h)<\infty$. For given $0<\zeta<1$ there exist bounds $\bar{h}>0$ and $\underline{\nu}$, such that (5.8) holds for all $\nu\in[\underline{\nu},\bar{\nu}(h))$ and all ℓ with $h_\ell\leq\bar{h}$, where the interval $[\underline{\nu},\bar{\nu}(h))$ is not empty (i.e. $\underline{\nu}<\bar{\nu}(h)$).

For the proof the theorems compare Hackbusch [20, Theorems 10.6.15-16].

5.5 Convergence of the Multi-Grid Iteration

In Theorem 3.2 the representation $M_\ell^{MGM}(\nu,0)=M_\ell^{TGM}(\nu,0)-\ldots$ of the multi-grid iterations matrix has been shown. We shall exploit the fact that the perturbation «...» is sufficiently small, so that the two-grid convergence implies the multi-grid convergence. Besides the smoothing and the approximation properties we require further conditions, which however are easy to satisfy. The first one is

(5.9) $\qquad \|S_\ell^\nu\|_2\leq C_S \qquad$ for all $\ell\geq1,\ 0<\nu<\bar{\nu}:=\min_{\ell\geq1}\bar{\nu}(h_\ell)$

with $\bar{\nu}(h_\ell)$ from (5.4c).

Remark 5.9 Assume $S_\ell:=S_\ell'+S_\ell''$. Let (5.9) hold for $S_\ell'^{\nu}$, and assume (5.5c): $\lim_{\ell\to\infty}\|S_\ell''\|_2=0$. Then, analogously to Theorem 7, (5.9) holds for S_ℓ.

Further, the comparability of $x_{\ell-1}$ and its prolongated (interpolated) value $px_{\ell-1}$ is requested:

(5.10) $\qquad \underline{C}_p^{-1}\|x_{\ell-1}\|_2\leq\|px_{\ell-1}\|_2\leq\bar{C}_p\|x_{\ell-1}\|_2 \qquad$ for all $x_{\ell-1}\in X_{\ell-1},\ \ell\geq1$.

$pA_{\ell-1}^{-1}rA_\ell S_\ell^\nu = S_\ell^\nu-[A_\ell^{-1}-pA_{\ell-1}^{-1}r]A_\ell S_\ell^\nu = S_\ell^\nu-M_\ell^{TGM}(\nu,0)$ implies

Lemma 5.10 Let (5.9) and (5.10) be valid. Then

(5.11) $\qquad \|A_{\ell-1}^{-1}rA_\ell S_\ell^\nu\|_2\leq\underline{C}_p(C_S+\|M_\ell^{TGM}(\nu,0)\|_2)$.

Assume $\nu=\nu_1>0,\ \nu_2=0$ as in §5.4. Using (5.10) and (5.11), one can estimate the multi-grid iterations matrix from (3.8a,b) by

(5.12a) $\qquad \|M_\ell^{MGM}(\nu,0)\|_2\leq\|M_\ell^{TGM}(\nu,0)\|_2+C^*\|M_\ell^{TGM}(\nu,0)\|_2^\gamma \qquad$ for $\ell\geq1$

(5.12b) $\qquad C^*:=\underline{C}_p\bar{C}_p(C_S+1)$.

Here, it is assumed that ν is chosen large enough so that $\|M_\ell^{TGM}(\nu,0)\|_2\leq1$ according to Theorem 7 or 8. Together with $M_0^{MGM}=O$ (cf. (3.8a)) inequality (5.12a) leads to the recursive inequalities (5.13a) for the quantities ζ_ℓ:

(5.12c) $\qquad \zeta_\ell:=\|M_\ell^{MGM}(\nu,0)\|_2$,

(5.13a) $\qquad \zeta_0:=0,\ \zeta_\ell\leq\zeta+C^*(\zeta_{\ell-1})^\gamma \qquad$ for $\ell\geq1$.

Here, ζ is the ℓ-independent bound for the two-grid convergence, whose existence is stated by Theorem 7 or 8:

(5.13b) $\qquad \|M_\ell^{TGM}(\nu,0)\|_2\leq\zeta$.

An analysis of the fixed point equation $x=\zeta+C^*x^\gamma$ yields

Lemma 5.11 Assume $\gamma\geq2$, $C^*\gamma>1$ and $\zeta\leq\frac{\gamma-1}{\gamma}/(C^*\gamma)^{1/(\gamma-1)}$. Then all solutions of the inequalities (5.13a) are bounded by

(5.14a) $\qquad \zeta_\ell \leqslant \zeta^* \leqslant \frac{\gamma}{\gamma-1}\zeta < 1 \qquad$ for all $\ell \geqslant 0$.

For the most interesting case $\gamma=2$ one has

(5.14b) $\qquad \zeta^* = 2\zeta/(1+\sqrt{1-4C^*\zeta}) \qquad$ for ζ^* from (5.14a).

Since, by (5.12c), ζ_ℓ are the contraction number bounds, we obtain

Theorem 5.12 Assume the smoothing and the approximation properties (5.4a-c) and (5.7), the conditions (5.9), (5.10), and in addition $\gamma \geqslant 2$. As in the Theorems 7/8 for every $0<\zeta'<1$ there are $\underline{\nu}$ and $\bar{h}>0$, such that

(5.15a) $\qquad \|M_\ell^{MGM}(\nu,0)\|_2 \leqslant \zeta' < 1 \qquad$ for $\underline{\nu} \leqslant \nu < \bar{\nu} := \min_{\ell\geqslant 1} \bar{\nu}(h_\ell)$,

provided $h_1 \leqslant \bar{h}$. $\underline{\nu} < \bar{\nu}$ holds. In the case of $\bar{\nu}(h)=\infty$ one may set $\bar{h} := \infty$ (i.e. the choice of the grid size is not restricted; cf. Hackbusch [20, Theorem 10.6.22]).

5.6 The Case of Weaker Regularity

The proof of the approximation property has made use of the full regularity, which in the case of Poisson's equation $A^{-1}=(-\Delta)^{-1}$ reads as $U=L^2(\Omega)=H^0(\Omega)\rightarrow H^2(\Omega)\cap H_0^1(\Omega)$. This assumption is true for convex domains, but it does not hold e.g. for domains with re-entrant corners. In the general case, one obtains statements of the form

$$A^{-1}:\ H^{-\sigma m}(\Omega) \rightarrow H^{(2-\sigma)m}(\Omega)\cap H_0^m(\Omega) \quad \text{for some } \sigma\in(0,1)$$

(cf. Hackbusch [15,§9.1]). Then, the approximation property (5.7) cannot be proved, but has to be formulated by means of other norms. The analysis shows that the contraction number behaves only as $O(1/\nu^{1-\sigma})$ with $0<\sigma<1$ measuring the degree of regularity, while $O(1/\nu)$ holds in the standard case (cf. (2.16)).

6. Symmetric Multi-Grid Methods

The analysis of the multi-grid iteration in §5 is performed for the general (non-symmetric) case in order to emphasize that multi-grid iterations are not restricted to symmetric or even only to positive definite problems. However, the symmetric case admits some stronger statements. Proofs of the following results can be found in Hackbusch [14] or [20,§10.7].

The required symmetry conditions are

(6.1a) $\qquad r=p^*$ and $A_\ell>0$ (i.e. A_ℓ positive definite) $\qquad$ for all $\ell\geqslant 0$,

(cf. (3.4)). Further, we use the variant (1.22b) with the pre-smoothing $\mathcal{S}_\ell$ and the post-smoothing $\hat{\mathcal{S}}_\ell$, where the post-smoothing is adjoint to $\mathcal{S}_\ell$:

(6.1b) $\qquad \hat{\mathcal{S}}_\ell = \mathcal{S}_\ell^* \qquad$ for all $\ell\geqslant 0$

(cf. Hackbusch [20,§4.8.4]). The number of pre- and post-smoothing should be equal:

(6.1c) $\qquad \nu_1=\nu_2=\nu/2>0$.

Occasionally, we need the Galerkin product property:

(6.1d) $\qquad A_{\ell-1} = r A_\ell p$.

If (6.1a-c), the two- and multi-grid iterations $\Phi_\ell^{TGM}(\frac{\nu}{2},\frac{\nu}{2})$ and $\Phi_\ell^{MGM}(\frac{\nu}{2},\frac{\nu}{2})$ are symmetric, i.e. their transformed iteration matrices $A_\ell^{1/2}M_\ell A_\ell^{-1/2}$ are Hermitian (here we use the abbreviations $M_\ell=M_\ell^{TGM}(\frac{\nu}{2},\frac{\nu}{2})$ and $M_\ell^{MGM}(\frac{\nu}{2},\frac{\nu}{2})$). If the iteration converges, it converges *monotonously with respect to the energy norm* $\|x_\ell\|_{A_\ell} := \|A_\ell^{1/2}x_\ell\|_2$.

The case $\nu_1=\nu>0$, $\nu_2=0$ has been treated in §5. The technique used there can also be applied to the general case $\nu_1\geq 0$, $\nu_2\geq 0$, $\nu:=\nu_1+\nu_2>0$ and specially to $\nu_1=\nu_2=\nu/2$.

Assume (1a,b). Then

(6.2a) $$\Phi_\ell^{TGM}(0,\nu_2)=\left(\Phi_\ell^{TGM}(\nu_2,0)\right)^*,\quad \Phi_\ell^{MGM}(0,\nu_2)=\left(\Phi_\ell^{MGM}(\nu_2,0)\right)^*,$$

(6.2b) $$\Phi_\ell^{TGM}(\nu_1,\nu_2) = \Phi_\ell^{TGM}(0,\nu_2)\circ\Phi_\ell^{TGM}(\nu_1,0) \quad \text{in the case (6.1d).}$$

Under the assumption (6.1d) the statements (6.2c,d) follow for the two-grid iteration matrices $M_\ell:=M_\ell^{TGM}$ from (6.2a,b):

(6.2c) $$M_\ell(\nu_1,\nu_2) = M_\ell(0,\nu_2)\,M_\ell(\nu_1,0) = A_\ell^{-1}M_\ell(\nu_2,0)\,A_\ell M_\ell(\nu_1,0),$$

(6.2d) $$A_\ell^{1/2}M_\ell(\nu_1,\nu_2)A_\ell^{-1/2}=\left(A_\ell^{1/2}M_\ell(\nu_2,0)A_\ell^{-1/2}\right)^H\left(A_\ell^{1/2}M_\ell(\nu_1,0)A_\ell^{-1/2}\right).$$

For estimating $A_\ell^{1/2}M_\ell(\nu,0)A_\ell^{-1/2}$ one may use the approximation property (6.3a) and the smoothing property (6.3b):

(6.3a) $$\| A_\ell^{1/2}(A_\ell^{-1} - pA_{\ell-1}^{-1}r)\|_2 \leq \sqrt{C_A/\|A_\ell\|_2}\,,$$

(6.3b) $$\|A_\ell S_\ell^\nu A_\ell^{-1/2}\|_2 \leq \sqrt{\eta(2\nu)\,\|A_\ell\|_2}\,.$$

Under the assumption (6.1d), inequality (6.3a) is equivalent to the approximation property (5.7). In the case of Richardson's iteration (6.1a,b), inequality (6.3b) holds because of

$$\|A_\ell S_\ell^\nu A_\ell^{-1/2}\|_2^2 = \|A_\ell^{1/2}S_\ell^\nu\|_2^2 = \|A_\ell S_\ell^{2\nu}\|_2 \leq \eta_0(2\nu)\,\|A_\ell\|_2$$

with $\eta(2\nu)=\eta_0(2\nu)$ (η_0 from (6.2b)). (6.3a,b) yield the estimate

(6.3c) $$\|M_\ell^{TGM}(\nu,0)\|_{A_\ell} = \|A_\ell^{1/2}M_\ell^{TGM}(\nu,0)A_\ell^{-1/2}\|_2 \leq \sqrt{\eta(2\nu)\,C_A}\,.$$

Using (6.2d), one finally proves the following convergence theorem:

Theorem 6.1 Assume (6.1a,b,d). The smoothing and approximation properties (6.3a,b) imply
$$\|M_\ell^{TGM}(\nu_1,\nu_2)\|_{A_\ell}\leq C_A\sqrt{\eta(2\nu_1)\eta(2\nu_2)},\quad \|M_\ell^{TGM}(\tfrac{\nu}{2},\tfrac{\nu}{2})\|_{A_\ell}\leq C_A\eta(\nu).$$
The two-grid convergence follows as in Theorems 5.7 or 5.8.

As in §5 the multi-grid convergence can be concluded from the two-grid convergence. However, it has to be emphasized that the proof technique of §5 requires $\gamma\geq 2$ and therefore excludes the V-cycle ($\gamma=1$).

The third normal form of the smoothing iteration $\mathcal{S}_\ell$ is

$$x_\ell^{m+1}=\mathcal{S}_\ell(x_\ell^m,b_\ell) \text{ is the solution of } W_\ell(x_\ell^m-x_\ell^{m+1})=A_\ell x_\ell-b_\ell.$$

To simplify the following considerations, the smoothing iteration $\mathcal{S}_\ell$ is assumed to satisfy the inequality $\gamma W_\ell\leq A_\ell\leq\Gamma W_\ell$ with $0\leq\gamma\leq\Gamma\leq 1$, although generalisations are possible. The assumptions are

(6.4) $$\hat{\mathcal{S}}_\ell = \mathcal{S}_\ell,\qquad \hat{S}_\ell = S_\ell = I-W_\ell^{-1}A_\ell,\qquad 0 < A_\ell \leq W_\ell.$$

The approximation property is assumed to hold in the form

(6.5a) $$\| W_\ell^{1/2}(A_\ell^{-1} - pA_{\ell-1}^{-1}r)W_\ell^{1/2}\|_2 \leq C_A \qquad \text{for all } \ell\geq 1.$$

Lemma 6.2 Assume (6.1a,d). The approximation property (6.5a) is equivalent to inequality (6:5b):

(6.5b) $$0 \leq A_\ell^{-1} - pA_{\ell-1}^{-1}r \leq C_A W_\ell^{-1} \qquad \text{for all } \ell \geq 1.$$

In contrast to Theorems 5.7-8, it is now possible to prove convergence for *all* $\nu > 0$.

Theorem 6.3 Assume (6.1a-d), (6.4), and the approximation property (6.5a). Then the two-grid iteration converges monotonously with respect to the energy norm $\|\cdot\|_{A_\ell}$:

(6.6) $$\rho(M_\ell^{TGM}(\tfrac{\nu}{2},\tfrac{\nu}{2})) \leq \begin{Bmatrix} C_A \eta_0(\nu) & \text{if } C_A \leq 1+\nu \\ (1-1/C_A)^\nu & \text{if } C_A > 1+\nu \end{Bmatrix} < 1.$$

The result also holds under the assumption $rA_\ell p \leq A_{\ell-1}$ instead of (6.1d).

It remains to discuss the approximation property (6.5a).

Remark 6.4 Assume the approximation property in the original form (5.7): $\|A_\ell^{-1} - pA_{\ell-1}^{-1}r\|_2 \leq C_A' / \|A_\ell\|_2$. Further, let (6.6b): $\|W_\ell\|_2 \leq C_W \|A_\ell\|_2$ be valid. Then (6.5a) is satisfied with $C_A := C_A' C_W$.

We transfer the technique of §5.4 to the multi-grid method.

Theorem 6.5 Under the same assumptions (6.1a-d, 4, 5a) as in Theorem 3, the V-cycle ($\gamma=1$) converges monotonously with respect to the energy norm $\|\cdot\|_{A_\ell}$ with the rate

(6.7) $$\rho(M_\ell^V(\tfrac{\nu}{2},\tfrac{\nu}{2})) = \|M_\ell^V(\tfrac{\nu}{2},\tfrac{\nu}{2})\|_{A_\ell} \leq \frac{C_A}{C_A+\nu} < 1.$$

Obviously, monotonous, h_ℓ-independent convergence can also be shown for the W-cycle (more general: $\gamma \geq 2$). For this case one finds (assuming $C_A \geq 1$) e.g. the estimate

(6.8) $$\|M_\ell^W(\tfrac{\nu}{2},\tfrac{\nu}{2})\|_{A_\ell} \leq \sqrt{C_A} / (\sqrt{C_A} + \nu).$$

In the case of weaker regularity and for $\gamma=2$ one can still prove $\|M_\ell^W(\tfrac{\nu}{2},\tfrac{\nu}{2})\|_{A_\ell} = O(\nu^{\sigma-1}) < 1$ for all $\nu > 0$. This result does not hold for the V-cycle. In the latter case the convergence rate ζ_ℓ proves to be ℓ-dependent.

One may weaken the approximation property, so that no more regularity is required (for an analysis cf. Hackbusch [14,§6.4.2]). Then the convergence statement covers only the case $\nu_1 = 1$, i.e. faster convergence for larger ν_1 does not follow.

7. The Nonlinear Multi-Grid Method

We denote the nonlinear boundary value problem by

(7.1) $$\mathcal{L}(u) = 0.$$

Some discretisation of (7.1) with step size parameter h leads to the discrete problem

(7.2) $$\mathcal{L}_h(u_h) = 0$$

for n_h nonlinear equations for $u_h \in \mathbb{R}^{n_h}$. By formal reasons problem (7.2) will be generalised to

(7.3) $$\mathcal{L}_h(u_h) = f_h$$

with small f_h. Assuming the existence of at least one solution u_h^* of (7.2) and the regularity of the Jacobian

(7.4) $$L_h(u_h) := \mathcal{L}_h'(u_h) = \partial \mathcal{L}_h(u_h) / \partial u_h$$

at $u_h = u_h^*$, we can assume the solvability of (7.3) for sufficiently small f_h.

For the multi-grid application we replace h by a sequence of step sizes: $h_\ell = 2^{-\ell} h_0$ $(\ell \geq 0)$. The notation $\mathcal{L}_{h_\ell}$, u_{h_ℓ}, f_{h_ℓ}, L_{h_ℓ} is replaced by $\mathcal{L}_\ell$, u_ℓ, f_ℓ, L_ℓ. Eq. (7.3) becomes

$$\mathcal{L}_\ell(u_\ell) = f_\ell. \tag{7.3}$$

The simplest multi-grid approach is the combination of the standard Newton iteration with (one or two steps of) the linear multi-grid method for solving the arising linear equations. However, this approach requires the Jacobian of $\mathcal{L}_\ell$. This is not necessary for the nonlinear multi-grid method NMGM, which will be described below. It is important to notice that NMGM is used in the form of a *nested iteration* (cf. §4 and Hackbusch [14, §5, §9.3.4]). The problems $\mathcal{L}_k(u_k) = 0$ are solved approximately for all levels $0 \leq k \leq \ell$ by

(7.4)
```
ũ₀ := approximation to the solution of 𝓛₀(u₀) = 0;
for k := 1 step 1 until ℓ do
begin ũ_k := p̃ ũ_{k-1};  f̃_{k-1} := 𝓛_{k-1}(ũ_{k-1});
    for j := 1 step 1 until α do NMGM(k, ũ_k, 0)
end;
```

The mapping $\tilde{p}$ is some prolongation into the fine-grid function, which may differ from p used in NMGM. Note that this nested iteration yields approximations $\tilde{u}_k$ and their defects $\tilde{f}_{k-1} = \mathcal{L}_{k-1}(\tilde{u}_{k-1})$ for all indices $1 \leq k \leq \ell$.

The ALGOL program for NMGM reads as

(7.5)
```
procedure NMGM(ℓ,u,f); integer ℓ; array u,f;
if ℓ=0 then u := approximate solution of 𝓛₀(u₀) = f
else
begin integer j; array v,d; real s;
    for j := 1 step 1 until ν do u := 𝒮_ℓ(u,f);
    d := r(𝓛_ℓ(u) - f);
    s := σ(ℓ-1,d);
    d := f̃_{ℓ-1} - s*d;
    v := ũ_{ℓ-1};
    for j := 1 step 1 until γ do NMGM(ℓ-1,v,d);
    u := u + p(v - ũ_{ℓ-1})/s
end;
```

For the motivation of the construction we refer to §9.3.2 of Hackbusch [14]. $\mathcal{S}_\ell$ is again the smoothing iteration. Note that the algorithm uses the approximate solution $\tilde{u}_{\ell-1}$ and its defect $\tilde{f}_{\ell-1}$ already obtained from the previous performance of the k-loop in (7.4). Furthermore, algorithm (7.5) uses a scaling factor s. The reason for this parameter is the following: Small values of s ensure that the right-hand side $d := \tilde{f}_{\ell-1} - s*d$ is small, provided we have $\tilde{f}_{\ell-1} \approx 0$ because of $\tilde{u}_{\ell-1} \approx u^*_{\ell-1}$. A simple choice of s is

$$s = \sigma(\ell-1, d) := \sigma/\|d\| \qquad \text{if } d \neq 0 \ \ (\sigma\text{: small constant}).$$

For linear problems $\mathcal{L}_\ell(u_\ell) = L_\ell u_\ell - f_\ell$ the nonlinear multi-grid method NMGM is algebraically equivalent to the linear algorithm from §3, whatever the values of s, $\tilde{u}_{\ell-1}$ and $\tilde{f}_{\ell-1} = \mathcal{L}_{\ell-1}(\tilde{u}_{\ell-1})$ are.

For the convergence analysis and variants like FAS or for continuation problems we refer to Hackbusch [14].

8. The Multi-Grid Method of the Second Kind

The discretisation of Fredholm integral equations of the second kind (cf. Hackbusch [16]) yields a system of linear equations of the form

(8.1) $\lambda x_\ell = K_\ell x_\ell - b_\ell.$

for all levels $\ell = 0, 1, 2, \ldots$ with $\lambda \neq 0$. Here, K_ℓ is a full matrix. Using the old iterand in the right-hand side of (3.1b), we obtain the *Picard iteration*

(8.2) $x_\ell^{old} \mapsto x_\ell^{new} := (K_\ell x_\ell^{old} + b_\ell)/\lambda.$

Although the Picard iteration converges if and only if the spectral radius satisfies $\rho(K_\ell) < |\lambda|$, we will use it as smoothing iteration in any case.

The matrix A_ℓ of the system (1.8a) is $A_\ell = I - K_\ell$. The multi-grid method of the second kind is the general algorithm (3.6) applied to the system (1.8a) with $A_\ell = I - K_\ell$, where the smoothing iteration is chosen as the Picard iteration. Furthermore, the analysis shows that (in almost all situations) it makes no sense to perform more than one smoothing step, i.e. the parameters ν_1 and ν_2 from (3.6b,f) are

(8.3) $\nu_1 = 1, \ \nu_2 = 0.$

The multi-grid algorithm contains the further parameter γ. From the analysis it turns out that the desired convergence results can be obtained only for $\gamma \geq 2$. But since $\gamma = 2$ is fully sufficient, we fix this value:

(8.4) $\gamma = 2$ (i.e. «W-cycle»)

and obtain the following «multi-grid algorithm of the second kind».

(8.5)	**Multi-Grid Procedure** *MGM* for solving $\lambda x_\ell = K_\ell x_\ell + b_\ell,\ \ell \geq 0$	
(8.5a)	procedure $MGM(\ell, x, b)$; integer ℓ; array x, b;	
(8.5b)	if $\ell = 0$ then $x := (\lambda I - K_0)^{-1} b$ else	
	begin array d, y; integer i;	
(8.5c)	$x := \frac{1}{\lambda}(b + K_\ell * x)$;	(Picard iteration)
(8.5d)	$d := r * (\lambda x - b - K_\ell * x)$;	(defect computation)
(8.5e$_1$)	$y := 0$;	(set starting value)
(8.5e$_2$)	for $i := 1$ to 2 do $MGM(\ell - 1, y, d)$;	(2 multi-grid calls)
(8.5f)	$x := x - p * y$	(coarse-grid correction)
	end;	

The multi-grid of the second kind behaves differently from the standard application to discrete elliptic problems (cf. §5). In the latter case simple iterative schemes have a convergence rate of $1 - O(h^{2m})$, if $2m$ is the order of the differential operator, whereas the multi-grid rate is independent of h: $O(1)$. In the integral equation context the simple Picard iteration has already the rate $O(1)$ provided $\rho(K_\ell) < |\lambda|$ holds (cf. Hackbusch [16, Folgerung 5.1.10]). Thus the multi-grid method is expected to be even faster convergent than $O(1)$. As illustration we start with the example

(8.6) $f(x) = g(x) + \int_0^1 \cos(\pi x y)\, f(y)\, dy.$

The *Nyström method* with the trapezoidal rule (cf. Hackbusch [16,§4.7]) is used as discretisation. The step sizes in the hierarchy (1.2) are $h_\ell = 2^{-\ell}$, $\ell \geq 0$. The appropriate prolongation p is the piecewise linear interpolation. The restriction is chosen as the trivial restriction, which is admitted in the context of integral equations. The errors of the iterands and the convergence speed given by the quotients of the error norms are shown in Table 8.1. Obviously, the rates are of the order $O(h^2)$.

The general result of the convergence analysis (e.g. presented in Hackbusch [14,§16] and [16,§5]) is

(8.7a) multi-grid convergence speed $= O(h^{\varkappa})$

with a *positive* exponent $\varkappa$. The size of $\varkappa$ is limited by

(8.7b) $\varkappa = \min\{\varkappa_{\text{regularity}}, \varkappa_{\text{consistency}}, \varkappa_{\text{interpolation}}\}$.

The trapezoidal rule has the consistency order $\varkappa_{\text{consistency}} = 2$. The expression $x_\ell - p r x_\ell$ can be understood as the «interpolation error» connected with p and r. For the piecewise linear interpolation p the error is of the order $\varkappa_{\text{interpolation}} = 2$. However, the most important property is the «discrete regularity» of K_ℓ. To understand this property we start with the continuous integral operator K from (2.1). Even for weakly singular kernel functions k the operator K can be a bounded mapping from the continuous function space $X = C^0(\Gamma)$ into the Hölder space $Y = C^\lambda(\Gamma)$ with $\lambda > 0$ (for integer λ Y is the space of λ-times differentiable functions). The finite dimensional space $\mathbf{X}_\ell$ can be endowed with discrete analogues $\|\cdot\|_X$ and $\|\cdot\|_Y$ of the $C^0(\Gamma)$-norm (maximum norm) and of the $C^\lambda(\Gamma)$-norm. The discrete regularity reads

(8.7c) $\|K_\ell x_\ell\|_Y \leq \text{const}\, \|x_\ell\|_X$ for all $x_\ell \in \mathbf{X}_\ell$.

The quantity $\varkappa_{\text{regularity}}$ from (8.7b) equals the number λ from above. For the example defined above one can prove $\varkappa_{\text{regularity}} = 2$. Hence, (8.7b) yields $\varkappa = 2$ in agreement with the numerical results from Table 8.1.

The inequality (8.7c) proves also why the Picard iteration may serve as smoothing iteration in the multi-grid method. The Picard iteration maps the error e_ℓ into $e_\ell' := K_\ell e_\ell / \lambda$. The smoothness of e_ℓ' is described by the estimate $\|e_\ell'\|_Y \leq \text{const}\, \|e_\ell\|_X / \lambda$.

An exact description and analysis of the multi-grid method of the second kind as well as many examples of application (also different from integral equations) can be found in Hackbusch [14,§16], [16], [13].

i	$h_1=1/2$		$h_2=1/4$		$h_3=1/8$		$h_4=1/16$		$h_5=1/32$	
0	2.21	$1.95_{10}-1$	2.04	$4.33_{10}-2$	2.37	$1.20_{10}-2$	2.46	$3.09_{10}-3$	2.47	$7.81_{10}-4$
1	$4.33_{10}-1$	$9.87_{10}-2$	$8.94_{10}-2$	$3.65_{10}-2$	$2.84_{10}-2$	$9.42_{10}-3$	$7.59_{10}-3$	$2.40_{10}-3$	$1.93_{10}-3$	$6.02_{10}-4$
2	$4.27_{10}-2$	$1.72_{10}-1$	$3.26_{10}-3$	$1.98_{10}-2$	$2.67_{10}-6$	$8.08_{10}-3$	$1.82_{10}-5$	$2.19_{10}-3$	$1.16_{10}-6$	$5.60_{10}-4$
3	$7.35_{10}-3$	$1.34_{10}-1$	$6.44_{10}-5$	$5.16_{10}-2$	$2.16_{10}-6$	$1.10_{10}-2$	$3.99_{10}-8$	$2.71_{10}-3$	$6.51_{10}-10$	
4	$9.81_{10}-4$	$1.38_{10}-1$	$3.33_{10}-6$	$1.86_{10}-2$	$2.38_{10}-8$	$8.39_{10}-3$	$1.08_{10}-10$		*	
5	$1.36_{10}-4$		$6.18_{10}-8$		$2.00_{10}-10$		*		*	

Table 8.1 Errors of the multi-grid iterands and convergence factors

9. Further Comments

In order to have a short presentation, other smoothers than the Richardson and the chequer-board Gauß-Seidel iterations have been mentioned only marginally. In applications to more complicated systems the problem of robustness arises: are those convergence rates which we known from the Poisson model problem uniformly valid in the larger class of problems? In the simplest case the equation $A(\varepsilon)x=b$ depends on one parameter $\varepsilon\in(0,\infty)$. If an iteration method achieves convergence rates not only of the form $\rho(\varepsilon)\leqslant 1-C(\varepsilon)h^{\varkappa}$, but with $C=C(\varepsilon)$ uniformly in $\varepsilon\in(0,\infty)$: $\rho(\varepsilon)\leqslant 1-Ch^{\varkappa}$, then this iteration is *robust* with respect to the class of problems. Robust multi-grid methods have to satisfy $\rho(\varepsilon)\leqslant\zeta<1$ (ζ h_ℓ- and ε-independent; Hackbusch [14,§10]).

Good experiences concerning robustness are made with incomplete LU-decompositions, which have been introduced by Wesseling [1,2] as multi-grid smoothers (cf. Kettler [1]). Robustness holds for conjugate gradient methods applied to the modified ILU iteration ($\omega=-1$) as well as for the use of point- or blockwise ILU iterations in the multi-grid method as smoother (then with $\omega=0$ or $\omega=1$; cf. Wittum [5], Kettler [1]).

Another approach is the *frequency decomposition multi-grid method* of Hackbusch [11], which uses not only one but several coarse-grid corrections with different coarse-grid equations. The prolongations from the different coarse grids into the fine grid are constructed in such a way that the corrections cover different frequency intervals.

The construction of the coarse-grid equation at level $\ell-1$ requires more information than given only by the system $A_\ell x_\ell=b_\ell$ for $\ell=\ell_{\max}$. This fact may lead to difficulties when the multi-grid method is wanted as blackbox solver. Therefore, it is remarkable that there are variants - so-called *algebraic multi-grid methods* - in which the coarse-grid matrix $A_{\ell-1}$ is constructed only by the information contained in the coefficients of A_ℓ (cf. Stüben [1]).

An essential characteristic of the multi-grid method - besides the use of a coarser grid - is the product form $\Phi_\ell^{\mathrm{GGK}}\circ\mathscr{S}_\ell^{\nu}$ with the different frequency intervals, in which the coarse-grid correction and the smoothing step are active. While many methods can be used as smoother, the question remains whether there exists an alternative to Φ_ℓ^{GGK}. It would be desirable to have a method filtering out the coarse frequencies and needing no hierarchy of grids. Such a method (the frequency filtering iteration) is due to Wittum [6,7] and is based on a sequence of partial steps Φ_ν reducing certain frequency intervals (cf. also Hackbusch [20,§10.9.4]; Axelsson/Polman [1]).

References

ASTRACHANCEV, G. P.: [1] An iterative method of solving elliptic net problems. *USSR Comput. Math. and Math. Phys.* ***11**,2* (1971) 171-182

AXELSSON, O. and B. POLMAN: [1] A robust preconditioner based on algebraic substructuring and two-level grids. In: Hackbusch [18] 1-26

BACHVALOV, N. S.: [1] On the convergence of a relaxation method with natural constraints on the elliptic operator. *USSR Comput. Math. Math. Phys.* ***6**,5* (1966) 101ff

BANK, R. E.: [1] A comparison of two multilevel iterative methods for nonsymmetric and indefinite elliptic finite element equations. *SIAM J. Numer. Anal.* ***18*** (1981) 724ff

BANK, R. E.: [2] PLTMG - user guide. Edition 5.0. Techn. Report, University of California at San Diego, 1988

BANK, R. E. and T. F. DUPONT: [1] Analysis of a two-level scheme for solving finite element equations. Report CNA-159, University of Texas at Austin, 1980

BANK, R. E. and T. F. DUPONT: [2] An optimal order process for solving elliptic finite element equations. *Math. Comp.* ***36*** (1981) 35-51

BANK, R. E., T. F. DUPONT and H. YSERENTANT: [1] The hierarchical basis multigrid method. *Numer. Math.* ***52*** (1988) 427-458

BRAESS, D.: [1] The contraction number of a multigrid method for solving the Poisson equation. *Numer. Math.* ***37*** (1981) 387-404

BRAESS, D., W. HACKBUSCH: [1] A new convergence proof for the multigrid method including the V-cycle. *SIAM J. Numer. Anal.* ***20*** (1983) 967-975

BRAESS, D., W. HACKBUSCH, U. TROTTENBERG (ed.): [1] *Advances in multi-grid methods*. Proceedings, Oberwolfach, Dec. 1984. Notes on Numerical Fluid Mechanics 11. Vieweg, Braunschweig 1985

BRAKHAGE, H.: [1] Über die numerische Behandlung von Integralgleichungen nach der Quadraturformelmethode. *Numer. Math.* **2** (1960) 183-196

BRAMBLE, J. H. and J. E. PASCIAK: [1] New convergence estimates for multigrid algorithms. *Math. Comp.* **49** (1987) 311-329

BRANDT, A.: [1] Multi-level adaptive solutions to boundary-value problems. *Math. Comp.* **31** (1977) 333-390

BRANDT, A.: [2] Guide to multigrid development. In: Hackbusch/Trottenberg [1] 220ff

BULIRSCH, R., R. D. GRIGORIEFF, J. SCHRÖDER (ed.): [1] *Numerical treatment of differential equations*. Proceedings, Oberwolfach, July 1976. Lecture Notes in Mathematics 631. Springer, Berlin 1978

D'JAKONOV, E. G.: [1] О шодимости однога итерационного процесса (On the convergence of an iterative process). *Usp. Mat. Nauk* **21** (1966) 179-182

FEDORENKO, R. P.: [1] A relaxation method for solving elliptic difference equations. *USSR Comput. Math. and Math. Phys.* **1**,*5* (1961) 1092-1096

FEDORENKO, R. P.: [2] The speed of convergence of one iterative process. *USSR Comput. Math. and Math. Phys.* **4**,*3* (1964) 227-235

HACKBUSCH, W.: [1] A fast iterative method solving Poisson's equation in a general region. In: Bulirsch / Grigorieff / Schröder [1] 51-62

HACKBUSCH, W.: [2] On the convergence of a multi-grid iteration applied to finite element equations. Report 77-8, Univ. zu Köln 1977

HACKBUSCH, W.: [3] On the multi-grid method applied to difference equations. *Computing* **20** (1978) 291-306

HACKBUSCH, W.: [4] Convergence of multi-grid iterations applied to difference equations. *Math. Comp.* **34** (1980) 425-440

HACKBUSCH, W.: [5] The fast numerical solution of very large elliptic difference schemes. *J. Inst. Maths Applics* **26** (1980) 119-132

HACKBUSCH, W.: [6] Die schnelle Auflösung der Fredholmschen Integralgleichung zweiter Art. *Beiträge Numer. Math.* **9** (1981) 47-62

HACKBUSCH, W.: [7] On the convergence of multi-grid iterations. *Beiträge Numer. Math.* **9** (1981) 213-239

HACKBUSCH, W.: [8] On the regularity of difference schemes. *Ark. Mat.* **19** (1981) 71ff

HACKBUSCH, W.: [9] On the regularity of difference schemes - part II: regularity estimates for linear and nonlinear problems. *Ark. Mat.* **21** (1983) 3-28

HACKBUSCH, W.: [10] Multi-grid convergence theory. In: Hackbusch / Trottenberg [1] 177-219

HACKBUSCH, W.: [11] The frequency decomposition multi-grid method. Part I: Application to anisotropic equations. *Numer. Math.* **56** (1989) 229-245

HACKBUSCH, W.: [12] A parallel variant of the conjugate gradient method. In: Hackbusch [19] 111-119

HACKBUSCH, W.: [13] The solution of large systems of BEM equations by the multi-grid and panel clustering method technique. In: Numerical Methods in Applied Science and Industry, Rend. Semin. Mat., Torino. Ed.: G. Monegato. Proceedings, Juni 1990, Turin.

HACKBUSCH, W.: [14] *Multi-grid methods and applications*. Springer, Berlin 1985

HACKBUSCH, W.: [15] *Theorie und Numerik elliptischer Differentialgleichungen*. Teubner, Stuttgart 1986

HACKBUSCH, W.: [16] *Integralgleichungen - Theorie und Numerik*. Teubner, Stuttgart 1989

HACKBUSCH, W. (ed.): [17] *Efficient solvers for elliptic systems*. Notes on numerical fluid mechanics ***10***. Vieweg, Braunschweig 1984

HACKBUSCH, W. (ed.): [18] *Robust Multi-Grid Methods*. Proceedings, Kiel, Jan. 1988. Notes on numerical fluid mechanics **23**. Vieweg, Braunschweig 1988

HACKBUSCH, W. (ed.): [19] *Parallel Algorithms for PDEs*. Proceedings, Kiel, Jan. 1990. To be published in the same series as [18]

HACKBUSCH, W.: [20] *Iterative Lösung großer schwachbesetzter Gleichungssysteme (Iterative solution of large sparse systems of equations).* Teubner, Stuttgart 1991 (engl. translation in preparation)

HACKBUSCH, W. and A. REUSKEN: [1] Analysis of a damped nonlinear multilevel method. *Numer. Math.* **55** (1989) 225-246

HACKBUSCH, W. and U. TROTTENBERG (ed.): [1] *Multi-grid methods.* Proceedings, Köln-Porz, Nov 1981. Lecture Notes in Mathematics **960**. Springer, Berlin 1982

HACKBUSCH, W. and U. TROTTENBERG (ed.): [2] *Multi-grid methods II.* Proceedings, Köln, Oct 1985. Lecture Notes in Mathematics **1228**. Springer, Berlin 1986

HEMKER, P. W.: [1] Mixed defect correction iteration for the accurate solution of the convection diffusion equation. In: Hackbusch / Trottenberg [2] 485-501

HEMKER, P. W. and H. SCHIPPERS: [1] Multiple grid methods for the solution of Fredholm integral equations of the second kind. *Math. Comp.* **36** (1981) 215-232

KETTLER, R.: [1] Analysis and comparison of relaxation schemes in robust multigrid and preconditioned conjugate gradient methods. In: Hackbusch/Trottenberg [1] 502ff

KRONSJÖ, L. and G. DAHLQUIST: [1] On the design of nested iterations for elliptic difference equations. *BIT* **11** (1971) 63-71

MAITRE, J.-F. and F. MUSY: [1] Multigrid methods: convergence theory in a variational framework. *SIAM J. Numer. Anal.* **21** (1984) 657-671

MANDEL, J.: [1] Multigrid convergence for nonsymmetric, indefinite variational problems and one smoothing step. *Appl. Math. Optim.* **19** (1986) 201-216

McCORMICK, S. (ed.): [1] *Multigrid methods.* SIAM, Philadelphia 1987

NICOLAIDES, R. A.: [1] On multiple grid and related techniques for solving discrete elliptic systems. *J. Comput. Phys.* **19** (1975) 418-431

NICOLAIDES, R. A.: [2] On the ℓ^2 convergence of an algorithm for solving finite element equations. *Math. Comp.* **31** (1977) 892-906

PADDON, D. J. and H. HOLSTEIN (eds.): [1] *Multigrid methods for integral and differential equations.* Proceedings, Bristol, Sept 1983. Clarendon Press, Oxford 1985

SONNEVELD, P., P. WESSELING, P. M. DE ZEEUW: [1] Multigrid and conjugate gradient methods as convergence acceleration techniques. In: Paddon/Holstein [1] 117ff

STÜBEN, K.: [1] Algebraic multigrid (AMG): experiences and comparisons. *Appl. Math. Comput.* **13** (1983) 419-451

WESSELING, P.: [1] Theoretical and practical aspects of a multigrid method. *SIAM J. Sci. Statist. Comput.* **3** (1982) 387-407

WESSELING, P.: [2] A robust and efficient multigrid method. In: Hackbusch / Trottenberg [1] 614-630

WITTUM, G.: [1] *Distributive Iterationen für indefinite Systeme als Glätter im Mehrgitterverfahren am Beispiel der Stokes- und Navier-Stokes-Gleichungen mit Schwerpunkt auf unvollständige Zerlegungen.* Dissertation, Kiel 1986

WITTUM, G.: [2] Multi-grid methods for Stokes and Navier-Stokes equations. Transforming smoothers: Algorithms and numerical results. *Numer. Math.* **54** (1989) 543ff

WITTUM, G.: [3] On the convergence of multi-grid methods with transforming smoothers. Theory with applications to the Navier-Stokes equations. *Numer. Math.* **57** (1989) 15-38

WITTUM, G.: [4] Linear iterations as smoothers in multigrid methods: Theory with applications to incomplete decompositions. *Impact of Computing in Science and Engineering* **1** (1989) 180-215

WITTUM, G.: [5] On the robustness of ILU smoothing. *SIAM J. Sci. Stat. Comput.* **10** (1989) 699-717

WITTUM, G.: [6] An ILU-based smoothing correction scheme. In: Hackbusch [19] 228ff

WITTUM, G.: [7] *Filternde Zerlegungen: Ein Beitrag zur schnellen Lösung großer Gleichungssysteme.* Habilitationsschrift, Univ. Heidelberg 1990

YSERENTANT, H.: [1] On the multi-level splitting of finite element spaces. *Numer. Math.* **49** (1986) 379-412

Two Multi-Level Methods for Nonuniformly Refined Grids

Harry Yserentant
Mathematisches Institut
Universität Tübingen
D-7400 Tübingen
Germany.

Abstract : The hierarchical basis method and the recent method of BRAMBLE, PASCIAK and XU are multi-level type methods which are especially well suited for nonuniformely refined grids. The paper gives a short summary of these methods.

Keywords : finite element discretizations, nonuniform grids, multi-level methods, hierarchical bases.

1 Introduction

The finite element discretization of partial differential equations is one of the main sources of large linear systems. In typical $2D$–applications one has many thousand unknowns, and in the three-dimensional case one million unknowns are not unusual. Of course, the solution of these systems requires specialized techniques which exploit their structure.

Although often used in engineering practice, a disadvantage of direct elimination techniques is that the amount of work and storage grows much faster than the number of unknowns. Iterative solvers like multigrid methods represent a major progress in this field because, for these methods, the growth is proportional to the number of unknowns. For a description of multigrid methods see Hackbusch's article in this volume or the monography [6].

A special problem are discretizations on highly nonuniform grids. An example of such a grid is shown in Figure 1.

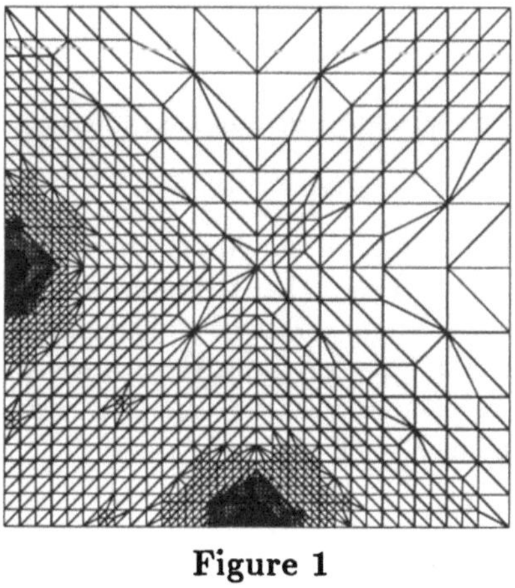

Figure 1

At one hand, such grids are necessary to approximate solutions with singularities arising from corners, cracks, interfaces or nonlinearities. On the other hand, if one applies classical

NATO ASI Series, Vol. F 77
Computer Algorithms for Solving Linear Algebraic Equations
Edited by E. Spedicato

multigrid methods to such problems, the amount of work and storage tends to increase too fast and, in addition, complications arise in the mathematical analysis.

This paper gives a short summary of two methods which are specialized to discretizations on nonuniform grids. These methods are the hierarchical basis method developed in [10] and the recent preconditioner of Xu [9] and Bramble, Pasciak and Xu [4]. Our discussion follows [11]. We do not treat another related method, the hierarchical basis multigrid method [2] which is the linear solver utilized in the widely distributed finite element package PLTMG [1].

2 A Finite Element Discretization

We describe the methods for a scalar two-dimensional model problem. Many of the ideas developed here can be extended to much more general problems including the elasticity equations, for example, and to other refinement schemes.

Let $\overline{\Omega} \subseteq \mathbb{R}^2$ be a bounded polygonal domain. As a model problem, we consider the differential equation

$$-\sum_{i,j=1}^{2} D_j(a_{ij}D_iu) = f \tag{1}$$

on Ω with homogeneous Dirichlet boundary conditions on the boundary piece Γ and homogeneous natural boundary conditions on the remaining part $\partial\Omega \setminus \Gamma$ of the boundary of Ω. We assume that Γ is composed of straight lines. The solution space of this boundary value problem is

$$\mathcal{H} = \{u \in H^1(\Omega) | u = 0 \text{ on } \Gamma\} \tag{2}$$

where the zero boundary conditions have to be understood in the sense of the trace operator. The weak formulation (see [7], for example) is to find a function $u \in \mathcal{H}$ satisfying

$$a(u,v) = \int_\Omega fv\,dx\ , \quad v \in \mathcal{H}\ , \tag{3}$$

where the bilinear form a is defined by

$$a(u,v) = \int_\Omega \sum_{i,j=1}^{2} a_{ij} D_i u D_j v\, dx\ . \tag{4}$$

We assume that the a_{ij} are measurable and bounded functions and that

$$a_{ij} = a_{ji}\ . \tag{5}$$

By a triangulation $\mathcal{T}$ of the polygonal domain $\overline{\Omega}$, we mean a set of triangles such that the union of these triangles is $\overline{\Omega}$ and such that the intersection of two such triangles either consists of a common side or a common vertex of both triangles or is empty. Here we start with an intentionally coarse triangulation $\mathcal{T}_0$ of $\overline{\Omega}$ with the property that the boundary piece Γ is composed of edges of triangles $T \in \mathcal{T}_0$. We assume that there are positive constants $0 < \delta \leq 1 \leq M$ and $\omega(T) > 0$ with

$$\delta\omega(T)\sum_{i=1}^{2}\eta_i^2 \leq \sum_{i,j=1}^{2} a_{ij}(x)\eta_i\eta_j \leq M\omega(T)\sum_{i=1}^{2}\eta_i^2 \tag{6}$$

for all $T \in \mathcal{T}_0$, almost all $x \in T$ and all $\eta \in \mathbb{R}^2$. The constants M and δ cover the local variation of the coefficient functions inside the triangles of the initial triangulation

$\mathcal{T}_0$, whereas the $\omega(T)$ have the task to catch the global behavior. For example, they can describe jumps of the coeffient functions across the boundaries of the triangles $T \in \mathcal{T}_0$.

By (5) and (6) a is a symmetric, bounded and coercive bilinear form on $\mathcal{H}$.

$$\|u\|^2 = a(u,u) \tag{7}$$

defines a norm on $\mathcal{H}$, the energy norm induced by the given boundary value problem. Since $\mathcal{H}$ is a Hilbert space under this norm, the Riesz representation theorem guarantees that the boundary value problem (3) has a unique solution.

In addition we use the weighted L_2–norm

$$\|u\|_0^2 = (u,u) \tag{8}$$

which is induced by the inner product

$$(u,v) = \sum_{T\in\mathcal{T}_0} \frac{\omega(T)}{h(T)^2} \int_T u(x)v(x)dx \ . \tag{9}$$

$h(T)$ denotes the diameter of the triangle T. Note that the norm (8) depends on the initial triangulation $\mathcal{T}_0$ and on the coefficients of the boundary value problem. It has to be distinguished from the usual L_2–norm

$$\|u\|_{0,2;\Omega}^2 = \int_\Omega |u(x)|^2 dx \ . \tag{10}$$

The weights $h(T)^{-2}$ in the inner product (9) are normalization factors which are necessary to make the performance of the methods and the mathematical estimates independent of the quasiuniformity of the initial triangulation. For two-dimensional applications as considered here, a natural modification of the inner product (9) is

$$(u,v) = \sum_{T\in\mathcal{T}_0} \frac{\omega(T)}{\text{area}(T)} \int_T u(x)v(x)dx. \tag{11}$$

With this modification, the preconditioners defined below would become independent of the degeneration of the triangles in the initial triangulation.

The triangulation $\mathcal{T}_0$ is refined several times, giving a family of nested triangulations $\mathcal{T}_0$, $\mathcal{T}_1$, $\mathcal{T}_2$, ... A triangle of $\mathcal{T}_{k+1}$ is either a triangle of $\mathcal{T}_k$ or is generated by subdividing a triangle of $\mathcal{T}_k$ into four congruent triangles or into two triangles by connecting one of its vertices with the midpoint of the opposite side. The first case is called a regular refinement and the resulting triangles as well as the triangles of the initial triangulation are regular triangles. The second case is an irregular refinement and results in two irregular triangles.

The irregular refinement is potentially dangerous because interior angles are reduced. Therefore, we add the rule that irregular triangles may not be further refined. This rule insures that every triangle of any triangulation $\mathcal{T}_k$ is geometrically similar to a triangle of the initial triangulation $\mathcal{T}_0$ or to an irregular refinement of a triangle in $\mathcal{T}_0$.

The triangles in $\mathcal{T}_0$ are level 0 elements, and the regular and irregular triangles created by the refinement of level $k-1$ elements are level k elements. It is important to recognize that not all elements in $\mathcal{T}_{k-1}$ need to be refined in creating $\mathcal{T}_k$. The mesh $\mathcal{T}_k$ may contain unrefined elements from all lower levels, and thus it may be a highly nonuniform mesh. We require that only level $k-1$ elements are refined in the construction of $\mathcal{T}_k$.

The described triangulations are meanwhile standard; we refer to [1] and to [3] . We remark that our levels usually do not reflect the dynamic refinement process in an adaptive algorithm, although the final triangulations can be decomposed a-posteriori as described above; see [5] for a detailed discussion. Due to the last rule, this decomposition is unique.

Corresponding to the triangulations $\mathcal{T}_k$, we have finite element spaces $\mathcal{S}_k$. $\mathcal{S}_k$ consists of all functions which are continuous on $\overline{\Omega}$ and linear on the triangles $T \in \mathcal{T}_k$ and which vanish on the boundary piece Γ. Clearly, $\mathcal{S}_k$ is a subspace of $\mathcal{S}_l$ for $l \geq k$.

Let $\mathcal{N}_k = \{x_1, \ldots, x_{n_k}\}$ be the set of vertices of the triangles in $\mathcal{T}_k$ not lying on the boundary piece Γ. Then $\mathcal{S}_k$ is spanned by the nodal basis functions $\psi_i^{(k)}$, $i = 1, \ldots, n_k$, which are defined by

$$\psi_i^{(k)}(x_l) = \delta_{il} \ , \ x_l \in \mathcal{N}_k \ . \tag{12}$$

The hierarchical basis functions are

$$\widehat{\psi}_i = \psi_i^{(0)} \ , \ x_i \in \mathcal{N}_0 \ , \tag{13}$$

and

$$\widehat{\psi}_i = \psi_i^{(k)} \ , \ x_i \in \mathcal{N}_k \setminus \mathcal{N}_{k-1} \ . \tag{14}$$

$\widehat{\psi}_i$, $i = 1, \ldots, n_k$, is the hierarchical basis of $\mathcal{S}_k$.

3 The Preconditioners

Introducing the selfadjoint and positive definite operators $A : \mathcal{S} \to \mathcal{S}$ by

$$(Au, v) = a(u, v) \ , \quad v \in \mathcal{S} \ , \tag{15}$$

and a vector $b \in \mathcal{S}$ by

$$(b, v) = \int_\Omega fv dx, \quad v \in \mathcal{S} \ , \tag{16}$$

(3) can be reformulated as

$$Au = b \ . \tag{17}$$

For the solution of (17), we consider iterations

$$u \leftarrow u + \omega C(b - Au) \tag{18}$$

with selfadjoint and positive definite operators $C : \mathcal{S} \to \mathcal{S}$ and with properly chosen constants $\omega > 0$, and conjugate gradient type accelerations of such iterations, respectively.

The speed of convergence of this iteration is governed by the spectral condition number $\kappa(C^{1/2}AC^{1/2})$, which is the quotient of the maximum and the minimum eigenvalue of the operator $C^{1/2}AC^{1/2}$.

(18) can be rewritten as

$$r \leftarrow b - Au \ , \quad u \leftarrow u + \omega Cr \ . \tag{19}$$

To realize (19) efficiently, the right representation of the vectors u, $r \in \mathcal{S}$ is essential. We store u by the values

$$u(x_i) \ , \quad i = 1, \ldots, n \ , \tag{20}$$

whereas r is represented by

$$(r, \psi_i) \ , \quad i = 1, \ldots, n \ , \tag{21}$$

where for simplicity $n = n_j$ and $\psi_i = \psi_i^{(j)}$. We get

$$(r, \psi_i) = f^*(\psi_i) - \sum_{l=1}^{n} a(\psi_i, \psi_l) u(x_l) \ . \tag{22}$$

Therefore only the usual residual has to be computed; neither an explicit representation of the operator A nor of the right–hand side b is needed. For the operators C below, $(Cr)(x_i)$, $i = 1, \ldots, n$, can be computed easily from the values (21).

Generalizing (15) we introduce selfadjoint, positive definite operators $A_k : \mathcal{S}_k \to \mathcal{S}_k$ by

$$(A_k u, v) = a(u, v), \quad v \in \mathcal{S}_k. \tag{23}$$

In its final form, the hierarchical basis preconditioner [10] in Xu's formulation [9] is

$$C_H r = A_0^{-1} Q_0 r + \sum_{k=1}^{j} 4^{-k} \sum_{x_i \in \mathcal{N}_k \setminus \mathcal{N}_{k-1}} \frac{(r, \widehat{\psi}_i)}{(1, \widehat{\psi}_i)} \widehat{\psi}_i \tag{24}$$

or, in terms of the nodal bases,

$$C_H r = A_0^{-1} Q_0 r + \sum_{k=1}^{j} 4^{-k} \sum_{i=n_k+1}^{n_k} \frac{(r, \psi_i^{(k)})}{(1, \psi_i^{(k)})} \psi_i^{(k)}. \tag{25}$$

$Q_0 r$ is the orthogonal projection of $r \in \mathcal{S}$ onto $\mathcal{S}_0$ given by $(Q_0 r, v) = (r, v)$, $v \in \mathcal{S}_0$.

Note that the values $(r, \widehat{\psi}_i)$ can be computed recursively beginning with the values $(r, \psi_i) = (r, \psi_i^{(j)})$, and that the summation of the single terms in (24),(25) can be formulated as a recursive process, too.

We remark that for a given $r \in \mathcal{S}$ the function

$$u_0 = A_0^{-1} Q_0 r \in \mathcal{S}_0 \tag{26}$$

satisfies

$$a(u_0, v) = (r, v)\,, \quad v \in \mathcal{S}_0\,. \tag{27}$$

To compute u_0 therefore one needs only $(r, \psi_i^{(0)})$, $i = 1, \ldots, n_0$, but not $Q_0 r$ itself , and one has to solve a linear system with the level 0 discretization matrix.

As the values $(r, \psi_i^{(j)}) = (r, \psi_i)$ are already known from (22) as the components of the matrix residual, the inner product (9) enters into the final algorithm only via the scaling factors $4^{-k}/(1, \psi_i^{(k)})$.

The original formulation [10] of the hierarchical basis preconditioner has been given in terms of the transformation matrix from the hierarchical basis- to the nodal basis representation. Mathematically and also algorithmically both formulations are equivalent.

The Bramble–Pasciak–Xu preconditioner

$$C_X r = A_0^{-1} Q_0 r + \sum_{k=1}^{j} 4^{-k} \sum_{i=1}^{n_k} \frac{(r, \psi_i^{(k)})}{(1, \psi_i^{(k)})} \psi_i^{(k)} \tag{28}$$

is quite similar to the hierarchical basis preconditioner (25). In the hierarchical basis method only the newly introduced nodes on the level k are taken into account in the inner sum whereas in the Bramble–Pasciak–Xu preconditioner all basis functions $\psi_i^{(k)}$ of $\mathcal{S}_k$ enter.

For a growing number j of refinement levels the crucial condition numbers behave like

$$\kappa(C_H^{1/2} A C_H^{1/2}) = O(j^2) \tag{29}$$

and

$$\kappa(C_X^{1/2} A C_X^{1/2}) = O(j^2). \tag{30}$$

If the acceleration parameter ω in (18) is chosen optimally, with every iteration step (18) the energy norm of the error is reduced at least by the factor

$$\frac{\kappa-1}{\kappa+1}. \tag{31}$$

For a *cg*–accelerated version this improves to

$$\frac{\sqrt{\kappa}-1}{\sqrt{\kappa}+1}. \tag{32}$$

This means that the number of iteration steps necessary to reach a fixed accuracy grows like $O(j^2)$ and $O(j)$, respectively. Usually j behaves like $\log n$. Thus the hierarchical basis- and the Bramble–Pasciak–Xu method are asymptotically nearly optimal. For both cases, the condition number estimates do not depend on regularity properties of the boundary value problem or on the distribution of the unknowns among the levels.

Note, however, that the double sum in (24), (25) contains only

$$\sum_{k=1}^{j}(n_k - n_{k-1}) = n_j - n_0 \tag{33}$$

terms whereas the double sum in (28) consists of

$$n_j \leq \sum_{k=1}^{j} n_k \leq j n_j \tag{34}$$

terms. Therefore, for the Bramble–Pasiak–Xu preconditioner, the amount of work per iteration step can grow a little bit faster than the number of unknowns (at least for a naive realization!).

The condition number estimate (29) for the hierarchical basis preconditioner is restricted to two-dimensional applications. In the three-dimensional case, κ grows like $O(2^j)$; see [8]. In contrast, the estimate (30) for the Bramble–Pasciak–Xu preconditioner is dimension independent. Therefore, for three-dimensional applications, the Bramble–Pasciak–Xu preconditioner is superior. In addition, it can take advantage of the regularity of the boundary value problem. In the extreme case (30) can be improved to $\kappa = O(j)$, see [4].

4 Theory

For a theoretical derivation of the condition number estimates (29) and (30) we refer to [11].

Here we give only a little improvement of the estimate in Lemma 5.1 of [11]. We show that for all functions $v \in \mathcal{S}_k$

$$\|v\|_0^2 \leq \sum_{i=1}^{n_k}(1,\psi_i^{(k)})|v(x_i)|^2 \leq 4\|v\|_0^2. \tag{35}$$

This simple estimate is the final step in the derivation of both preconditioners.

To show (35), let T be a triangle of $\mathcal{T}_k$ with vertices x_1, x_2, x_3 and associated basis funtions ψ_1, ψ_2, ψ_3. Let v be a linear function with values v_1, v_2, v_3 at these vertices. Then

$$\|v\|_{0,2;T}^2 = \frac{\text{area}(T)}{12}((v_1^2+v_2^2+v_3^2)+(v_1+v_2+v_3)^2)$$

$$\sum_{i=1}^{3}(1,\psi_i)_{0,2;T}|v(x_i)|^2 = 4\frac{\text{area}(T)}{12}(v_1^2+v_2^2+v_3^2).$$

(35) follows immediately.

References

[1] Bank, R. E.: *PLTMG: A Software Package for Solving Elliptic Partial Differential Equations.* Philadelphia: SIAM (1990).

[2] Bank, R. E., Dupont, T., Yserentant, H.: *The Hierarchical Basis Multigrid Method.* Numer. Math. **52**, pp. 427–458 (1988).

[3] Bank, R. E., Sherman, A. H., Weiser, A.: *Refinement Algorithms and Data Structures for Regular Local Mesh Refinement.* In: Scientific Computing (eds.: R. Stepleman et al.), Amsterdam: IMACS/North Holland, pp. 3–17 (1983).

[4] Bramble, J. H, Pasciak, J. E., Xu, J.: *Parallel Multilevel Preconditioners.* Math. Comp. **55**, pp. 1-22 (1990).

[5] Deuflhard, P., Leinen, P., Yserentant, H.: *Concepts of an Adaptive Hierarchical Finite Element Code.* IMPACT of Computing in Science and Engineering **1** , pp. 3–35 (1989).

[6] Hackbusch, W.: *Multigrid Methods and Applications.* Berlin, Heidelberg, New York: Springer (1985).

[7] Hackbusch, W.: *Theorie und Numerik elliptischer Differentialgleichungen.* Stuttgart: Teubner (1986). (English translation in preparation).

[8] Ong, M. E. G.: *Hierarchical Basis Preconditioners for Second Order Elliptic Problems in Three Dimensions.* Technical Report No. 89–3, Department of Applied Mathematics, University of Washington, Seattle (1989).

[9] Xu, J.: *Theory of Multilevel Methods.* Report No. AM48, Department of Mathematics, Pennsylvania State University (1989).

[10] Yserentant, H.: *On the Multi-Level Splitting of Finite Element Spaces.* Numer. Math **49**, pp. 379–412 (1986).

[11] Yserentant, H.: *Two Preconditioners Based on the Multi-Level Splitting of Finite Element Spaces.* Numer. Math. **58**, 163-184 (1990).

Preconditioning methods for block H-matrices

Owe Axelsson
Faculty of Mathematics and Informatics
University of Nijmegen
Nijmegen, The Netherlands

ABSTRACT

Consider a matrix $A = [A_{i,j}]$, partitioned in matrix blocks $A_{i,j}$, and assume that A is a block H-matrix. A relaxed incomplete LU-factorization method without pivoting is presented and a short proof of its existence for this class of matrices is given. Different approximations of the inverse of the pivot matrix can be used in different block matrix columns.

The present existence result generalizes previous results known for M-matrices and pointwise H-matrices (i.e. for generalized diagonally dominant matrices), in particular for the special case of a full LU-factorization.

Also it is shown that a certain explicit (multiplicative) preconditioner exists and provides a convergent splitting.

1. INTRODUCTION

Consider the solution of a linear system $A\hat{\mathbf{x}} = \mathbf{b}$, with solution $\hat{\mathbf{x}}$, by iterative methods, such as

(1.1) $$C\delta^\ell = \tau_\ell \mathbf{r}^\ell, \quad \mathbf{x}^{\ell+1} = \mathbf{x}^\ell + \delta^\ell$$

or

(1.2) $$\mathbf{x}^{\ell+1} = \mathbf{x}^\ell + \tau_\ell G \mathbf{r}^\ell,$$

where $\mathbf{r}^\ell = \mathbf{b} - A\mathbf{x}^\ell$ is the residual. Here $\tau_\ell = 1$ in a basic iterative method or the sequence $\{\tau_\ell\}$ is chosen to accelerate the convergence, such as in a Chebyshev iterative method. A similar iterative method is the (generalized) conjugate gradient method. C and G are preconditioners, the latter on a multiplicative form, and the most efficient choice of such a preconditioner is a balance between the cost per iteration and the number of iteration steps required to find an approximation $\mathbf{x}^\ell$ of $\hat{\mathbf{x}}$, for which, say $\|\mathbf{r}^\ell\| \leq \varepsilon$, for some small positive number ε (usually ε depends on the condition number of A.) Frequently A is a sparse matrix and for efficiency reasons, it is then important to choose C and G on some sparse form also.

There exists two principal classes of preconditioners:

(i) explicit preconditioners, here called G-methods

NATO ASI Series, Vol. F 77
Computer Algorithms for Solving Linear Algebraic Equations
Edited by E. Spedicato

(ii) implicit preconditioners, here called C-methods
(G stands for a (discrete) Green's function.)

In an explicit method, one computes a preconditioner G, which approximates the inverse of A, in the sense that GA is close to the identity matrix in some measure.

In an implicit method, one computes first some approximation of A, frequently as a product of (block) triangular matrices and each application of the preconditioner in (1.1) requires then a solution of a linear system with matrix C, while the preconditioner in (1.2) requires only matrix-vector multiplications.

However, for elliptic difference type matrices for instance, the implicit class frequently can reduce the condition number of the matrix, and increase the rate of convergence of the iterative method, more efficiently for a given sparsity pattern than explicit methods can. The reason for this is that when A is sparse, usually its inverse is a full matrix and an accurate explicit approximation needs to include many non-zero entries, i.e., must be less sparse, than implicit preconditioners, because as it turns out, for the latter sparsity in A can be more easily preserved in the construction of C. As is well known, for difference matrices, A^{-1} is typically a discrete Greens function.

On the other hand, the explicit methods can be more easily implemented in an efficient way on parallel vector computers. In fact, for certain explicit methods, even the computation of the preconditioner can be done in parallel, while the computation of an implicit preconditioner usually requires some recursion, i.e., cannot be done fully in parallel.

The best choice of a particular explicit or implicit preconditioner is a function of computer architecture and the relative speed of various computer operations. However, this topic is outside the scope of the present paper. Instead we want to analyze the existence of such preconditioners for the class of block H-matrices. To define this class we recall first the definition of an M matrix.

Consider then so-called $\mathbf{Z}$-matrices, i.e., real square matrices $A = [a_{i,j}]$, for which $a_{i,j} \leq 0$, $i \neq j$. Note that any $\mathbf{Z}$-matrix can be written in the form $A = aI - B$, where $B \geq 0$ (i.e. B is non-negative, that is, each entry of B is non-negative). A $\mathbf{Z}$-matrix, A is said to be an M-matrix if A can be written in the form $A = aI - B$ with $\alpha > \rho(B)$, the spectral radius of B. For any $A \in \mathbf{Z}$, the following are equivalent (see Ostrowski (1937), Fiedler and P̂tak (1962) and Berman and Plemmons (1979)):

(i) A is an M-matrix
(ii) A is a monotone, i.e. $A^{-1} \geq 0$
(iii) There exists a positive vector $\mathbf{x}$, such that $A\mathbf{x} > 0$
(iv) There exists a positive vector $\mathbf{x}$ such that $A^T\mathbf{x} > 0$, or equivalently, such that $\mathbf{x}^T A > 0$.

In practice, the most useful conditions when proving existence are the last two.

In order to extend this definition to rectangular matrices, we define here A, of order $m \times n$, $m \geq n$ to be an M-matrix if $A \in \mathbf{Z}$ and there exists a positive vector $\mathbf{x}$, $\mathbf{x} \in \mathbf{R}^m$, such that $\mathbf{x}^T A \geq 0$ ($\mathbf{x}^T A \in \mathbf{R}^n$), i.e., the weighted column sums of A are positive.
Let $\mathcal{M}(A)$, called comparison matrix to A, be defined as

$$\{\mathcal{M}(A)\}_{i,j} = \begin{cases} |a_{i,i}|, & \text{for } i = j \\ -|a_{i,j}|, & \text{for } i \neq j \end{cases}.$$

Then A is said to be an H-matrix if $\mathcal{M}(A)$ is an M-matrix. Clearly, if A is an M-matrix, then $\mathcal{M}(A) = A$.

A perhaps more known class of matrices is the class of generalized diagonally dominant matrices: $A = [a_{i,j}]$ is said to be generalized diagonally dominant if there exists a positive vector $\mathbf{x}$ such that

$$x_j|a_{j,j}| > \sum_{\substack{i=1 \\ i\neq j}} x_i|a_{i,j}|, \quad j = 1, 2, \ldots, n$$

This means that the weighted column sums of

$$A = [b_{i,j}], \quad b_{i,j} = \begin{cases} |a_{j,j}|, & i = j \\ -|a_{i,j}|, & i \neq j \end{cases},$$

are positive. (Alternatively, we could have defined generalized diagonal dominance using weighted rowsums.)
The class of H-matrices and the class of generalized diagonally dominant matrices are in fact equivalent.

Lemma 1.1. A is an H-matrix if and only if A is generalized diagonally dominant.

Proof. If A is generalized diagonally dominant then by definition, there exists a vector $\mathbf{x} > \mathbf{0}$ such that

$$x_j|a_{j,j}| > \sum_{i\neq j} x_i|a_{i,j}|$$

or equivalently

$$\mathbf{x}^T\mathcal{M}(A) > 0 \text{ for a vector } \mathbf{x} > \mathbf{0}.$$

But this is equivalent with $\mathcal{M}(A)$ being an M-matrix, which, by definition, is equivalent with A being an H-matrix. □

Consider now a matrix A partitioned in block-matrix form, $A = [A_{i,j}]$, $1 \le i \le m$, $1 \le j \le n$, where $m \ge n$ and where $A_{i,i}$ are square, nonsingular matrices of order n_i. For such block-matrices there exists various extensions of the H-matrix concept. Varga and Feingold (1962) defines the *block comparison matrix* $\mathcal{M}_b(A) = [b_{i,j}]$ by

$$b_{i,j} = \begin{cases} \|A_{ii}^{-1}\|^{-1}, & \text{for } i = j \\ -\|A_{i,j}\|, & \text{for } i \neq j \end{cases} \tag{1.3}$$

and we call A a block H-matrix if $\mathcal{M}_b(A)$ is an M-matrix. Ostrowski (1961) used a similar definition. In the case $m = n$, Robert (1969) calls A a block H-matrix, if the diagonal block matrix part, D_A of A is nonsingular and $\mathcal{M}_b\left(D_A^{-1}A\right)$ is an M-matrix. Alternatively, we can call A an H-matrix if $\mathcal{M}_b\left(AD_A^{-1}\right)$ is an M-matrix. More generally, Polman (1987) defines a matrix to be a block H-matrix if $\mathcal{M}_b(DAE)$ is an $\mathcal{M}$-matrix, where D and E are nonsingular block diagonal matrices. For the purpose of this presentation, if suffices to choose D and E identity matrices, i.e., we call A an block H-matrix if D_A is nonsingular and $\mathcal{M}_b(A)$ is an M-matrix. This definition is then valid also for rectangular matrices.

Note that the class of block H-matrices can be significantly larger than the class of H-matrices.

To check if a matrix is a block H-matrix we need among other things to compute the norms of $A_{i,i}^{-1}$. If $A_{i,i}$ are monotone, i.e. $A_{i,i}^{-1} \ge 0$, in particular if $A_{i,i}$ are M-matrices then this can readily be done using the infinity norm, and the identity

$$\|A_{i.i}^{-1}\|_\infty = \|A_{i,i}^{-1}\mathbf{e}\|_\infty, \quad i = 1, 2, \ldots, m,$$

i.e. by solving linear systems with matrices $A_{i,i}$. An interesting method for general matrices to compute the ℓ_1-norm is discussed in Hager (1984), see also the references in that report. This method gives only approximations of the norm but they seem to be quite accurate after only few solutions of certain systems with $A_{i,i}$ and $A_{i,i}^T$.

We shall consider incomplete factorization methods. We recall that an incomplete factorization method proceeds as a Gaussian elimination method, but with deletion of entries outside a certain sparsity pattern. This sparsity pattern is defined either *a priori* or during the factorization itself. In addition, for matrices partitioned in block form we may need to approximate the inverses of the pivot block matrix entries.

Existence, i.e., the existence of non-zero pivot entries, of pointwise incomplete factorization methods for M-matrices was first shown by Meijerink and van der Vorst (1977) and for pointwise H-matrices, by Varga, Saff, and Mehrman (1980). Existence of incomplete factorization methods for M-matrices on block form, was shown in Axelsson, et al. (1984) for block tridiagonal matrices, in Axelsson (1986) for general block matrices and in Axelsson and Polman (1986), for relaxed versions of such methods.

For square block H-matrices, Polman (1987) shows existence of an incomplete block-matrix factorization method but without deletion of fill-in and with no relaxation. Kolotilina (1989) shows existence of convergent splittings for block H-matrices.

In the present paper we consider block H-matrices for rectangular matrices and show existence of relaxed forms of the incomplete block matrix factorization algorithm. In addition, we show that the approximations of the inverses of the pivot block matrix entries need to satisfy a condition, which is even weaker than the condition assumed by Axelsson (1986) and Beauwens and Ben Bouzid (1987) for block M-matrices, and furthermore, that this approximation can vary between matrix block columns. This latter property can have significant advantages in practical applications of the methods. The existence results generalizes also previous existence results for full LU-factorization methods shown for diagonally dominant matrices, see Wilkinson (1965), for instance. We show also that a certain explicit factorization method exists and provides a convergent splitting for the block H-matrix class of matrices.

The present report has appeared in an earlier version in Axelsson (1990).

2. AN IMPLICIT PRECONDITIONER: THE RELAXED INCOMPLETE FACTORIZATION METHOD FOR MATRICES PARTITIONED IN BLOCK FORM

Consider a matrix $A = [A_{i,j}]$, partitioned in block-matrix form with $m \times n$ blocks and with nonsingular diagonal blocks of order n_i. We assume that $m \geq n$. The basic step in any incomplete factorization method is the typical Gaussian elimination step at the rth stage,

$$A_{ij}^{(r+1)} := A_{ij}^{(r)} - A_{ir}^{(r)} A_{rr}^{(r)^{-1}} A_{rj}^{(r)}, \tag{2.1}$$

where we assume that the pivot matrices $A_{rr}^{(r)}$ are nonsingular. As is well known, Gaussian elimination and the factorization of A in lower and upper block triangular (LU) factors, are closely related.

If $A_{ij}^{(r)} = 0$ but $A_{ir}^{(r)} A_{rr}^{(r)^{-1}} A_{rj}^{(r)} \neq 0$, we call the latter a fill-in entry. In a full factorization method, usually many fill-in entries occur, in particular since fill-in entries at a stage will usually cause new fill-in entries at later stages, so there is a tendency of a rapid growth, even "exponential" of the number of fill-in entries as the factorization proceeds, at least for general sparse and unstructured matrices.

In an incomplete factorization method, factors L and U will be computed, where the product LU only approximates A. We have then a sparsity pattern S, which defines the positions where entries will be accepted during the factorization. Hence, the entries $A_{i,j}^{(r+1)}$ in (2.1) will only be used if the index pair (i,j) belongs to the set S. In this way, since not accepted fill-in entries cannot cause new fill-in entries at later stages, the sparsity pattern of L and U can be controlled, contrary to the case for the full factorization.

The sparsity set S can be defined *a priori*, i.e., before the outset of the factorization, or during the factorization, where in the latter case, an entry will be accepted only if

$$\|A_{ij}\| \geq c\{\|A_{ii}\|\;\|A_{jj}\|\}^{\frac{1}{2}},$$

for some c, $0 < c < 1$. Here $c = 0$ will give the full factorization, and with the Euclidean norm, $\|\cdot\|_2$, $c = 1$ will give a diagonal matrix if A is symmetric and positive definite. It is also possible to combine the two approaches. For definiteness we consider here only the case where S is chosen *a priori*.

In addition to confining nonzero entries to a sparse set S, when the individual blocks $A_{i,j}$ are sparse themselves, we need to approximate the inverses of the pivot entries, ${A_{rr}^{(r)}}^{-1}$, which occur during the factorization. Assume then that D_r is some sparse approximation of ${A_{rr}^{(r)}}^{-1}$. In fact, it will be seen that we can use different approximations for the entries in the different columns, i.e., we can use $D_r^{(j)} A_{rj}^{(r)}$ as approximation of ${A_{rr}^{(r)}}^{-1} A_{rj}^{(r)}$. This added freedom in the choice of approximations can increase the efficiency of the method significantly, because for blocks $A_{rj}^{(r)}$ with small entries we can use less accurate, and hence less costly approximations $D_r^{(j)}$ than for blocks with larger entries, without any major deterioration of the quality of the preconditioner. We present in section 3 various methods to compute such approximations.

Furthermore, instead of simply neglecting the block matrix entries which are deleted during the incomplete factorization (because they fall outside the sparsity pattern S) we can use some information from them, by adding certain entries to the diagonal of the diagonal matrix block in the same block matrix column. (In this presentation, we have chosen a column oriented algorithm and storage of matrices. A row oriented algorithm would have been equally applicable.) This addition is controlled by the components of a positive vector $\mathbf{x}$ (of order m) such that $\mathbf{x}^T \mathcal{M}_b(A) > \mathbf{0}$ and by the vector $\mathbf{e}_i = (1,1,\ldots,1)^t$, of order n_i, so that we compute a vector

$$\mathbf{z}_i^{(j)} := x_i \mathbf{e}_i^T \left(A_{ij}^{(r)} - A_{ir}^{(r)} {A_{rr}^{(r)}}^{-1} A_{rj}^{(r)} \right), \tag{2.2}$$

if $(i,j) \neq S$, and add the components of the entries of this vector to the diagonal.

Also, we can correct for approximations of the inverses in a similar way when we compute the approximations $D_r^{(j)}$. More generally, in *the relaxed method* we add the components of a diagonal matrix $W^{(i)}$ times the correction vector, where $W^{(i)} = \text{diag}\left(w_k^{(i)}\right)$, with entries $w_k^{(i)}$, $0 \leq w_k^{(i)} \leq 1$, $k = 1,2,\ldots$

Let the sparsity pattern, defining where the block matrix entries will be accepted, be S and let S^c be the complementary set, i.e.

$$S^c = \{(i,j);\; 1 \leq i \leq m,\; 1 \leq j \leq n\} \backslash S.$$

S will always include at least the set $\{(i,i); 1 \leq i \leq m\}$, i.e., $(i,i) \neq S^c$. In fact, we shall assume for simplicity that S contains at least the set of index pairs defined by the sparsity pattern of

A itself. This means that (2.2) (with the weight matrix $W^{(i)}$) takes the simpler form

(2.2′) $$\mathbf{z}_i^{(j)} = -x_i \mathbf{e}_i^T W^{(i)} A_{ir}^{(r)} A_{rr}^{(r)^{-1}} A_{rj}^{(r)}.$$

Then the incomplete factorization method for block matrices (presented in a pseudocode) takes the form:

ALGORITHM (IBFC)
[C stands for column oriented]

```
for r := 1 step 1 until m − 1 do
begin
  for j := r+1 step 1 until n do
  begin
  compute approximation(s) D_r^(j) of A(r,r)^(−1);
  if (r,j) ∈ S then
    begin F := A(r,j)); E := D_r^(j) × F;
      for i := (r + 1) step 1 until m do
      if (i,r) ∈ S then
        begin if (i,j) ∈ S then
          A(i,j) := A(i,j) − A(i,r) × E;
          else
        diag(A(j,j)) := diag(A(j,j)) − (x_i/x_j) e(i)^T W(i) × A(i,r) × A(r,r)^(−1) F
          [comment: here corrections to the diagonal for the deleted fill in entries
          are computed, so that the weighted column sums will be preserved,
          if w_k^(i) = 1; the computation is done by vector matrix multiplications
          and solution of a linear system with matrix A^T(r,r). If there occurs
          several fill-in entries in a row, it is advisable to first sum up the
          i-dependent part of the correction vectors, so that it suffices with
          one solution with A^T(r,r)]
        end
      end (loop in i)
    end (loop in j)
  end (loop in r)
```

Remark 2.1. Note that if we let $S^c = \emptyset$, and $D_r = A_{rr}^{(r)^{-1}}$, then this algorithm will compute the full LU factorization of A. Hence, the full factorization is a special case of the IBFC algorithm. Note also that we could have used row or column permutations but that this would have changed the sparsity pattern, so we have not considered this here. However, we can permit permutations if the permuted matrix will remain a block H-matrix.

The block matrix entries can be stored in the positions originally occupied by the entries of A. (Naturally, for the iterative method we need to have stored another copy of A.) The entries, except the unit diagonal blocks of U, of the block triangular matrices L and U are found in the lower and upper triangular part of this matrix, when the algorithm is completed. If we use a compact storage of the entries of A, i.e., if we store only entries in positions defined by the sparsity pattern of S, the entries of L and U will be found in corresponding positions.

Obviously we are interested in knowing if the entries $A_{rr}^{(r)}$, which occupy the main diagonal of L, are nonsingular. We shall show that this is so for the class of H-matrices.

Theorem 2.2. Let A be a block H-matrix and let $\mathbf{x}$ be a positive vector such that $\mathbf{x}^T\mathcal{M}_b(A)>0$, where the block diagonal part of A is nonsingular and $\mathcal{M}_b$ is the comparison matrix, defined in (1.3). Then the IBFC algorithm exists, i.e., each pivot entry $A_{rr}^{(r)}$ is nonsingular, if the approximations $D_r^{(j)}$ of $A_{rr}^{(r)^{-1}}$, which can differ in different block matrix columns (j), satisfy

$$\|D_r^{(j)}A_{rj}^{(r)}\| \le \|A_{rr}^{(r)^{-1}}A_{rj}^{(r)}\|, \quad (r,j)\in S, \ j\ge r+1. \tag{2.3}$$

In addition, at every stage (r) of the method, the remaining part $A^{(r)}$ of the matrix is an H-matrix.

Proof (by induction). Let $A^{(1)}=A$, and for $r=1,2,\dots,n$, let

$$s_j^{(r)} = x_j - \|A_{jj}^{(r)^{-1}}\| \sum_{i=r,i\neq j}^{m} x_i\|A_{ij}^{(r)}\|, \quad j=r,\dots,n, \tag{2.4}$$

and note that $s_j^{(1)}$ is positive, since by assumption $A^{(1)}$ is a block H-matrix for the vector x. First we need to show that $A_{jj}^{(r+1)}$ is nonsingular. To this end, using the inverse triangle inequality, and noting that $w_k^{(i)}\ge 0$, we find for $j\ge r+1$,

$$\begin{aligned}
&x_j\|A_{jj}^{(r+1)}\| - \sum_{\substack{i=r+1\\ i\neq j,(i,j)\in S}}^{m} x_i\|A_{ij}^{(r+1)}\| \\
&\ge x_j\|A_{jj}^{(r)}\| - x_j\|A_{jr}^{(r)}D_r^{(j)}A_{rj}^{(r)}\| \\
&\quad - \sum_{\substack{i=r+1\\ i\neq j,(i,j)\in S}}^{m} \left\{x_i\|A_{ij}^{(r)}\| + x_i\|A_{ir}^{(r)}D_r^{(j)}A_{rj}^{(r)}\|\right\} \\
&\quad - \max_{k,i} w_k^{(i)} \sum_{\substack{i=r+1\\ (i,j)\in S^c}}^{m} x_i\|A_{ir}^{(r)}A_{rr}^{(r)^{-1}}A_{rj}^{(r)}\| \\
&\ge x_j\|A_{jj}^{(r)}\| - \sum_{\substack{i=r+1\\ i\neq j}}^{m} x_i\|A_{ij}^{(r)}\| \\
&\quad - \|A_{rr}^{(r)^{-1}}A_{rj}^{(r)}\| \sum_{i=r+1}^{m} x_i\|A_{ir}^{(r)}\| + t_j^{(r)},
\end{aligned} \tag{2.5}$$

where

$$t_j^{(r)} = \left(1-\max_{k,i} w_k^{(i)}\right)\|A_{rr}^{(r)^{-1}}A_{rj}^{(r)}\| \sum_{\substack{i=r+1\\ (i,j)\in S^c}}^{m} x_i\|A_{ir}^{(r)}\|,$$

and where we have used (2.3). Note that $t_j^{(r)}\ge 0$.

For later use let

$$F_{ij}^{(r)} = A_{ir}^{(r)}A_{rr}^{(r)^{-1}}A_{rj}^{(r)}.$$

Since for any nonsingular matrix A, $\|A\| \geq 1/\|A^{-1}\|$, we find using (2.5),

$$x_j\|A_{jj}^{(r+1)}\| - \sum_{\substack{i=r+1\\ i\neq j,(i,j)\in S}}^{m} x_i\|A_{ij}^{(r+1)}\|$$

$$\geq x_j\|A_{jj}^{(r)^{-1}}\|^{-1} - \sum_{\substack{i=r\\ i\neq j}}^{m} x_i\|A_{ij}^{(r)}\| + x_r\|A_{rj}^{(r)}\|$$

$$- \|A_{rj}^{(r)}\|\,\|A_{rr}^{(r)^{-1}}\| \sum_{i=r+1}^{m} x_i\|A_{ir}^{(r)}\| + t_j^{(r)}$$

$$= \|A_{jj}^{(r)^{-1}}\|^{-1}s_j^{(r)} + \|A_{rj}^{(r)}\|s_r^{(r)} + t_j^{(r)} > 0.$$

Similarly, multiplying all matrix terms by a vector $\mathbf{v}$, prior to taking norms in (2.5), yields

$$\|A_{jj}^{(r+1)}\mathbf{v}\| > \frac{1}{x_j} \sum_{\substack{i=r+1\\ i\neq j,(i,j)\in S}} x_i\|A_{ij}^{(r+1)}\mathbf{v}\| \geq 0, \quad i \leq j \leq n \tag{2.6}$$

so

$$\inf_{\mathbf{v};\|\mathbf{v}\|=1} \|A_{jj}^{(r+1)}\mathbf{v}\| > 0,$$

which shows that $A_{jj}^{(r+1)}$ is nonsingular, $1 \leq j \leq n$.

Using the relation $\|(I-B)^{-1}\|^{-1} \geq 1 - \|B\|$, which is valid for any matrix with $\|B\| < 1$, we find

$$x_j/\|A_{jj}^{(r+1)^{-1}}\| = x_j\Bigg/\left\|\left[A_{jj}^{(r)} - A_{jr}^{(r)}D_r^{(r)}A_{rj}^{(r)} - \sum_{\substack{i=r+1\\ i\neq j,(i,j)\in S^c}}^{m} \operatorname{diag}\left(\frac{x_i}{x_j}W^{(i)}e(i)^T F_{ij}\right)\right]^{-1}\right\|$$

$$\geq \left(x_j/\|A_{jj}^{(r)^{-1}}\|\right)\left\|\left[I - A_{jr}^{(r)}D_r^{(r)}A_{rj}^{(r)}A_{jj}^{(r)^{-1}} - \sum_{\substack{i=r+1\\ i\neq j,(i,j)\in S^c}}^{m} \operatorname{diag}\left(\frac{x_i}{x_j}e(i)^T W^{(i)}F_{ij}^{(r)}\right)A_{jj}^{(r)^{-1}}\right]^{-1}\right\|$$

$$\geq \frac{x_j}{\|A_{jj}^{(r)^{-1}}\|}\left(1 - \left[\|A_{jr}^{(r)}\|\,\|A_{rr}^{(r)^{-1}}A_{rj}^{(r)}\| + \sum_{\substack{i=r+1\\ i\neq j,(i,j)\in S^c}}^{m} \max_{k,i} w_k^{(i)}\frac{x_i}{x_j}\|A_{ir}^{(r)}\|\,\|A_{rr}^{(r)^{-1}}A_{rj}^{(r)}\|\right]\|A_{jj}^{(r)^{-1}}\|\right)$$

$$= \frac{x_j}{\|A_{jj}^{(r)^{-1}}\|} - \left[x_j\|A_{jr}^{(r)}\| + \sum_{\substack{i=r+1\\ i\neq j,(i,j)\in S^c}}^{m} \max_{k,i} w_k^{(i)}x_i\|A_{ir}^{(r)}\|\right]\|A_{rr}^{(r)^{-1}}A_{rj}^{(r)}\|,$$

and a derivation similar to the one in (2.5) reveals that

$$\frac{1}{\|A_{jj}^{(r+1)^{-1}}\|}s_j^{(r+1)} = \frac{x_j}{\|A_{jj}^{(r+1)^{-1}}\|} - \sum_{\substack{i=r+1\\ i\neq j,(i,j)\in S}}^{m} x_i\|A_{jj}^{(r+1)}\|$$

$$\geq \frac{x_j}{\|A_{jj}^{(r)^{-1}}\|} - \sum_{\substack{i=r+1\\ i\neq j,(i,j)\in S}}^{m} x_i\|A_{ij}^{(r)}\| + \left(x_r\|A_{rj}^{(r)}\| - x_r\|A_{rj}^{(r)}\|\right) - \|A_{rr}^{(r)^{-1}}A_{rj}^{(r)}\| \sum_{i=r+1}^{m} x_i\|A_{ir}^{(r)}\| + t_j^{(r)}$$

$$\geq \|A_{jj}^{(r)^{-1}}\|^{-1}s_j^{(r)} + \|A_{rj}^{(r)}\|s_r^{(r)} + t_j^{(r)} > 0, \quad r+1 \leq j \leq n,$$

which shows that $\mathcal{M}_b\left(A^{(r+1)}\right)$ is an H-matrix.

Corollary 2.3. The exact (full) block matrix factorization of a rectangular block H-matrix exists. In addition, the diagonal blocks of U are nonsingular.

Proof. $S^c = \emptyset$, the empty set and $D_r^{(j)} = A_{rr}^{(r)^{-1}}$, produces the exact factorization and Theorem 2.2 is valid for this choice of S^c and $D_r^{(j)}$ also. The diagonal blocks of U are $A_{rr}^{(r)}$, which by the proof of Theorem 2.2 are nonsingular.

Remark 2.4. Clearly, the above includes pointwise versions of the algorithm as well. For block M-matrices the proof of existence becomes particularly simple, see Axelsson (1986), Axelsson and Polman (1986) and Beauwens and Ben Bouzid (1987). In these papers it was assumed that $D_r^{(j)} = D_r \leq A_{rr}^{(r)^{-1}}$. Clearly, if in addition $A_{rj}^{(r)} \leq 0$, $j \neq r$ then $\|D_r A_{rj}^{(r)}\| \leq \|A_{rr}^{(r)^{-1}} A_{rj}^{(r)}\|$ for the infinity norm.

Remark 2.5 (choice of norms). The computation of $\mathcal{M}_b(A)$ to check if A is an H-matrix can be troublesome in practice. However, if A_{jj}^{-1} is nonnegative and each $A_{i,j}$, $i \neq j$ is either nonpositive or nonnegative, then we have

$$\|A_{jj}^{-1}\|_\infty = \|A_{jj}^{-1}\mathbf{e}\|_\infty \quad \text{and} \quad \|A_{ij}\|_\infty = \|A_{ij}\mathbf{e}\|_\infty,$$

so the norms can be computed simply by solving a linear system with A_{jj} and using matrix-vector multiplications for $A_{i,j}, i \neq j$. If A_{jj} is symmetric and positive definite, we have

$$\|A_{jj}^{-1}\|_2 = [\min \lambda(A_{jj})]^{-1}$$

for the Euclidean norm. Since $\|A\|_2 \leq \{\|A\|_1 \|A\|_\infty\}^{\frac{1}{2}}$ we may still use the practically easier norms $\|\cdot\|_1$ and $\|\cdot\|_\infty$ for the estimate of the norms of the off-diagonal blocks.
As has already been remarked we can alternatively use the method presented in Hager (1984) to estimate these norms, $\|D_r A_{rj}^{(r)}\|$ and $\|A_{rr}^{(r)^{-1}} A_{rj}^{(r)}\|$.

Finally, it is readily seen from the proof that an alternative condition to (2.3) is

$$\|D_r^{(j)}\| \leq \|A_{rr}^{(r)^{-1}}\|.$$

Note that we can use simpler (i.e. more sparse) approximation in columns where $\|A_{rj}^{(r)}\|$ is small than in other columns where $\|A_{rj}^{(r)}\|$ is bigger.

Remark 2.6 (diagonal modification). We have seen that in the algorithm $IBFC$ we may correct for the neglected fill-in entries. If we choose $w_k^{(i)} = 1$ for all k, i then it follows that we have generalized (weighted) columnsums preserved, i.e. a form of mass-balance.

We can also correct (modify) for the sparse matrix approximation $D_r^{(j)}$ of the inverses, $A_{rr}^{(r)^{-1}}$ of the pivot matrices.

Let then $G_r^{(j)}$ be an initial approximation of $A_{rr}^{(r)^{-1}}$, for instance computed by one of the methods to be presented in the next section.

Let

$$D_r^{(j)} = G_r^{(j)} + \hat{D}_r^{(j)},$$

where $\hat{D}_r^{(j)}$ is a diagonal matrix such that

$$\left(G_r^{(j)} + \hat{D}_r^{(j)}\right)\mathbf{u} = A_{rr}^{(r)^{-1}}\mathbf{u} \tag{2.7}$$

and where $\mathbf{u}$ is a positive matrix. For instance, if $A_{rj}^{(r)} \leq 0$ and there is no zero row in $A_{rj}^{(r)}$, we may choose $\mathbf{u} = \mathbf{u}^{(j)} = -A_{rj}^{(r)}\mathbf{e}$, where $\mathbf{e} = (1, 1, \ldots, 1)^t$. Then (2.7) shows that

$$D_r^{(j)} A_{rj}^{(r)} \mathbf{e} = A_{rr}^{(r)^{-1}} A_{rj}^{(r)} \mathbf{e},$$

i.e.

$$\|D_r^{(j)} A_{rj}^{(r)}\|_\infty = \|A_{rr}^{(r)^{-1}} A_{rj}^{(r)}\|_\infty,$$

if $D_r^{(j)} \geq 0$ and $A_{rr}^{(r)}$ is monotone.

Remark 2.7. As for M-matrices, we can define *singular H-matrices.* We define a singular block H-matrix as a matrix for which its comparison matrix $\mathcal{M}_b(A)$ is a singular M-matrix. For a singular M-matrix A, there exists no positive vector $\mathbf{x}$, for which $\mathbf{x}^T A > 0$ (because otherwise A would be a nonsingular M-matrix). Let $\mathbf{x} > 0$ be such that $\mathbf{x}^T \mathcal{M}_b(A) = 0$. Then it is readily seen even if we choose $W^{(i)} = I$, that (2.6) shows that $A_{jj}^{(r+1)}$ remains nonsingular as long as there exists at least one nonzero block $A_{ij}^{(r+1)}$, $i > j$. If the sparsity pattern S equals the sparsity pattern of A, and if there exists at least one nonzero block A_{ij}, $i > j$ for each $j = 1, 2, \ldots, n-1$ in the original matrix A, then it is readily seen that $A^{(r+1)}$ also will contain at least one such nonzero block for each $r \leq n-1$. This is so because the outermost entries in the envelope (skyscraper) structure of A will in fact not be changed during the factorization. Hence, the only singular diagonal block of U for such a singular block H-matrix, will occur in the last diagonal block, i.e., only $A_{n,n}^{(n)}$ will be singular.

Remark 2.8. Note that (2.5) indicates that the "stability" of the factorization increases as the amount of relaxation decreases, i.e. the norms of the pivot entries gets larger as $w_k^{(i)}$ gets smaller.

3. EXPLICIT PRECONDITIONERS

Let A be a square matrix of order n partitioned in block matrix form and define a sparsity pattern S as in Section 2 where $(i,i) \notin S^c$, $i = 1, 2, \ldots, n$.

Consider the following explicit preconditioner G, which is partitioned in the same form as A and which has sparsity pattern S, i.e. $G_{i,j} = 0$ if $(i,j) \in S^c$ and which satisfies

$$(GA)_{i,j} = \delta_{i,j}, \quad (i,j) \in S, \tag{3.1}$$

where

$$\delta_{i,j} = \begin{cases} 1, & \text{for } j = i \\ 0, & \text{for } j \neq i. \end{cases}$$

Note that due to the sparsity pattern of G, the ith row of (3.1) takes the form

$$\sum_{\substack{k=1 \\ (i,k) \in S}}^{n} G_{i,k} A_{k,j} = \delta_{i,j}, \quad (i,j) \in S. \tag{3.1'}$$

If we consider the matrix graph of G we see that in the summation in (3.1'), only entries $A_{k,j}$ will appear where both vertices k and j are directly connected to vertex i.
Let $|S|$ be the cardinality of S. Note that G contains $|S|$ entries and that (3.1) provides the same number of equations. We shall show that if A is a block H-matrix, then such a matrix G exists and is unique. Note that we have not imposed G to be symmetric, for instance. Such a result has previously been shown by Kolotilina (1989), but the proof below is constructive and different. For a recent presentation of symmetric versions of such methods, see Kolotilina and Yeremin (1990).

We show first the practically important result that the entries of each block row (i) of G can be computed independently of the entries of the other block rows, i.e. the block rows can be computed in parallel. Let then $\alpha^{(i)}$ be a matrix partitioned in the same block matrix form as A and let $\alpha^{(i)} \otimes A$ be a Hadamard (pairwise entry) product, where

$$(3.2) \qquad \alpha^{(i)}_{k,j} = \begin{cases} 1 & \text{(all entries in this block are equal to 1), for all } k,j \\ & \text{such that } (i,k) \in S \wedge (i,j) \in S \\ 1, & \text{for all } k = j \\ 0, & \text{otherwise,} \end{cases}$$

$i = 1, 2, \dots, n$ and let $G^{(i)}$, partitioned as G, be the *exact inverse* of $\alpha^{(i)} \otimes A$ (it will be shown below that this inverse exists).
The equation for the ith row of $G^{(i)}$ is then

$$(3.3) \qquad \sum_{k=1}^{n} G^{(i)}_{i,k} \left(\alpha^{(i)} \otimes A\right)_{k,j} = \delta_{i,j}.$$

By (3.2), here all entries $(\alpha^{(i)} \otimes A)_{k,j} = 0$ for $(i,k) \in S^c$ or $(i,j) \in S^c$ that is, we have

$$\sum_{\substack{k=1 \\ (i,k) \in S}}^{n} G^{(i)}_{i,k} A_{k,j} = \delta_{i,j}, \quad i \text{ fixed}, (i,j) \in S.$$

As this is the same as (3.1') we see that the ith row of $G^{(i)}$ and G are identical.

Corollary 3.1. The block rows of G, defined by (3.1) and the sparsity set S, can be computed in parallel.
Proof. (3.3) shows how the ith row of G can be computed independently of the other rows. □

The next Lemma is an immediate consequence of the definition of a block H-matrix.
Lemma 3.2. Let A be a block H-matrix. Then $\alpha^{(i)} \otimes A$, where $\alpha^{(i)}$ is defined by (3.2), is also a block H-matrix.

We show now that the inverse of a block H-matrix exists and is unique.
Theorem 3.3. Let A, partitioned in block matrix form of square order n, be a block H-matrix. Then the inverse of A exists.
Proof. Corollary 2.3 shows that the LU factorization of A exists, where L and U are triangular block-matrices of order $n \times n$. Since the diagonal blocks of L and U are nonsingular, it follows that their inverses, and hence $A^{-1} = U^{-1}L^{-1}$, exists. □

(3.3), Lemma 3.2 and Theorem 3.3 show now:

Corollary 3.4. Let A be a block H-matrix. Then there exists a matrix G, with sparsity pattern $\mathcal{S}$ such that

$$(GA)_{i,j} = \delta_{i,j}, \quad (i,j) \in S.$$

The above method was one of two methods proposed in Axelsson, et al. (1984), to compute approximate inverses. A similar method was proposed earlier by Benson (1973) and Frederickson (1975). An alternative way is to compute G as a sparse part of the exact inverse A^{-1}. In particular, if we compute a band part of A^{-1}, this can be done efficiently as shown in [2]. It is also interesting to note that, as Kolotilina and Yeremin (1986) show, the above approximate inverse methods for symmetric matrices are best approximations to A^{-1} in the following sense. Let G have a sparsity pattern defined by S and let

$$F_W(G) = tr\{(I - GA)^T W(I - GA)\},$$

where W is symmetric and positive definite and tr stands for the trace. Then one finds that the minimum is taken for a matrix G which satisfies

$$(GAWA^T)_{i,j} = (WA^T)_{i,j}, \quad (i,j) \in S. \tag{3.4}$$

For $W = A^{-1}$ where A is symmetric and positive definite one finds that the minimum of $F_W(G)$ is taken for a matrix G satisfying

$$(GA)_{i,j} = \delta_{i,j}, \quad (i,j) \in S. \tag{3.5}$$

When $W = A^T A$ (in this case A does not have to be symmetric, but it must be nonsingular, of course) one finds that the minimum of G is taken for a matrix G satisfying

$$G_{i,j} = (A^{-1})_{i,j}, \quad (i,j) \in S \tag{3.6}$$

The explicit preconditioning methods have the advantage that the implementation of them requires only matrix vector multiplications which can be implemented easily on parallel computers. As we have seen, the computation of G satisfying (3.4) can also occur in parallel between rows. The computation of G satisfying (3.5) on the other hand can not be done so efficiently in parallel. For instance, if we want to compute a band part of A^{-1} and if A itself is a band matrix, then an efficient algorithm uses first the factorization of A in triangular factors and the band part of A^{-1} can now be computed using a similar recursion as is used for the factorization of A (see [2]).

It is readily seen that the preconditioner in (3.4) does not preserve symmetry and/or positive definiteness of A in general. However when using an iterative method, such as the conjugate gradient method it is of interest to preserve symmetry and positive definiteness, i.e. to use a symmetric, and positive definite preconditioning matrix when A is symmetric and positive definite. The preconditioner in (3.5) will clearly be symmetric if S is a symmetric sparsity set, but it may be indefinite, even if A is positive definite. In order to compute an s.p.d. preconditioner to an s.p.d. matrix A we can proceed as follows.

Assume first that we have factored A as $A = L_A D^{-1} L_A^T$ where D is diagonal and where $L_A = I - \tilde{L}$ and $\tilde{L}$ is strictly lower triangular. We compute then an approximate inverse L_G of L_A. This can be done using formulas corresponding to (3.4). We get

$$(L_G L_A)_{i,j} = \delta_{i,j}, \quad (i,j) \in S_L \tag{3.5'}$$

and

$$(3.6') \qquad (L_G)_{i,j} = (L_A^{-1})_{i,j}, \quad (i,j) \in S_L$$

respectively, where S_L denotes the "lower triangular part" of S, i.e. $S_L = \{(i,j) \in S, \ j \leq i\}$. The preconditioner $G = L_G^T D L_G$ to be used in a multiplicative preconditioner is now symmetric and positive definite.

If a factorization of A is not known, for instance would be too expensive to compute, we can use the following method, again based on minimizing the matrix approximation error in a (weighted) Frobenius norm.

To derive this method we let $A = L_A D^{-1} L_A^T$ but it is our intention to find a method where L_A (i.e. $\tilde{L}$) actually is not needed. We want to minimize

$$F_W(L_G) = tr\{(I - L_G L_A)^T W (I - L_G L_A)\}.$$

Then for $W = (L_A^T L_A)^{-1}$, (3.4), shows that

$$(L_G L_A (L_A^T L_A)^{-1} L_A^T)_{i,j} = (L_A^T L_A)^{-1} L_A^T)_{i,j}, \quad (i,j) \in S_L$$

or

$$(L_G)_{i,j} = (L_A^{-1})_{i,j}, \quad (i,j) \in S_L$$

which is (3.6$'$). Hence this choice of W requires that L_A has been computed, but the above characterizes the solution of (3.6$'$) as an optimal solution in the corresponding norm.

We try now $W = D^{-1}$. Then (3.4) shows that

$$(L_G L_A D^{-1} L_A^T)_{i,j} = (D^{-1} L_A^T)_{i,j}, \quad (i,j) \in S_L,$$

or

$$(3.7) \qquad (L_G A)_{i,j} = \{(L_A D^{-1})^T\}_{i,j}, \quad (i,j) \in S_L.$$

Since S_L is a lower triangular set but the right matrix is upper triangular, (3.7) takes the form

$$(L_G A)_{i,j} = \begin{cases} 0, & i \neq j \\ D_i^{-1}, & i = j \end{cases}.$$

Hence we do not need to know L_A!

If D is also not known, Yeremin and Kolotilina (1990) suggested to use a diagonal scaling. Compute then first $\hat{L}_G$ such that

$$(\hat{L}_G A)_{i,j} = \delta_{i,j}, \quad (i,j) \in S_L.$$

Then set $L_G = \hat{D}^{\frac{1}{2}} \hat{L}_G$, where $\hat{D}$ is a diagonal matrix to be defined below and let $G = L_G^T L_G = \hat{L}_G^T \hat{D} \hat{L}_G$ be the preconditioner. $\hat{D}$ will be chosen so that

$$(L_G A L_G^T)_{i,i} = 1$$

i.e.

$$(\hat{D}^{\frac{1}{2}} \hat{L}_G A \hat{L}_G^T \hat{D}^{\frac{1}{2}})_{i,i} = 1$$

or, with $\hat{D} = \mathrm{diag}(\hat{d}_1, \ldots, \hat{d}_n)$, we find

$$\sum_{j=1}^{n} \hat{d}_i^{\frac{1}{2}} (\hat{L}_G A)_{i,j} (\hat{D}^{\frac{1}{2}} \hat{L}_G)_{ji}^T = 1$$

so, since $(\hat{L}_G A)_{i,j} = \delta_{i,j}$,

$$\hat{d}_i (\hat{L}_G)_{ii} = 1$$

that is

$$\hat{D} = [\mathrm{diag}(\hat{L}_G)]^{-1}.$$

Hence the preconditioner is

$$L_G^T L_G = \hat{L}_G^T \hat{D} \hat{L}_G = \hat{L}_G^T [\mathrm{diag}(\hat{L}_G)]^{-1} \hat{L}_G.$$

4. A HYBRID PRECONDITIONING METHOD

It turns out that while explicit preconditioners are readily implemented in an iterative method, they frequently require $|S|$ to be relatively large to get an accurate approximation of the inverse. In particular, this is the case when A has a large condition number. On the other hand, implicit preconditioners can be accurate approximations of A if A is sparse, even when $|S|$ is relatively small.

In some problems, a combination of explicit and implicit preconditioners in a two-level way can be efficient. As has been shown by Concus, et al. (1985) and Axelsson, et al. (1984) such combinations can be efficient when the condition numbers of the pivot matrices $A_{rr}^{(r)}$, which occur in the algorithm to compute the implicit preconditioner of a matrix, partitioned in block form, are relatively small. For the approximation of the inverses of these matrices, we can then use one of the explicit methods (3.5) or (3.6). Frequently, we will then use a sparsity pattern in the form of a band matrix. Note then that it can be efficient to use different bandwidths for approximations which occur in the different columns of the pivot row. For instance, we can use diagonal approximations when the entries in some columns are small, while using more accurate approximations in other columns. For other related approximation methods and preconditioners, see [1].

There are other possible ways to combine explicit and implicit preconditioners. For instance, we can compute first an explicit preconditioner G to A and then an incomplete factorization C to $I - E$, where $E = I - GA$. Hence, to solve $A\mathbf{x} = \mathbf{b}$, we use the preconditioner $C^{-1}G$, which requires a matrix vector multiplication with G and a solution of a linear system with C at each iteration step. Since the computation and application of G can occur in parallel, it costs little to use such a matrix, at least on a parallel computer. The accuracy of the preconditioner $C^{-1}G$ can be expected to be much better than for an implicit preconditioner C_0 alone, computed from A involving a sparsity structure S_0, with about the same cardinality ($|S|$) as the sparsity pattern S, when S is used for C, the preconditioner of $I - E$.

We shall now prove that $GA = I - E$ is a block H-matrix. To this end we show first the corresponding result for M-matrices.

Lemma 4.1. Let $\tilde{A}$ be a M-matrix and $\tilde{G}$ defined by $(\tilde{G}\tilde{A})_{i,j} = \delta_{i,j}$, $(i,j) \in S$. Then $\tilde{G} \geq 0$, $\tilde{G}\tilde{A}$ is an M-matrix, $I - \tilde{G}\tilde{A} \geq 0$, and there exists a positive vector $\mathbf{x}$, such that for each i, either $\{(I - \tilde{G}\tilde{A})\mathbf{x}\}_i > 0$, or if $\{(I - \tilde{G}\tilde{A})\mathbf{x}\}_i = 0$ for some i, then the ith row of $I - \tilde{G}\tilde{A}$ is zero.

Proof. We show first that $\tilde{G}\tilde{A}$ is a Z-matrix. For $(i,j) \in S$, we have

$$(\tilde{G}\tilde{A})_{i,j} = \delta_{i,j} \tag{4.1}$$

and for $(i,j) \in S^c$, we find that

$$(\tilde{G}\tilde{A})_{i,j} = \sum_{k,(i,k)\in S} \tilde{g}_{i,k}\tilde{a}_{k,j} = \sum_{k,k\neq j,(i,k)\in S} \tilde{g}_{i,k}\tilde{a}_{k,j}, \tag{4.2}$$

because $\tilde{g}_{i,j} = 0$, for $(i,j) \in S^c$. Since, as we have seen in section 3, each row of $\tilde{G}$ is the ith row in the exact inverse of an M-matrix, we have $\tilde{g}_{ik} \geq 0$. Hence, since $\tilde{A}$ is a Z-matrix, (4.2) shows that $(\tilde{G}\tilde{A})_{i,j} \leq 0$ for $(i,j) \in S^c$. This together with (4.1) show that $\tilde{G}\tilde{A}$ is a Z-matrix. Since $\tilde{A}$ is an M-matrix, there exists a positive vector $\mathbf{x}$, such that $\tilde{A}\mathbf{x} > 0$. Hence, noting that there is no zero row in $\tilde{G}$, $\tilde{G}\tilde{A}\mathbf{x} > 0$, so $\tilde{G}\tilde{A}$ is also an M-matrix.

This and (4.1) show that $\tilde{G}\tilde{A} = I - \tilde{E}$, where $\tilde{E} \geq 0$, so $(I - \tilde{G}\tilde{A})\mathbf{x} = \tilde{E}\mathbf{x} \geq 0$. If $(\tilde{E}\mathbf{x})_i = 0$ for some i, then $(\tilde{G}\tilde{A}\mathbf{x})_i = x_i$, but since $(\tilde{G}\tilde{A})_{i,i} = 1$, we have then $x_i = (\tilde{G}\tilde{A}\mathbf{x})_i = \sum_j (\tilde{G}\tilde{A})_{i,j}x_j = x_i + \sum_{j\neq i}(\tilde{G}\tilde{A})_{i,j}x_j$, so this together with $(\tilde{G}\tilde{A})_{i,j} \leq 0$ for $(i,j) \in S^c$, show that $(\tilde{G}\tilde{A})_{i,j} = 0$, $j \neq i$, i.e., the ith row of $\tilde{E}$ is zero. □

Let now A be a block H-matrix and let G be defined by $(GA)_{i,j} = \delta_{i,j}$, $(i,j) \in S$. Further, let $\tilde{G} = [\tilde{g}_{ik}]$ be the approximate inverse of the comparison matrix $\mathcal{M}_b(A)$, i.e. let

$$\{\tilde{G}\mathcal{M}_b(A)\}_{i,j} = \delta_{i,j}, \quad (i,j) \in S.$$

Let $E = I - GA$ and $\tilde{E} = I - \tilde{G}\mathcal{M}_b(A)$. Then $E_{i,j} = 0$, $(i,j) \in S$ and $\tilde{e}_{i,j} = 0$, $(i,j) \in S$. (Note that $E_{i,j}$ are block matrices and $\tilde{e}_{i,j}$ are scalar entries.)

A theorem by Ostrowski (1961) shows that

$$\|G_{i,j}\| \leq |\tilde{g}_{i,j}|, \quad (i,j) \in S$$

and shows also that

$$\|E_{i,j}\| \leq |\tilde{e}_{i,j}|, \quad (i,j) \in S^c.$$

Let $\mathbf{x}$ be a positive vector, such that $\mathcal{M}_b(A)\mathbf{x} > 0$. Then

$$\sum_{j,j\neq i} \|E_{i,j}\|x_j \leq \sum_{j,j\neq i} |\tilde{e}_{i,j}|x_j.$$

But the proof of Lemma 4.1 shows that $\tilde{e}_{i,j} \geq 0$ and $0 \leq \tilde{E}\mathbf{x} = (I - \tilde{G}\mathcal{M}_b(A))\mathbf{x} < \mathbf{x}$.
Hence

$$\sum_{j,j\neq i} \|E_{i,j}\|x_j < x_i. \tag{4.3}$$

Noting that the block diagonal part of $I - E$ is simply I, (4.3) shows that $I - E$ is a block H-matrix. Hence, Theorem 2.2 shows that the incomplete block-matrix factorization $C = LU$ of $I - E$ exists, whose results we state in the final theorem:

Theorem 4.2. Let A be a block H-matrix and let G be defined by $(GA)_{i,j} = \delta_{i,j}$, $(i,j) \in S$. Then $I - E$ is a block H-matrix, where $E = I - GA$, the incomplete block-matrix factorization $C(IBFC)$ of $I - E$ exists, and therefore, the combined preconditioner $C^{-1}G$ exists.

REFERENCES

1. O. Axelsson (1985), A survey of preconditioned iterative methods for linear systems of algebraic equations, *BIT* 25:166-187.
2. O. Axelsson (1985), Incomplete block matrix factorization preconditioning methods. The ultimate answer? *J. Comp. Appl. Math.* 12&13:3-18.
3. O. Axelsson (1986), A general incomplete block-matrix factorization method, *Linear Algebra Appl.* 74:179-190.
4. O. Axelsson (1990), Preconditioning methods for block H-matrices, FSU-SCRI-90-06, Supercomputer Computations Research Institute, The Florida State University, Tallahassee, Florida.
5. O. Axelsson, S. Brinkkemper and V.P. Il'in (1984), On some versions of incomplete block-matrix factorization iterative methods, *Linear Algebra Appl.* 58:3-15.
6. O. Axelsson and B. Polman (1986), On approximate factorization methods for block matrices suitable for vector and parallel processors, *Linear Algebra Appl.* 77:3-26.
7. R. Beauwens and M. Ben Bouzid (1987), On sparse block factorization, iterative method, *SIAM J. Numer. Anal.* 24:1066-1076.
8. M.W. Benson (1973), Iterative solution of large scale linear systems, Thesis, Lakehead University, Thunder Bay, Canada.
9. A. Berman and R.J. Plemmons (1979), Nonnegative matrices in the mathematical sciences, Academic Press, New York.
10. P. Concus, G.H. Golub and G. Meurant (1985), Block preconditioning for the conjugate gradient method, *SIAM J. Sci. Stat. Comput.* 6:220-252.
11. D.G. Feingold and R.S. Varga (1962), Block diagonally dominant matrices and generalizations of the Gershgorin circle theorem, *Pacific J. Math.* 12:1241-1250.
12. M. Fiedler and V. P̂tak (1962), On matrices with nonpositive off-diagonal elements and positive principal minors, *Czech. Math. J.* 12:283-400.
13. P.O. Frederickson (1975), Fast approximate inversion of large sparse linear systems, Math. Report 7, Lakehead University, Thunder Bay, Canada.
14. W.W. Hager (1984), Condition estimates, *SIAM J. Sci. Stat. Comput.* 5:311-316.
15. L.Yu. Kolotilina (1989), On approximate inverses of block H-matrices, in Numerical Analysis and Mathematical Modelling, Moscow (in Russian).
16. L.Yu. Kolotilina and A. Yu. Yeremin (1986), On a family of two-level preconditionings of the incomplete block factorization type, *Sov. J. Numer. Anal. Math. Modelling* 1:293-320.
17. L. Yu Kolotilina and A.Yu. Yeremin (1990), Factorized sparse approximate inverse preconditionings, manuscript in progress.
18. J.A. Meyerink and H.A. van der Vorst (1977), An iterative solution method for linear systems of which the coefficient matrix is a symmetric M-matrix, *Math. Comp.* 31:148-162.
19. A.M. Ostrowski (1937), Über die Determinanten mit überwiegender Hauptdiagonale, *Comment. Math. Helv.* 10:69-96.
20. A.M. Ostrowski (1961), On some metrical properties of operator matrices and matrices partitioned into blocks, *J. Math. Anal. Appl.* 2:161-209.
21. B. Polman (1987), Incomplete blockwise factorizations of (block) H-matrices, *Linear Algebra Appl.* 90:119-132.
22. F. Robert (1969), Blocs H-matrices and convergence des méthodes iteratives classiques par blocs, *Linear Algebra Appl.* 2:223-265.
23. R.S. Varga, E.B. Saff, and V. Mehrman (1980), Incomplete factorization of matrices and connections with H-matrices, *SIAM J. Numer. Anal.* 17:787-793.
24. Wilkinson (1965), The Algebraic Eigenvalue Problem, Clarendon Press, Oxford.

Domain Decomposition methods for solving large sparse linear systems

Gérard Meurant

CEA, Centre d'Etudes de Limeil-Valenton,

94195 Villeneuve St Georges

France

Abstract

This paper reviews some domain decomposition techniques for solving symmetric sparse linear systems. These techniques are used to construct efficient parallel preconditioners for the conjugate gradient method. We describe different kinds of methods with overlapping and non overlapping subdomains as well as methods for finite elements approximations and also purely algebraic preconditioners.

Keywords Linear systems / Domain decomposition / Conjugate gradient / Preconditioners /

1 Introduction

In this paper we review some domain decomposition (DD) techniques for solving symmetric linear systems. Domain decomposition has been used for a long time, but have received recently a renewal of interest due to the development of parallel computers. Today, there are commercially available parallel computers like, for instance, supercomputers with a few powerful vector processors, the prototypes of which in 1990 are the Cray machines. It is a challenge for numerical analysts to be able to find algorithms that can be efficiently used on such computers.

Due to the lack of space and the rapid growth in the area during last years, we will not be able to review all the techniques that have been proposed for domain decomposition. So, we will only display a selection of algorithms, not necessarily the most efficient ones as these were chosen mainly to give the reader a broad view of the field. We will concentrate on problems arising from the discretization of elliptic or parabolic partial differential equations although these techniques can be applied to a larger class of problems. More detailed information can be found in the proceedings of the annual domain decomposition conferences that have been organized since 1987, [15], [6], [7], [16].

The modern perspective on DD is to use these techniques to construct preconditioners that will be used in some iterative methods.

As a matter of introduction to the field and as our main goal is the efficient solution of problems, Section 2 will introduce the conjugate gradient method and will recall some facts about widely used efficient preconditioners and the modifications that are needed to use them on vector and eventually parallel computers. Section 3 is devoted to the Schwarz algorithm which is a method that originated in the last century and use overlapping subdomains. Section 4 describes a method by Bramble, Pasciak and Schatz that defines a preconditioner for finite element approximations. Section 5 studies some algebraic preconditioners developed for block tridiagonal matrices. Finally we describe a method specially tuned for parabolic problems and we draw some conclusions.

NATO ASI Series, Vol. F 77
Computer Algorithms for Solving Linear Algebraic Equations
Edited by E. Spedicato

2 Some classical preconditioners and their vectorizable variants

The problem we are considering is solving a linear system

$$Ax = b,$$

where A is a large sparse symmetric positive definite matrix of order n.

To do this, we use the conjugate gradient method (CG) and we will mainly concentrate on the design of preconditioners. Most of the preconditioners we will described can be easily extended to non symmetric matrices. Of course, then CG cannot be used anymore and we must rely on other methods like Orthomin, GMRES or CGS, [36].

The conjugate gradient method which was introduced in the 50's can be described in many different ways. It can be considered as a minimization method, as an acceleration method or as an orthogonalization method.

To introduce CG as an acceleration method, let us first recall the first order Richardson iterative method :

$$x^{k+1} = x^k + \alpha(b - Ax^k),$$

x^0 being given and α being a parameter to be chosen. The method converges if $\alpha < \frac{2}{\lambda_{min}}$ where λ_{min} is the smallest eigenvalue of A.

A straightforward generalization of Richardson's method leads to :

$$Mx^{k+1} = Mx^k + \alpha_k(b - Ax^k),$$

where M is a (non singular) symmetric preconditioning matrix as, in fact, instead of solving $Ax = b$, we are solving

$$M^{-1}Ax = M^{-1}b.$$

If we try to apply acceleration algorithms to the previous method, we formally get, see Golub–Meurant [18]

$$x^{k+1} = \omega_{k+1}(\alpha_k z^k + x^k - x^{k-1}) + x^{k-1}$$

$$Mz^k = r^k = b - Ax^k$$

Then, we can compute the coefficients α_k, ω_k as in an orthogonalization method. We choose α_k and ω_k such that the generalized residuals z^i are mutually orthogonal in the scalar product defined by M as we now suppose M to be positive definite

$$(z^i, Mz^j) = 0, \; i \neq j$$

This property gives $r^n = 0$ and the final termination of the algorithm in at most n steps. Unfortunately, in finite precision, this is not always true. But, we will see that we can use the algorithm as an iterative method.

To obtain the orthogonalization property we have to choose

$$\alpha_k = \frac{(z^k, Mz^k)}{(z^k, Az^k)}$$

$$\omega_{k+1} = \frac{1}{1 - \frac{\alpha_k (z^k, Mz^k)}{\omega_k \alpha_{k-1} (z^{k-1}, Mz^{k-1})}}.$$

It is relatively easy to see that the $(k+1)^{th}$ iterate is a polynomial in x^0. More precisely,

$$x^{k+1} = x^0 + Q_k(M^{-1}A)z^0$$

where Q_k is a polynomial of degree less or equal to k. Between methods that can be written that way, CG is the one which, at each iteration, minimizes $E(x^k)$

$$E(x^k) = (A(x - x^k), x - x^k),$$

which is a measure of the error in the A–norm.

As a consequence (see Golub–Meurant [18]), it can be proven using Chebychev polynomials that

$$E(x^k) \leq 4\left(\frac{\sqrt{\kappa} - 1}{\sqrt{\kappa} + 1}\right)^{2k} E(x^0)$$

where $\kappa = \frac{\lambda_{max}}{\lambda_{min}}$ is the condition number of $M^{-1}A$. This shows that if κ is not too large, the method can be efficiently used as an iterative method. This fact was recognized by Reid [33] and led to the revival of CG in the 70's.

It can also be easily seen that this method is equivalent to the "usual" form of CG :

$$r^0 = b - Ax^0,$$

for $k = 0, 1, \ldots$ until convergence,

$$Mz^k = r^k,$$

$$\beta_k = \frac{(r^k, z^k)}{(r^{k-1}, z^{k-1})}, \; \beta_0 = 0,$$

$$p^k = z^k + \beta_k p^{k-1},$$

$$\alpha_k = \frac{(r^k, z^k)}{(Ap^k, p^k)},$$

$$x^{k+1} = x^k + \alpha_k p^k,$$

$$r^{k+1} = r^k - \alpha_k Ap^k.$$

If we put aside the problem of finding a good preconditioner M, it is clear that CG is a good method for vector computers as, on these machines, there exist efficient implementations of the basic CG operations.

Vector computing can be viewed as a form of low level (loop) parallelism exploiting pipelined functional units [19]. Unfortunately, CG is not well suited for parallel computers that need a higher level parallelism. One way of introducing more parallelism in CG, is to take the "usual" form and to notice that (in exact arithmetic) we have

$$(Mz^i, z^j) = 0, i \neq j.$$

Then, it follows that

$$(r^{k+1}, z^{k+1}) = \alpha_k^2(M^{-1}Ap^k, Ap^k) - (r^k, z^k).$$

The modified algorithm is (see Meurant [27]) :

$$x^0 \text{ given}, r^0 = b - Ax^0, Mz^0 = r^0, p^0 = z^0.$$

For each k until convergence,

$$Mv^k = Ap^k,$$

$$(v^k, Ap^k), \qquad (p^k, Ap^k), \qquad (r^k, z^k)$$

$$\alpha_k = \frac{(r^k, z^k)}{(p^k, Ap^k)}$$

$$s_{k+1} = \alpha_k^2(v^k, Ap^k) - (r^k, z^k)$$

$$\beta_{k+1} = \frac{s_{k+1}}{(r^k, z^k)}$$

$$x^{k+1} = x^k + \alpha_k p^k$$

$$r^{k+1} = r^k - \alpha_k Ap^k$$

$$z^{k+1} = z^k - \alpha_k v^k$$

$$p^{k+1} = (z^k - \alpha_k v^k) + \beta_{k+1} p^k$$

In this algorithm a predictor–corrector is used for the dot product (r^{k+1}, z^{k+1}). First, a value is predicted to be able to compute β_{k+1}. Then, the value is corrected after the new vectors are computed, to improve the stability of the algorithm.

With this new form of the CG algorithm, we can compute in parallel the 3 dot products and after a small scalar section, all the vectors involved. Unfortunately, there is one more scalar product to be computed but the number of operations is the same as with the three terms form.

In preconditioned CG, the most important part of the work is usually solving $Mz^k = r^k$, see Meurant [29]. So, preconditioners adapted to parallel computation must be developed. Of course, this is not too much of a problem (consider, for instance, the Jacobi or diagonal preconditioner). The difficult task is to have both parallelism and efficiency (that is a small number of iterations). The properties that are requested for the preconditioner M are :

- M symmetric positive definite
- M sparse
- M easy to construct
- M allows for a fast solve of $Mz = r$ on a vector or parallel computer

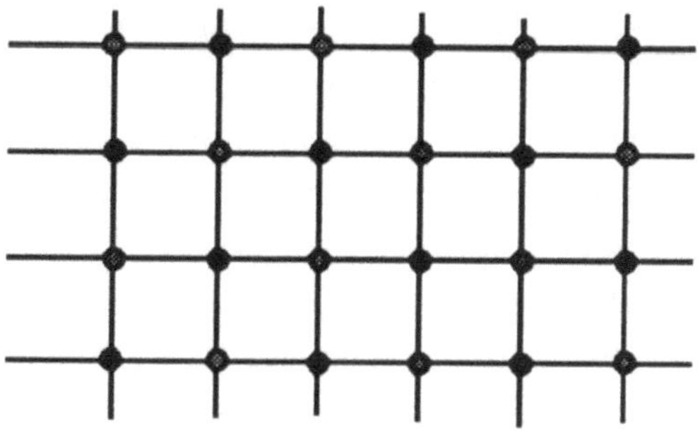

Figure 1
The Red–Black ordering

- "good" distribution of the eigenvalues of $M^{-1}A$ or a small condition number.

Below, some well known preconditioners are described as well as the modifications to be made for their use on vector and parallel computers.

As it was said before, the simplest preconditioner is the diagonal one (DIAG),

$$M = diag(A).$$

Unfortunately, this preconditioner, even though it is perfectly parallel, is not efficient regarding the rate of convergence. A more interesting method is the SSOR algorithm that can be used as a preconditioner because the corresponding iteration matrix is symmetric. It is defined as

$$M = \frac{1}{\omega(2-\omega)}(D+\omega L)D^{-1}(D+\omega L^T)$$

where we have the usual splitting $A = D + L + L^T$, D being diagonal and L strictly lower triangular.

If the model problem of Poisson equation in a unit square domain is discretized with the standard five point scheme, it is known that with the row-wise (or natural) ordering of the mesh points,

$$\kappa(A) = O(\frac{1}{h^2})$$

but there exists an optimal value of ω $(0 < \omega < 2)$ such that

$$\kappa(M^{-1}A) = O(\frac{1}{h}),$$

see, Axelsson [1].

Unfortunately, the actual value of ω is not easy to find for general problems and moreover, as there are some recurrences in the solves with the triangular factors, the SSOR method is not well suited for parallel computing. So, one way to use this algorithm is to change the ordering of the unknowns. One possibility is the Red–Black ordering for the 5 point finite difference scheme, see Figure 1.

Suppose the Red nodes are ordered first (in a row-wise fashion) and then the black nodes. With the 5 point scheme, a Red (resp. Black) node is only related to Black (resp. Red) nodes. So, in the SSOR sweep the Red nodes, as they only depends on components of the previous iterates, can be computed in parallel. Then, during the sweep on the Black nodes, we need only components corresponding to the Red nodes that are known from the first half sweep. Of course, this method is specific of the 5 point scheme. With a 9 point scheme or finite element matrices, more colors must be used. The problem is that for the Red–Black ordering the optimal value of ω is 1 and the condition number is still $O(\frac{1}{h^2})$.

The most well known and used preconditioner is probably the Incomplete Choleski decomposition (IC), see Meijerink and Van der Vorst [25]. If we do the standard Choleski decomposition of A, during the elimination we get some fill ins (see Duff [12]), some elements that are zero in A become non zero in the triangular factors. For the problems under consideration, it can be shown that, within the band of the matrix, the fill ins decrease in magnitude as we go towards the middle of the band, see Duff–Meurant [11] for examples. The idea of the Incomplete decomposition is to neglect some of the fill ins.

For a five diagonals model problem with the row-wise ordering and the most simplest method, things can be formulated very easily in the following way. Let $M = L\ D^{-1}\ L^T$, where L has the same sparsity pattern as the lower triangular part of A. Suppose the five non zero terms on a row of A are

$$A = (c_i, b_i, a_i, b_{i+1}, c_{i+m})$$

and rows of L and D are given by

$$L = (\bar{c}_i, \bar{b}_i, d_i, 0, 0),\ \ D = (0, 0, d_i, 0, 0).$$

By inspection, we can easily see that

$$\bar{c}_i = c_i$$

$$\bar{b}_i = b_i$$

$$d_i = \frac{1}{a_i - b_{i-1}^2 d_{i-1} - c_{i-m}^2 d_{i-m}}.$$

Elements with negative indices are taken to be zero. This method is called IC(1,1). For the Poisson equation, we still have $\kappa(A) = O(\frac{1}{h^2})$ but a better distribution of the eigenvalues than with DIAG. As before, there are some troubles solving the triangular systems on vector and parallel computers. Hence, several modifications to the preconditioner were devised to solve that problem.

One possibility is to compute the unknowns diagonal-wise but, some troubles arise, short vector lengths, non constant strides, etc... The advantage is that, as the same system as in the basic method is solved, the convergence rate is unchanged.

Another possibility is to modify the preconditioner such that solving triangular systems is avoided, see Van der Vorst [37]. This can be done in the following way. First L is symmetrically scaled to put 1 on the diagonal, the system to solve is then,

$$(D^{-1/2} L D^{-1/2})(D^{-1/2} L^T D^{-1/2}) D^{1/2} z = D^{-1/2} r.$$

The first half of the solve is

$$D^{-1/2}LD^{-1/2}y = D^{-1/2}r.$$

We consider the matrix in block form (each block corresponding to a mesh line)

$$D^{-1/2}LD^{-1/2} = \begin{pmatrix} E_1 & & & \\ B_2 & E_2 & & \\ & \ddots & \ddots & \\ & & B_n & E_n \end{pmatrix}$$

Then the solve reduces to

$$E_1 y_1 = D_1^{-1/2} r_1$$

$$E_i y_i = -B_i y_{i-1} + D_i^{-1/2} r_i$$

E_i can be written as

$$E_i = I + F_i$$

with F_i having the following structure (x denoting non zero entries)

$$F_i = \begin{pmatrix} 0 & & & & & \\ x & 0 & & & & \\ & x & 0 & & & \\ & & x & 0 & & \\ & & & x & 0 & \\ & & & & x & 0 \end{pmatrix}$$

$E_i^{-1} = (I + F_i)^{-1}$ is approximated by a truncated Neumann series. It is usual to use three terms in the series, see Van der Vorst [37] for a justification.

$$E_i^{-1} \approx I - F_i + F_i^2 - F_i^3 = (I - F_i)(I + F_i^2)$$

F_i^2 is a matrix with just one non zero block diagonal which is easily computed,

$$F_i^2 = \begin{pmatrix} 0 & & & & & \\ 0 & 0 & & & & \\ x & 0 & 0 & & & \\ & x & 0 & 0 & & \\ & & x & 0 & 0 & \\ & & & x & 0 & 0 \end{pmatrix}.$$

The interest of this modification is that the solves with bidiagonal matrices are replaced by matrix multiplications that can easily be done in vector or parallel mode. We call this method ICVDV. However, it must be understood that we modified the preconditioner. So, the problem is to know what the convergence rate of the new method is. With most problems, the number of iterations for ICVDV is within 2 or 3 the same as for IC(1,1). However, examples can be constructed (see Meurant [26]) where the matrices involved are not enough diagonally dominant, so the Neumann series converge very slowly and give bad approximations to the inverses. This has the consequence that the number of iterations is much higher for the modified method.

These ways of parallelizing are specific of the problems we considered. For more general examples, several methods can be considered. More or less, they are all based on re–ordering the unknowns. Two things can be done: the first one is to compute the Incomplete Decomposition with a "natural" ordering and to change the ordering only for the solve of $Mz = r$. This approach has been recently used by Filippone and Radicati [14] using a front-wise ordering that reduced to a diagonal ordering for the model problem. Another approach is to first reorder the matrix A according to some numbering scheme and then, do the Incomplete Decomposition. This has been done by Duff and Meurant [11], not only for the sake of parallelism but, to try to understand the effect of orderings on IC. Some theory has been subsequently developed by Eijkhout [13] and Doi [9].

Now, we consider block preconditioners which, on the problems they are suitable for, give better results than point methods like IC. Let us suppose A has a block tridiagonal structure

$$A = \begin{pmatrix} D_1 & A_2^T & & & \\ A_2 & D_2 & A_3^T & & \\ & \ddots & \ddots & \ddots & \\ & & A_{m-1} & D_{m-1} & A_n^T \\ & & & A_m & D_m \end{pmatrix}.$$

Suppose D_i is a point tridiagonal strictly diagonally dominant matrix and A_i is diagonal. Let L denote the block lower triangular part of A

$$L = \begin{pmatrix} 0 & & & \\ A_2 & 0 & & \\ & \ddots & \ddots & \\ & & A_m & 0 \end{pmatrix}.$$

The (complete) block Choleski decomposition of A can be written as

$$A = (\Sigma + L)\, \Sigma^{-1}\, (\Sigma + L^T)$$

with Σ a block diagonal matrix

$$\Sigma = \begin{pmatrix} \Sigma_1 & & & \\ & \ddots & & \\ & & \ddots & \\ & & & \Sigma_m \end{pmatrix}.$$

By inspection, matrices Σ_i are given by

$$\begin{cases} \Sigma_1 = D_1, \\ \Sigma_i = D_i - A_i\, \Sigma_{i-1}^{-1}\, A_i{}^T. \end{cases}$$

The matrices Σ_i are dense but, as the D_is are diagonally dominant, the magnitude of the elements of the Σ_is decay away from the diagonal. The idea introduced in Concus, Golub and Meurant [8] is to approximate these dense matrices with tridiagonal matrices. Let B a matrix and $trid(B)$, a triagonal matrix whose non zero elements are the same as the corresponding ones of B. The preconditioner INV is defined as

$$A = (\Delta + L)\, \Delta^{-1}\, (\Delta + L^T)$$

with

$$\Delta = \begin{pmatrix} \Delta_1 & & & \\ & \ddots & & \\ & & \ddots & \\ & & & \Delta_n \end{pmatrix},$$

and

$$\begin{cases} \Delta_1 = D_1, \\ \Delta_i = D_i - A_i\ trid(\Delta_{i-1}^{-1})\ A_i^T. \end{cases}$$

It has been shown with numerical experiments that this method is very efficient and robust, see Concus, Golub and Meurant [8]. INV is not well suited for vector and parallel machines for two reasons. First of all, in solving $Mz = r$, there is a block recurrence. Secondly, for each block we have to solve a tridiagonal system like

$$\Delta_i w = c.$$

A way to vectorize is to replace Δ_i^{-1} by a banded approximation, as was done before. However, instead of using a Neumann series, a better way is to use some diagonals of the inverse Δ_i^{-1}. Taking a 7 diagonals approximation is fine with most problems (with enough diagonal dominance), see Meurant [26]. Then, everything is in vector mode within a block. We denote this method by INVV.

The last class of preconditioners we consider in this section is polynomial preconditioners. M^{-1} is directly approximated with

$$M^{-1} = P_k(A) = \sum \alpha_i A^i.$$

This idea may seem strange because we have seen that the conjugate gradient method generates an optimal polynomial and, in that respect, the CG polynomial after $k+m$ iterations with no preconditioner is better than the polynomial generated by k iterations with a polynomial preconditioner of degree m. But the latter can be of some use on vector machines as there are less scalar products to be computed and a lot of parallelism.

There are many ways to choose the polynomial coefficients. A robust one is to consider least square polynomials, see Saad [35]. Suppose estimates a and b of the minimum and maximum eigenvalues of A are known. The ideal would be to have the polynomial $P_k(\lambda)\lambda$ equal to 1 on the set of eigenvalues of A. Unfortunately, we do not know the eigenvalues, so it is natural to ask for the polynomial that minimizes

$$\int_a^b (1 - \lambda P_k(\lambda))^2 w(\lambda)\ d\lambda,$$

where w is a weigth function which is usually chosen as

$$w(\lambda) = (b - \lambda)^{-1/2}(\lambda - a)^{-1/2}$$

With this choice, it is relatively easy to construct the polynomial coefficients. It can be notice that we need approximations to the extreme eigenvalues and that this is not always easy to get. Moreover, if such approximations are known, we possibly can use other methods like Chebyshev acceleration. The number of iterations depends on the polynomial degree. From experiments it is seen that, usually, when the degree of the polynomial is increased, the number of iterations drops rapidly in the beginning and then there is a slow down. Hence, looking at the computer time, a low order polynomial is good enough for most problems.

As we have seen, some of the previous method that were devised for vector computers may also be used on parallel machines with a loop level parallelism. However, it is desirable to devise specific methods that can involve higher level parallelism. Domain decomposition methods offer such a possibility.

For constructing algorithms for parallel computers, a good principle is to divide the problem into smaller pieces, solve the subproblems in parallel and then to paste the results together. This has often been called "divide and conquer". DD methods proceed in a similar way. First of all, the domain Ω (or preferably the problem) is split into subdomains (or subproblems), a problem is solved on each subdomain in parallel and then the partial solutions are glued together to get the global solution. Originally DD algorithms were devised to use existing software for solving problems where separation of variables can be used and also for computers with small memories, as the subproblems can fit on these machines. Today, memories are much larger and cheaper and these methods are used to introduce parallelism in very strongly coupled problems. In the next Section, we begin our study of some examples of such algorithms.

3 The Schwarz method

For simplicity, we consider the domain Ω split into two overlapping subdomains Ω_1 and Ω_2. Let Γ_i, $i = 1, 2$, be the part of the boundary of Ω_i enclosed in Ω (see Figure 2).

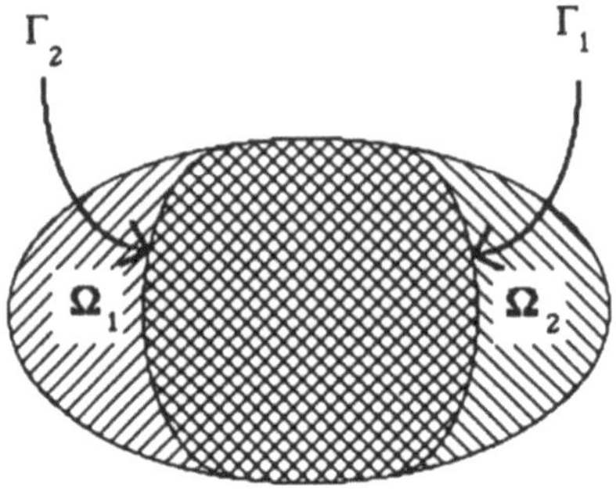

Figure 2

Overlapping subdomains

Roughly speaking, the Schwarz method is the following : guess a value for the unknowns on the inner boundary Γ_1, solve the problem exactly in Ω_1, use the computed values on the inner boundary Γ_2 to solve exactly in Ω_2 and repeat the process until convergence. This simple method "almost always" converges. A general theory has been given by P.L. Lions [22],[23], [24], in terms of projections in Hilbert spaces. Of course, the rate of convergence will depend on the extent of overlapping. The larger the overlapping, the faster the convergence. This is expected since, when $\Omega_1 = \Omega_2 = \Omega$, the method converges in 1 iteration. The following exposition is taken from Lions [22]. The model problem we are solving is

$$-\Delta u = f \text{ in } \Omega, \quad u|_{\partial\Omega} = 0$$

The Schwarz algorithm can be formulated as

$$-\Delta u^{2k} = f \text{ in } \Omega_1, \quad u^{2k}|_{\Gamma_1} = u^{2k-1}|_{\Gamma_1}$$

$$-\Delta u^{2k+1} = f \text{ in } \Omega_2, \quad u^{2k+1}|_{\Gamma_2} = u^{2k}|_{\Gamma_2}$$

Let us define a bilinear form a as

$$a(u,v) = \int_\Omega \nabla u \cdot \nabla v \, dx$$

The model problem can be written as

$$a(u,v) = (f,v), \quad \forall v \in H_0^1(\Omega).$$

Let $V_1 = H_0^1(\Omega_1)$ and $V_2 = H_0^1(\Omega_2)$ and the projectors defined by

$$a(P_i v, w) = a(v,w), \ \forall w \in V_i$$

The functions defined only on subdomains are extended by 0 to $H_0^1(\Omega)$. Then we have

$$a(u^{2k} - u, v_1) = 0, \ \forall v_1 \in V_1, \ u^{2k} - u^{2k-1} \in V_1$$

$$a(u^{2k+1} - u, v_2) = 0, \ \forall v_2 \in V_2, \ u^{2k+1} - u^{2k} \in V_2$$

It is easy to see that

$$u - u^{2k} = (I - P_1)(u - u^{2k-1})$$
$$u - u^{2k+1} = (I - P_2)(u - u^{2k})$$

Therefore,

$$u - u^{2k+1} = (I - P_2)(I - P_1)(u - u^{2k-1})$$

Therefore, Schwarz can be viewed as an iterative method for solving

$$-(I - P_2)(I - P_1)u = (P_1 + P_2 - P_1 P_2)u = g$$

The mathematical formulation of the problem for studying convergence is

$$v^0 \in V, \quad v^{2k} = (I - P_1)v^{2k-1}, \quad v^{2k+1} = (I - P_2)v^{2k}$$

The following result shows that the method converges.

Theorem (P.L. Lions)

If $V = \overline{V_1 + V_2}$ then $v^k \longrightarrow 0$

If $V = V_1 + V_2$ then

$$\|(I - P_2)(I - P_1)\| \leq c, \quad c < 1$$

Schwarz method can also be viewed as applying the block Gauss–Seidel algorithm to a larger enhanced problem, [34]. Note that there is no parallelism in the previous algorithm as we must solve for the subdomains sequentially. This can be fixed in the following way. First, the Schwarz method can be generalized to k subdomains with $k > 2$, the subproblems being solved in sequential order.

Unfortunately, the rate of convergence is slower for a large number of subdomains and small overlaps. This method is still not parallel, but the same devices as for the SSOR preconditioner can be used, namely a red–black ordering of the subdomains. For a domain divided into stripes, every other stripe is black, and red stripes alternate with black stripes. This means that a subdomain of one color is only related to subdomains of the other color, so we can compute at once, in parallel, all red subdomains say, and then all black subdomains.

Now the method is parallel, the degree of parallelism being half the number of subdomains. The drawback is, of course, that for a large number of processors the method is slowly converging but a nice feature is that it is very easy to implement.

Another way to parallelize is to use the Additive Schwarz method (Dryja–Widlund [10]).

Recall that "multiplicative" Schwarz is solving

$$(P_1 + P_2 - P_1 P_2)u = g$$

The non parallel part comes from the term $P_1 P_2$. The idea is to find g' such that

$$(P_1 + P_2)u = g'$$

But

$$a(P_i u, w) = a(u, w) = (f, w), \ \forall w \in V_i$$

Therefore the right hand sides can be constructed by solving problems for all V_is and adding the results. Moreover as $P_1 + P_2$ is symmetric and positive definite, we can use the conjugate gradient method to accelerate the convergence, then the Schwarz method can be considered as a preconditioner.

4 The Bramble–Pasciak–Schatz method

In this section, we consider a preconditioner that has been proposed by Bramble, Pasciak and Schatz [3]for a finite element problem in two dimensions. This is only one of a series of algorithms that have been published during the last years and have had an important influence on the development of DD methods. The method relies on deriving another simpler finite element approximation of the problem.

Suppose, we would like to solve

$$a(u, v) = (f, v)$$

where a is a coercive bilinear form like the one of the previous Section. The aim of the algorithm is to construct another bilinear form $b(u, v)$ such that

$$\lambda_0 b(v, v) \leq a(v, v) \leq \lambda_1 b(v, v), \ \forall v \in V$$

and to use b as a preconditioner. To do so, Ω is divided into non overlapping subdomains Ω_k as in Figure 3. Therefore $a(u, v)$ can be written as

$$a(u, v) = \sum_{i,j=1}^{N} \int_{\Omega} a_{ij} \frac{\partial u}{\partial x_i} \frac{\partial v}{\partial x_j} \, dx$$

Another intermediate form is introduced to allow for average coefficients,

$$\tilde{a}(u,v) = \sum_k \sum_{i,j} \int_{\Omega_k} a_{ij}^k \frac{\partial u}{\partial x_i} \frac{\partial v}{\partial x_j} \, dx = \sum_k \tilde{a}_k(u,v)$$

The method then separates interior, edges and vertices unknowns in the following way :

$$u = u_P + u_H$$

where u_P is in $\sum_{\oplus} V^0(\Omega_k)$ where functions in $V^0(\Omega_k)$ have homogeneous Dirichlet boundary conditions and

$$u_P = 0 \text{ on } \Gamma_{lm}.$$

u_P is defined by

$$\tilde{a}_k(u_P, \phi) = \tilde{a}_k(u, \phi), \ \forall \phi \in V^0(\Omega_k)$$

and u_H by

$$\tilde{a}_k(u_H, \phi) = 0 \ \forall \phi \in V^0(\Omega_k)$$

The method goes one step further and decomposes u_H as

$$u_H = u_E + u_V, \quad u_E, u_V \in V(\Omega_k),$$

where u_E stands for edge unknowns, u_V for vertices unknowns, $u_V(v_j) = u(v_j)$ and $u_V|_{\Gamma_{ij}}$ is linear, $u_E(v_j) = 0$.

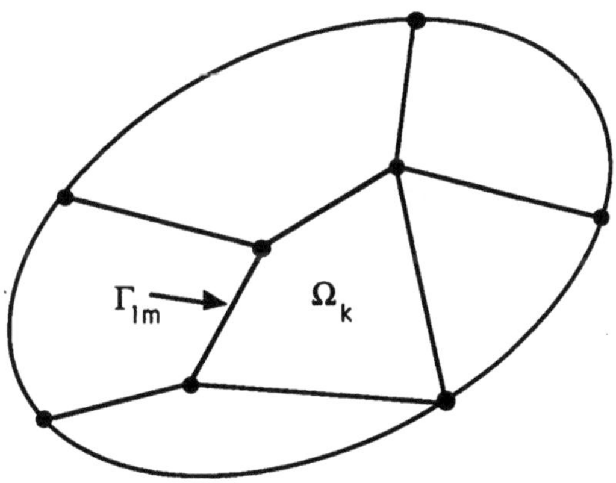

Figure 3

The coarse mesh

An operator on the edges $l_0 : V^0(\Gamma_{ij}) \rightarrow V^0(\Gamma_{ij})$ is defined by

$$\int_{\Gamma_{ij}} c^{-1} l_0(w)\phi = \int_{\Gamma_{ij}} c w'\phi', \; \forall \phi \in V^0(\Gamma_{ij})$$

where c is piecewise constant. In more trivial words, this defines something which behaves like the one dimensional Laplace operator.

Now, everything is set to define b as

$$\begin{aligned} b(w,\phi) =& \tilde{a}(u_P, \phi_P) \\ &+ \sum_{\Gamma_{ij}} \int_{\Gamma_{ij}} \alpha_{ij} a^{-1} l_0^{1/2}(u_E)\phi_E \\ &+ \sum_{\Gamma_{ij}} (u_V(v_i) - u_V(v_j))(\phi_V(v_i) - \phi_V(v_j)) \end{aligned}$$

However this is not completely obvious, this preconditioner amounts to perform the following steps :

solve Dirichlet problems in parallel $\longrightarrow u_P$

solve 1D edge equations in parallel $\longrightarrow u_E$

solve a coarse mesh system on vertices $\longrightarrow u_V$

From u_E and u_V we know the boundary values of u_H, therefore the last step is

solve Dirichlet problems in parallel $\longrightarrow u_H$

The final solution of the preconditioner is $u_P + u_H$. In their paper, Bramble, Pasciak and Schatz proved that under suitable hypothesis, the following is true for the preconditioned system in two dimensions

$$\kappa \leq C\left(1 + \ln^2\left(\frac{H}{h}\right)\right)$$

where H is the coarse mesh size. Therefore, there is only a slight h dependency in the condition number and this gives a good CG rate of convergence. In subsequent papers, some other methods were derived including one for two dimensional problems that gives a condition number bounded by a constant. It can be seen also that these methods have a good degree of parallelism, the only "sequential" part being the solve for the vertices unknowns on the coarse mesh.

Recently, preconditioners that seem as efficient and conceptually simpler, the so-called multilevel methods were developed by Bramble, Pasciak and Xu [4].

5 Algebraic Schur complement preconditioners

In this Section, we consider algebraic methods that rely only on the structure of the matrix without exploiting too much knowledge about the continuous problem. They are also methods without overlapping of the subdomains, therefore the domain Ω is divided into two different sets, the interior of the subdomains and the interfaces. The first step is to number first the interior unknowns followed by the interface unknowns.

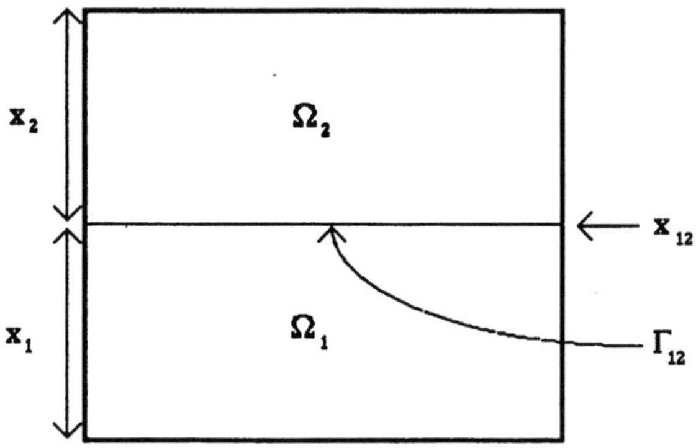

Figure 4

Partitioning with two non overlapping subdomains

Consider, for instance, the 2 subdomains case, then the permuted system is

$$\begin{pmatrix} B_1 & 0 & C \\ 0 & B_2 & E \\ C^T & E^T & B_{1,2} \end{pmatrix} \begin{pmatrix} x_1 \\ x_2 \\ x_{1,2} \end{pmatrix} = \begin{pmatrix} b_1 \\ b_2 \\ b_{1,2} \end{pmatrix}.$$

where x_1 and x_2 are the subdomains unknowns and $x_{1,2}$ is related to the interface. B_1 (resp. B_2) represents the coupling between the unknowns in Ω_1 (resp. Ω_2).

Then the next step is to eliminate x_1 and x_2 to get an equation on the interface

$$\begin{aligned} Sx_{12} &= (B_{12} - C^T B_1^{-1} C - E^T B_2^{-1} E)\, x_{12} \\ &= b_{12} - C^T B_1^{-1} b_1 - E^T B_2^{-1} b_2 \end{aligned},$$

S is the so-called Schur complement.

Now, there are many ways for solving the problem. We have to take decisions concerning solving for the subdomains and for the interfaces. For example, we can

1) use a direct solver for subdomains

2) use an approximate solver for subdomains.

In the first case, it can be better to iterate (with CG) on the interface unknowns. Each iteration needs one solve on each subdomain to form the matrix vector product

$$Sx = (B_{12} - C^T B_1^{-1} C - E^T B_2^{-1} E)\, x.$$

The main problem to speed up the iterative method is to find a good preconditioner for S.

In the second case, it seems better to iterate on all the domain forming a global preconditioner. This can be done taking the preconditioner to be

$$M = L \begin{pmatrix} M_1^{-1} & 0 & 0 \\ 0 & M_2^{-1} & 0 \\ 0 & 0 & M_{1,2}^{-1} \end{pmatrix} L^T,$$

with

$$L = \begin{pmatrix} M_1 & 0 & 0 \\ 0 & M_2 & 0 \\ C^T & E^T & M_{1,2} \end{pmatrix}.$$

Then, by multiplication, one gets

$$M = \begin{pmatrix} M_1 & 0 & C \\ 0 & M_2 & E \\ C^T & E^T & M_{1,2}^* \end{pmatrix}$$

where,

$$M_{1,2}^* = M_{1,2} + C^T M_1^{-1} C + E^T M_2^{-1} E.$$

Therefore it is natural to take M_i as an approximation to B_i and M_{12} as an approximation to

$$B_{12} - C^T M_1^{-1} C - E^T M_2^{-1} E.$$

This approximates the Schur complement

$$S = B_{12} - C^T B_1^{-1} C - E^T B_2^{-1} E.$$

Hence, once again, we are faced with the problem of constructing an approximation to the Schur complement (or to an approximate Schur complement). There are many ways to do this. Let us consider two algebraic approximations that rely on the specificities of the problem to be solved.

The first one is the Probing method (Chan [5]). It has been observed experimentally that for problems arising from elliptic partial differential equation, the elements of S decay rapidly away from the main diagonal (property of the inverses of subdomain matrices).

Therefore, the method looks for a banded approximation to S. More precisely, the problem is to find a banded matrix C with bandwith $2k+1$ such that

$$Cv = (B_{12} - C^T M_1^{-1} C - E^T M_2^{-1} E)v$$

for some chosen vectors v

The choice for v depends on S being symmetric or not. If S is non symmetric, Chan suggested to choose the $2k+1$ vectors as

$$\begin{aligned} k = 0 : v_1 &= [1,1,1,1,1,1,\ldots]^T \\ k = 1 : v_1 &= [1,0,0,1,0,0,\ldots]^T \\ v_2 &= [0,1,0,0,1,0,\ldots]^T \\ v_3 &= [0,0,1,0,0,1,\ldots]^T \\ k = 2 : v_1 &= [1,0,0,1,0,0,\ldots]^T \\ v_2 &= [0,1,0,0,1,0,\ldots]^T \\ v_3 &= [0,0,1,0,0,1,\ldots]^T \\ v_4 &= [0,0,0,1,0,0,\ldots]^T \\ &\vdots \end{aligned}$$

If S is symmetric then $k+1$ vectors are chosen as

$$\begin{aligned}
k=0 : v_1 &= [1,1,1,1,1,1,\dots]^T \\
k=1 : v_1 &= [1,0,0,1,0,0,\dots]^T \\
v_2 &= [0,1,0,0,1,0,\dots]^T \\
k=2 : v_1 &= [1,0,0,1,0,0,\dots]^T \\
v_2 &= [0,1,0,0,1,0,\dots]^T \\
v_3 &= [0,0,1,0,0,1,\dots]^T \\
v_4 &= [0,0,0,1,0,0,\dots]^T \\
&\vdots
\end{aligned}$$

Notice that computation of C implies (parallel) solves on each domain and that vectors v are multiple eigenvectors of $C^{-1}S$ with eigenvalue 1. Other choices for the probing vectors exist (Axelsson [2]).

The second method considers block preconditioners (Meurant [28]). This method is designed for finite difference schemes where, with a natural ordering, the matrices for each (stripe) subdomain can be written as

$$B_i = \begin{pmatrix} D_i^1 & (A_i^2)^T & & & \\ A_i^2 & D_i^2 & (A_i^3)^T & & \\ & \ddots & \ddots & \ddots & \\ & & A_i^{m_i-1} & D_i^{m_i-1} & (A_i^{m_i})^T \\ & & & A_i^{m_i} & D_i^{m_i} \end{pmatrix}$$

where D_i^j and $B_{i,j}$ are point tridiagonal matrices, m_i is the number of mesh lines in Ω_i. The matrices C and E have a very special structure, of which we will take advantage of in our methods, namely

$$C = \begin{pmatrix} 0 \\ \vdots \\ 0 \\ C^{m_1} \end{pmatrix}, \qquad E = \begin{pmatrix} E^1 \\ 0 \\ \vdots \\ 0 \end{pmatrix}.$$

Let $trid(B)$ be a tridiagonal matrix whose non zero elements are the same as the corresponding ones from the three main diagonals of B.

In our particular case with two subdomains, we can take for M_1 a block LU INV approximation :

$$M_1 = (\Delta_1 + L_1)\, \Delta_1^{-1}\, (\Delta_1 + L_1^T),$$

where L_1 is the block lower triangular part of B_1, Δ_1 is a block diagonal matrix whose blocks are given by,

$$\begin{cases} \Delta_1^1 = D_1^1, \\ \Delta_1^j = D_1^j - A_1^j\; trid((\Delta_1^{j-1})^{-1})\; (A_1^j)^T \end{cases}$$

The same can be done for M_2 but, because we need to compute $M_2^{-1}E$ and E has a special sparse block structure, it is more interesting to take for M_2 a block UL INV approximation :

$$M_2 = (\Sigma_2 + L_2^T)\, \Sigma_2^{-1}\, (\Sigma_2 + L_2),$$

L_2 is the block lower triangular part of B_2, Σ_2 is a block diagonal matrix,

$$\begin{cases} \Sigma_2^{m_2} = D_2^{m_2}, \\ \Sigma_2^j = D_2^j - (A_2^{j+1})^T \; trid((\Sigma_2^{j+1})^{-1}) \; A_2^{j+1}, \quad j = m_2 - 1, \ldots, 1. \end{cases}$$

Then, we define M_{12} as an approximation to the Schur complement,

$$M_{12} = B_{1,2} - (C^{m_1})^T \; trid((\Delta_1^{m_1})^{-1}) \; C^{m_1} - (E^1)^T \; trid((\Sigma_2^1)^{-1}) \; E^1.$$

With suitable hypothesis and the same techniques as in Concus, Golub & Meurant [8] , it is possible to show that the preconditioner M just constructed is positive definite.

It is fairly easy to generalize the method to more than 2 subdomains, [28]. At each iteration of CG, a linear system with matrix M has to be solved. It is obvious that the subdomains can be handled in parallel. There is a potential problem for the interface solve when there is more than two subdomains, although in some cases a completely parallel preconditioner can be constructed, see Meurant [28] for details. However, the number of interfaces is the number of subdomains minus one, so there will be a slight unbalance in the computation, one processor being idle during the interface solves. This not much of a problem as very large computer speeds and speed ups can be achieved with these methods on parallel vector computers like the Cray Y-MP, see Meurant [29].

6 A domain decomposition method for parabolic problems

In this section, we consider a method specially designed for parabolic partial differential equations like

$$\frac{\partial u}{\partial t} - \frac{\partial}{\partial x}(a(x,y)\frac{\partial u}{\partial x}) - \frac{\partial}{\partial y}(b(x,y)\frac{\partial u}{\partial y}) = f$$
$$\text{in } \Omega \subset R^2,$$

$$u \mid_{\partial\Omega} = 0, \quad u(x,0) = u_0(x).$$

The model problem that is considered is the heat equation :

$$\frac{\partial u}{\partial t} - \Delta u = f,$$

with Dirichlet boundary conditions, Ω being the unit square.

For stability and efficiency an implicit Crank–Nicolson scheme is used that can be described in the following framework :

consider the general parabolic partial differential equation,

$$\frac{\partial u}{\partial t} + Lu = f,$$

with L being a second order self–adjoint linear elliptic operator. The (space) operator L is discretized with a standard finite difference scheme with $n + 1$ points on each side of the square domain Ω. The mesh size h is given by

$$h = \frac{1}{n+1}.$$

Then we obtain a sparse matrix A (of order $n^2 \times n^2$) from L,

$$L \Longrightarrow \frac{1}{h^2}A$$

where A is a block tridiagonal matrix (corresponding to the steady state problem $Lu = f$). With the standard five point schemes, A is a diagonally dominant M–matrix.

Time is discretized with the usual centered scheme : with $t \in [0, T]$, $k = \Delta t$ being the time step, m refering to the values of unknowns at time mk, we have

$$\frac{u^{m+1} - u^m}{k} + \frac{1}{2h^2}(Au^{m+1} + Au^m) = \frac{1}{2}(f^{m+1} + f^m),$$

or

$$2\frac{h^2}{k}(u^{m+1} - u^m) + Au^{m+1} + Au^m = h^2(f^{m+1} + f^m).$$

At every time step this gives a linear system to solve

$$(2\frac{h^2}{k}I + A)u^{m+1} = 2\frac{h^2}{k}u^m - Au^m + h^2(f^{m+1} + f^m).$$

The advantages of the implicit scheme are its second order accuracy (which means that the truncation error is $O(k^2)+O(h^2)$) and unconditional stability that allows larger time steps than more straightforward explicit schemes. In practice, one chooses to solve the problem in the form :

let $\theta = 2\frac{h^2}{k}$, and $A_t = \theta I + A$,

$$A_t x = h^2(f^{m+1} + f^m) - 2Au^m,$$

then,

$$u^{m+1} = x + u^m.$$

Remark that for our problem, A_t is a symmetric strictly diagonally dominant M–matrix.

Although domain decomposition preconditioners for elliptic problems can also be used in our case, we will use the specificities of the time dependent problems to derive a preconditioner that is both efficient and very easy to implement. This method was inspired by an algorithm of Y. Kuznetsov [21]. However, contrary to Kuznetsov, we use CG to solve the linear system at each time step and domain decomposition, only to provide a preconditioner.

The domain Ω is divided into non–overlapping subdomains Ω_i, $i = 1, \ldots, l$. For the sake of the numerical experiments on the unit square, the Ω_is will be either stripes or squares (also called boxes), and it will be supposed that their boundaries are parallel to the rows and columns in the mesh but not mesh lines. For more general problems, the subdomains can have any shape.

Each Ω_i is extended to a domain $\widehat{\Omega}_i$ that overlaps the neighbours of Ω_i (restricted of course to Ω), see Figure 5, where Ω_i corresponds to the gray area and $\widehat{\Omega}_i$ to the same plus the stripped area.

To solve $Mz = r$, the following steps are performed : for each subdomain $\widehat{\Omega}_i$, $i = 1, \ldots, l$, let $\widehat{A_i}$ be the $n_i \times n_i$ matrix arising from the discretization of the problem on $\widehat{\Omega}_i$ with homogeneous Dirichlet boundary conditions. Let $\widehat{r_i}$ be the vector of length n_i that will be the right hand side on $\widehat{\Omega}_i$, whose entries are equal to those of r for components corresponding to mesh points in Ω_i and 0 elsewhere. Then let $\widehat{z}_i$ be defined by solving a problem on $\widehat{\Omega}_i$:

$$\widehat{A_i}\widehat{z_i} = \widehat{r_i},$$

then, we extend $\widehat{z_i}$ to a vector z_i on the whole Ω with 0 components corresponding to mesh points outside $\widehat{\Omega}_i$. The solution z is simply defined as

$$z = \sum_{i=1}^{l} \widehat{z}_i.$$

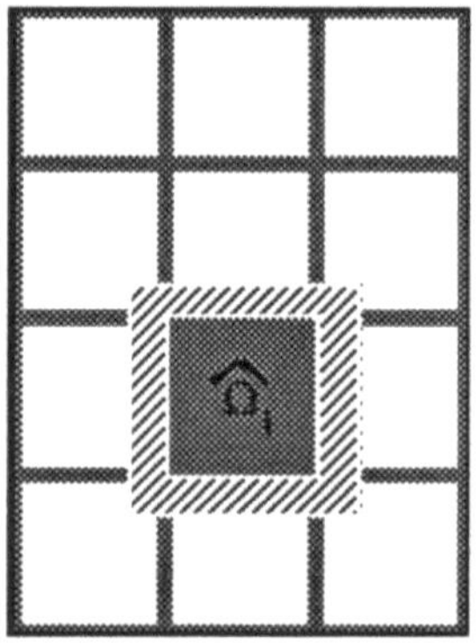

Figure 5
Subdomain $\widehat{\Omega}_i$

The main problem is to know where to put the artificial boundaries, that is the definition of $\widehat{\Omega}_i$. Some hints and some bounds are given by the study of the decay of elements of the inverse A_i^{-1}, Meurant [30]. However, usually a moderate amount of overlapping gives good results, see [31], [32]. This example shows that to develop efficient methods, one has to take into account the specific nature of the problem to be solved, rather than using more general methods.

7 Conclusions

In the previous sections, we described some domain decomposition methods for different kinds of applications. All of them introduce some high level parallelism and therefore can be efficiently used on parallel vector computers with a small (up to the order of 64) number of powerful processors. There are much more methods that have appeared in the literature. For instance, a lot of work was done on applying domain decomposition to Fluid Mechanics, see Glowinski, Périaux [17] using different physical models in subdomains (let say Potential flow and full Navier–Stokes equations). For a while, studies were restricted to model problems that are easier to analyze but, more and more researchers are applying domain decomposition techniques to real world problems, see Keyes [20].

It remains to be seen what can be done for the emerging class of massively parallel computers that offer very high potential performances but for which all the techniques we have developed in this review do not really apply.

References

[1] O. AXELSSON, *A class of iterative methods for finite element equations.* Comp. Meth. Appl. Mech. Eng. v 9, (1976) pp 123–137.

[2] O. AXELSSON, *On inverse free factorization methods suitable for vector and parallel processors.* Report 8431, University of Nijmegen, (1984).

[3] J.H. BRAMBLE, J.E. PASCIAK & A.H. SCHATZ, *The construction of preconditioners for elliptic problems by substructuring I.* Math. Comp. v 47, n 175, (1986) pp 103–104.

[4] J.H. BRAMBLE, J.E. PASCIAK & J. XU, *Parallel multilevel preconditioners* Math. Comp. v 55, n 191, (1990) pp 1–22.

[5] T. CHAN, *Boundary probe preconditioners for fourth order elliptic problems.* In [6] pp 160–167.

[6] T. CHAN, R. GLOWINSKI, J. PERIAUX & O. WIDLUND Eds., *Proceedings of the second international symposium on domain decomposition methods for partial differential equations.* SIAM (1989).

[7] T. CHAN, R. GLOWINSKI, J. PERIAUX & O. WIDLUND Eds., *Proceedings of the third international symposium on domain decomposition methods for partial differential equations.* SIAM (1990).

[8] P. CONCUS, G.H. GOLUB & G. MEURANT, *Block preconditioning for the conjugate gradient method.* SIAM J. Sci. Stat. Comp., v 6, (1985) pp 220–252.

[9] S. DOI, *Ordering effects for convergence of some preconditioned iterative methods* submitted to SIAM J. of Numerical Analysis (1990).

[10] M. DRYJA & O. WIDLUND, *Towards a unified theory of domain decomposition algorithms for elliptic problems.* In [7] pp 3- 21.

[11] I.S. DUFF & G. MEURANT, *The effect of ordering on preconditioned conjugate gradients.* BIT v 29, (1989) pp 635–657.

[12] I.S. DUFF, *Direct solvers.* These proceedings.

[13] V. EIJKHOUT, *Analysis of parallel incomplete point factorizations.* Ph.D. Thesis University of Nijmegen (Netherlands) 1990.

[14] S. FILIPPONE & G. RADICATI, *Vectorized ILU preconditioners for general sparsity patterns.* Proceedings of the International Meeting on Parallel Computing, Verona, September 28–30 1988

[15] R. GLOWINSKI, G.H. GOLUB, G. MEURANT & J. PERIAUX Eds., *Proceedings of the first international symposium on domain decomposition methods for partial differential equations.* SIAM (1988).

[16] R. GLOWINSKI, Y. KUZNETSOV, G. MEURANT & J. PERIAUX & O. WIDLUND Eds., *Proceedings of the fourth international symposium on domain decomposition methods for partial differential equations.* SIAM (1991).

[17] R. GLOWINSKI, J. PERIAUX & G. TERRASSON, *On the coupling of viscous and inviscid models for compressible fluid flows via domain decomposition.* In [7] pp 64–97.

[18] G.H. GOLUB & G. MEURANT, *Résolution numérique des grands systèmes linéaires.* Eyrolles, Paris (1983).

[19] R. HOCKNEY & C.JESSHOPE, *Parallel Computers 2.* Adam Hilger (1988).

[20] D. KEYES, *Domain decomposition methods for the parallel computation of reacting flows.* Comput. Phys. Comm. v 53 (1989) pp 181–200.

[21] Y. KUZNETSOV, *New algorithms for approximate realization of implicit difference schemes.* Sov. J. Numer. Anal. Math. Modelling v 3 n^0 2 (1988).

[22] P.L. LIONS, *On the Schwarz alternating method I.* in [15] (1988) pp 1–42.

[23] P.L. LIONS, *On the Schwarz alternating method II : stochastic interpretation and order properties.* in [6] pp 47–70.

[24] P.L. LIONS, *On the Schwarz alternating method III : a variant for non overlapping subdomains.* in [7] pp 202–223.

[25] J.A. MEIJERINK & H.A. VAN DER VORST, *An iterative solution method for linear systems of which the coefficient matrix is a symmetric M-matrix.* Math. Comp. 31 (1977) pp 148–162.

[26] G. MEURANT, *The block preconditioned conjugate gradient method on vector computers.* BIT v 24 (1984) pp 623–633.

[27] G. MEURANT, *Multitasking the conjugate gradient method on the CRAY X-MP/48.* Parallel Computing 5 (1987) pp 267–280.

[28] G. MEURANT, *Domain Decomposition Preconditioners for the Conjugate Gradient Method.* Calcolo v 25 n 1–2 (1988) pp 103–119.

[29] G. MEURANT, *Practical use of the conjugate gradient method on parallel supercomputers.* Comput. Phys. Comm. v 53 (1989) pp 467–477.

[30] G. MEURANT, *A review on the inverse of symmetric tridiagonal and block tridiagonal matrices.* accepted for publication in SIAM J. Matrix Anal. Appl. (1990).

[31] G. MEURANT, *Numerical experiments with a domain decomposition method for parabolic problems on parallel computers.* to appear in [16], Moscow, May 1990.

[32] G. MEURANT, *A domain decomposition method for parabolic equations.* to appear in Applied Numerical Mathematics (1990).

[33] J.K. REID, *On the method of conjugate gradients for the solution of large sparse systems of linear equations.* In "Large sparse sets of linear equations", J.K. Reid ed., Academic Press (1971).

[34] G. RODRIGUE & J. SIMON *A generalization of the numerical Schwarz algorithm.* In "Computing methods in applied sciences and engineering VI", R. Glowinski & J.L. Lions Eds, North-Holland (1984), pp 273–283.

[35] Y. SAAD, *Practical use of polynomial preconditioning for the conjugate gradient method.* SIAM J. Sci. Stat. Comput. v 6 n 4 (1985) pp 865–881.

[36] Y. SAAD & M.H. SCHULTZ, *GMRES : a generalized residual algorithm for solving nonsymmetric linear systems.* SIAM J. Sci. Stat. Comput. v 7 (1986) pp 856–869.

[37] H.A. VAN DER VORST, *A vectorizable variant of some ICCG methods.* SIAM J. Sci. Stat. Comput. v 3 (1982) pp 86–92.

A Splitting Method for Large Sparse Linear Systems on a Multivector Computer

Ilio Galligani
Department of Mathematics
University of Bologna
Italy

Abstract: This paper is concerned with the analysis and implementation on a computer consisting of two vector processors of the Preconditioned Simultaneous Displacements method with an additive preconditioner for solving large sparse systems of linear equations. This method converges for systems with coefficient matrices that are symmetric positive definite or irreducible L-matrices with a strong diagonal dominance.

In the case in which the coefficient matrix is symmetric positive definite, «good» estimates for the iterative parameters have been obtained. In this case, also, a polynomial acceleration of the method has been considered. The results of some computational experiments on CRAY Y-MP have been reported.

1. The basic iterative method

We shall consider the numerical solution of very large but sparse systems of linear algebraic equations

$$(1) \qquad Ax = b$$

where A is a given (n x n) real sparse matrix, **b** is the (n x 1) right hand side vector and **x** is the unknown vector to be determined.
We assume throughout that the matrix A is non-singular and moreover that its diagonal entries a_{ii} are all non zero numbers: A has no specific structure.

To approximately solve (1), we consider the iterative method of first degree

$$(2) \qquad B\,\frac{u_{k+1} - u_k}{\tau} + Au_k = b \qquad k = 0,1,\ldots$$

where $\mathbf{u}_0$ is an arbitrary initial vector approximation to the exact solution $\mathbf{x}^*$ of (1), B is a non-singular matrix and τ is an iterative parameter ($\tau > 0$). The matrix B is taken to be an easily invertible matrix, i.e., for any vector **r** it is relatively easy to solve the system B**w** = **r**.

NATO ASI Series, Vol. F 77
Computer Algorithms for Solving Linear Algebraic Equations
Edited by E. Spedicato

An alternative form of (2) is given by

(3) $$u_{k+1} = u_k + \tau B^{-1} (b - A u_k)$$

or

(4) $$u_{k+1} = T u_k + c$$

where $T = I - \tau G$, $G = B^{-1} A$ and $\mathbf{c} = \tau B^{-1} \mathbf{b}$.

If A has the splitting

(5) $$A = L + D + U$$

where $D = \text{diag}\{a_{11}, a_{22}, \ldots, a_{nn}\}$ is the diagonal matrix with the same diagonal elements of A and L and U are strictly lower and upper triangular matrices, respectively, then a suitable form of B is

(6) $$B^{-1} = \frac{1}{2}\left(\left(\frac{1}{\omega}D + L\right)^{-1} + \left(\frac{1}{\omega}D + U\right)^{-1}\right) \qquad 0<\omega<2$$

It is interesting to note that the iterative method (2) with the choice (6) of B^{-1} is characterized by having within its overall mathematical structure certain well-defined substructures that can be executed simulteneously during each iteration *k*. This feature makes the method (3)-(6) ideally suited for implementation on a multiprocessor system with two or more vector processors. The lower triangular system $(\frac{1}{\omega} D+L)\, \mathbf{w}_k^{(1)} = \mathbf{r}_k$ and the upper triangular system $(\frac{1}{\omega} D+U)\, \mathbf{w}_k^{(2)} = \mathbf{r}_k$, where $\mathbf{r}_k = \mathbf{b} - A\mathbf{u}_k$, can be solved *simultaneously* on two different vector processors. Hence, the effectiveness of the method (3)-(6) depends on the availability of an efficient algorithm for solving these triangular systems.

In literature (see, e.g. [5]) a widely used form of B in (2) is

(7) $$B^{-1} = \omega(2-\omega)\, (D+\omega U)^{-1}\, D\, (D+\omega L)^{-1} \qquad 0<\omega<2$$

In this case, during each iteration *k*, the two triangular systems $(D + \omega L)\, \mathbf{w}_k^{(1)} = \mathbf{r}_k$ and $(D + \omega U)\, \mathbf{w}_k^{(2)} = D\, \mathbf{w}_k^{(1)}$ are solved *sequentially*.

In (6) B^{-1} is formed by considering the two splittings of A: $A = H_1 - K_1$ and $A = H_2 - K_2$ where $H_1 = \frac{1}{\omega}D + L$, $K_1 = (\frac{1}{\omega} - 1)D - U$, $H_2 = \frac{1}{\omega}D + U$ and $K_2 = (\frac{1}{\omega} - 1)D - L$.
In the paper [11] B^{-1} is associated with a multisplitting of A. Multisplittings are most prevalent when A has a specific structure (for example, A is a block-biangular matrix).

If we set $R = \frac{1}{2}(L - U)$, the matrix B of formula (6) has the expression

(8) $$B = \frac{1}{2}\left(A + \frac{2-\omega}{\omega}D\right) - R\left(\frac{1}{2}\left(A + \frac{2-\omega}{\omega}D\right)\right)^{-1} R$$

We assume that $A + \frac{2-\omega}{\omega} D = \frac{2}{\omega} D + L + U$ is non-singular. Indeed, we have

$$2B^{-1} = \left(\frac{1}{\omega}D+L\right)^{-1} + \left(\frac{1}{\omega}D-U\right)^{-1} =$$

$$= \left(\left(\frac{1}{\omega}D+U\right)\left(\left(\frac{1}{\omega}D+L\right)+\left(\frac{1}{\omega}D+U\right)\right)^{-1}\left(\frac{1}{\omega}D+L\right)\right)^{-1} =$$

$$= \left(\left(\frac{1}{\omega}D+U\right)\left(\frac{2}{\omega}D+L+U\right)^{-1}\left(\frac{1}{\omega}D+L\right)\right)^{-1} =$$

$$= \left(\frac{1}{2}\left(\frac{2}{\omega}D+L+U-(L-U)\right)\left(\frac{2}{\omega}D+L+U\right)^{-1}\frac{1}{2}\left(\frac{2}{\omega}D+L+U-(U-L)\right)\right)^{-1} =$$

$$= \left(\frac{1}{4}\left(\frac{2}{\omega}D+L+U\right)-R\left(\frac{2}{\omega}D+L+U\right)^{-1}R\right)^{-1}$$

Hence

$$B^{-1} = 2\left(\left(\frac{2-\omega}{\omega}D+A\right) - 4R\left(\frac{2-\omega}{\omega}D+A\right)^{-1}R\right)^{-1} =$$

$$= \left(\frac{1}{2}\left(\frac{2-\omega}{\omega}D+A\right)-R\left(\frac{1}{2}\left(\frac{2-\omega}{\omega}D+A\right)\right)^{-1}R\right)^{-1}$$

so that, clearly, (8) holds.

If A is symmetric and positive definite, the matrix B of (6), with $0 < \omega < 2$, is symmetric and positive definite.

This result follows from the following lemma [9].

Lemma Let A, H_1, K_1, H_2, K_2 be (n x n) matrices and assume that H_1 and H_2 are non singular, $A = H_1 - K_1 = H_2 - K_2$ and A is symmetric and positive definite. If the matrices H_1+K_1 and H_2+K_2 are positive real, then the matrix $\rho_1 H_1^{-1} K_1 + \rho_2 H_2^{-1} K_2$, with $\rho_1 > 0$, $\rho_2 > 0$, $\rho_1 + \rho_2 = 1$, is convergent.

(A matrix S is positive real if the symmetric matrix $\frac{1}{2}(S+S^T)$ is positive definite).

2. *Convergence analysis*

A study of the convergence properties of the method (2)-(6) is given under the assumption that A is symmetric and positive definite or is an irreducible L-matrix with strong diagonal dominance [12, pg 42].

We consider the iteration matrix Q of the arithmetic mean method [10]

$$Q = \frac{1}{2}\left(H_1^{-1}K_1 + H_2^{-1}K_2\right) =$$

(9)

$$= \frac{1}{2}\left(\left(\frac{1}{\omega}D + L\right)^{-1}\left(\left(\frac{1}{\omega} - 1\right)D - U\right) + \left(\frac{1}{\omega}D + U\right)^{-1}\left(\left(\frac{1}{\omega} - 1\right)D - L\right)\right)$$

Under the above assumptions on the matrix A, Q is a convergent matrix for any ω in well definite ranges.
Since

$$G = B^{-1}A = \frac{1}{2}\left(\left(\frac{1}{\omega}D + L\right)^{-1} + \left(\frac{1}{\omega}D + U\right)^{-1}\right)(L + D + U) =$$

$$= \frac{1}{2}\left(\left(\frac{1}{\omega}D + L\right)^{-1} + \left(\frac{1}{\omega}D + U\right)^{-1}\right)\left(\left(\frac{1}{\omega}D + L\right) + \left(\frac{1}{\omega}D + U\right) - \frac{2-\omega}{\omega}D\right) =$$

$$= I + \frac{1}{2}\left(\left(\frac{1}{\omega}D + L\right)^{-1}\left(\left(1 - \frac{1}{\omega}\right)D + U\right) + \left(\frac{1}{\omega}D + U\right)^{-1}\left(\left(1 - \frac{1}{\omega}\right)D + L\right)\right) =$$

$$= I - Q$$

we have

(10) $$T = I - \tau G = (1 - \tau)I + \tau Q$$

Thus, the method (2)-(6) is convergent if $0 < \tau \leq 1$ [12, pg 107].
Therefore, we have the following general
Theorem If the matrix A is an irreducible L-matrix with strong diagonal dominance, then the iterative method (2), where B has the form (6), is convergent for $0 < \omega < 1$ and $0 < \tau \leq 1$. If A is symmetric and positive definite, then the iterative method (2)-(6) is convergent for $0 < \omega < 2$ and $0 < \tau \leq 1$. If, in addition, A is «2-cyclic» and $\tau = 1$, it is possible to derive an *optimal* value of ω [2].

In order to evaluate the *effectiveness* of the iterative method (2)-(6) some computational experiments have been carried out on the multivector computer CRAY Y-MP, using a Fortran program which uses CMIC$ microtasking directives. In these experiments the matrix A is obtained from the finite-difference method applied to the elliptic partial differential equation

$$-\phi_{xx} - \phi_{yy} + \alpha\phi_x + \beta\phi_y + \gamma\phi = f \qquad in\ \Omega$$

with boundary conditions ϕ=g on $\partial\Omega$; here Ω = (0,1) x (0,1). This is a test-problem which allows us to study the influence of asymmetry of A on the solution behaviour of the iterative methods.

Under certain conditions on the coefficients α, ß, γ and the mesh-spacing it is possible to generate matrices A for which the hypotheses of the above convergence theorem hold.

The method (2)-(6) has been compared with the biconjugate gradient method [6] and the generalized conjugate residual (GCR) method [3]. An interesting result has been obtained [10]: the method (2)-(6) produces a significantly better convergence behaviour than the biconjugate gradient and GCR methods for strongly asymmetric matrices A. Method (2)-(6) is less efficient than the biconjugate gradient method for nearly symmetric matrices.

The problem of determining the optimal parameters ω^* and τ^* for ω and τ, respectively, such that the spectral radius $\rho(T)$ of $T = I-\tau B^{-1}A$ is minimized is of paramount importance.

«Good» estimates for ω and τ of the method (2)-(6) can be developed when A is symmetric and positive definite. In this case $U = L^T$ and $a_{ii} > 0$ (i = 1,2,...,n).

For any ω in the range $0 < \omega < 2$ the following well-known theorem gives a «good» choice of τ in order for $\rho(T)$ to attain its minimum value.

Theorem. If A and B are symmetric and positive definite, then the iterative method (3) is convergent if, and only if, $0 < \tau < 2/\lambda_{max}(G)$ and

(11) $$\tau_* = 2/\left(\lambda_{max}(G) + \lambda_{min}(G)\right)$$

is the best choice of the acceleration parameter τ in (3). Here, $\lambda_{max}(G)$ and $\lambda_{min}(G)$ are the largest and smallest eigenvalues of $G = B^{-1}A$, respectively.

With the choice (11) of τ, it is possible to compare the condition numbers of two matrices $B_1^{-1}A$ and $B_2^{-1}A$. Indeed, it is easy to prove the following

Theorem. Let A, B_1 and B_2 be symmetric and positive definite matrices satisfying the condition $\rho(I-\tau_1B_1^{-1}A) < \rho(I-\tau_2B_2^{-1}A)$, where $\tau_i=2/(\lambda_{max}(B_i^{-1}A) + \lambda_{min}(B_i^{-1}A))$, i=1,2, then $K(B_1^{-1}A) < K(B_2^{-1}A)$. Here, $K(B_i^{-1}A)$ is the condition number of the matrix $B_i^{-1}A$.

Since the iterative method (2) can be regarded as belonging to a generalized class referred to as the method of Simultaneous Displacements with some form of «preconditioning» to the original system (1), a «good» choice of ω is obtained by minimizing the condition number $K(B^{-1}A)$ of the matrix $B^{-1}A$.

The following relation, proved in [7], gives an upper bound for $K(B^{-1}A)$, when B^{-1} has the expression (6), which can be studied as a function of ω:

(12) $$K(B^{-1}A) \le \theta(\zeta)$$

where

$$\theta(\zeta) = K(D^{-1}A)\left(\frac{\zeta + \lambda_{min}(D^{-1}A)}{\zeta + \lambda_{max}(D^{-1}A)} + \frac{4\,\|RR^T\|\,\lambda_{min}(D^{-1}A)}{\lambda_{min}(A)\,(\zeta + \lambda_{max}(D^{-1}A))\,(\zeta\,\lambda_{min}(D) + \lambda_{min}(A))}\right)$$

and $\zeta = (2 - \omega)/\omega > 0$.

In figure we plot $K(B^{-1}A)$ and $\theta(\zeta)$ as a function of ω for the matrix A of order n=1024 obtained from the five-point finite-difference equations related to the Laplace equation in the unit square region $\Omega = (0,1) \times (0,1)$ with Dirichlet boundary conditions; the net superimposed upon Ω is uniform.

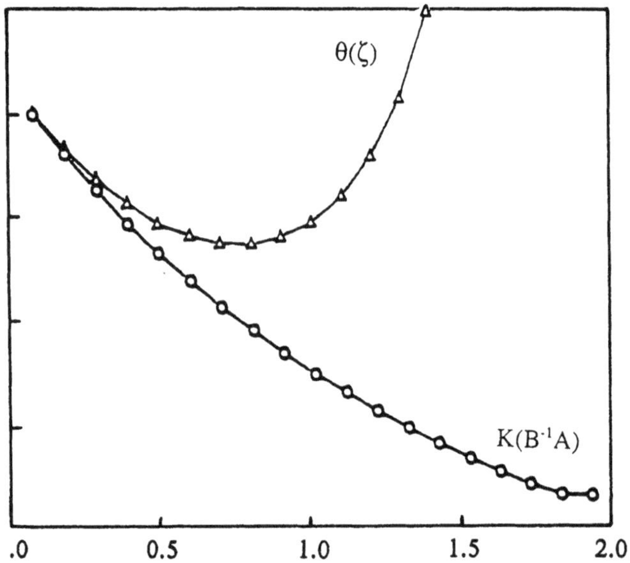

3. *Polynomial acceleration*

When A is symmetric and positive definite and B has the form (6), the basic iterative method (4) is symmetrizable. Thus it is possible to define an iterative procedure with iterates $\mathbf{v}_k$ given by ($k \geq 1$).

$$(13)\quad \begin{aligned} v_{k+1} &= \rho_{k+1}\,(\gamma_{k+1}\,(Tv_k + c) + (1 - \gamma_{k+1})\,v_k) + (1 - \rho_{k+1})\,v_{k-1} \\ v_1 &= \gamma_1\,(Tu_0 + c) + (1 - \gamma_1)\,u_0 \end{aligned}$$

which is a *polynomial acceleration method* of the method (2)-(6) [8].

If the parameters γ_{k+1} and ρ_{k+1} in (13) have the form ($k \geq 1$)

(14)
$$\gamma_{k\,1} = \frac{r_{k-1}^T \, B^{-1} \, r_{k-1}}{(B^{-1} \, r_{k-1})^T \, A \, (B^{-1} \, r_{k-1})}$$

$$\rho_{k+1} = \left(1 - \frac{\gamma_{k+1}}{\gamma_k} \, \frac{r_k^T \, B^{-1} \, r_k}{r_{k-1}^T \, B^{-1} \, r_{k-1}} \, \frac{1}{\rho_k} \right)^{-1} \qquad \rho_1 = 1$$

with $\mathbf{r}_k = \mathbf{b}\text{-}\mathbf{A}\mathbf{v}_k$, formula (13) is the three-term recursion formula of the preconditioned conjugate gradient (PCG) method applied to the system (1) with preconditioner (6) (see, e.g. [4]).

Therefore, the PCG method is one of many approaches that may be used to accelerate the convergence of the basic iterative method (2)-(6) with τ=1. Also this acceleration procedure is well suited to multivector computers.

In [7] a comparison between the additive preconditioner (6) and the SSOR preconditioner (7) for the PCG method has been carried out on different test-matrices. Also the *m-step* preconditioners generated by these two preconditioners ([2], [1]) have been compared. For the test-matrix arising from the discretization of the diffusion problem on a rectangular domain Ω:

$$-\operatorname{div}\,(p \text{ grad } \phi) + q\,\phi = f \qquad \textit{in } \Omega$$

$$\phi = g \qquad \text{on } \partial\Omega$$

where the coefficients p(x y) and q(x y) are positive piecewise continuous functions, we have the results of the following table, as ω varies. In this table, k^* is the number of iterations for an *error* $\|\mathbf{v}_{k^*} - \mathbf{x}^*\|_\infty$ and *time* indicates the computer-time, expressed in seconds, for solving the system (1) on CRAY Y-MP. For the preconditioner (7) only one vector processor is used; for the preconditioner (8) two vector processors are used. The triangular systems which appear in (6) and (7) are solved with the block forward-elimination algorithm or the block back-substitution algorithm combined with the cyclic reduction algorithm. In table, for ω=0, we have the results of the basic conjugate gradient (CG) method. K(A) is the condition number of the test-matrix A. The notation 1.6(-11), for instance, means $1.6\cdot10^{-11}$.

order of A	ω	preconditioner (6)			preconditioner (7)		
		k*	*time*	*error*	k*	*time*	*error*
n = 1024	0	3380	.30	1.7(-11)		.41	
K(A)=	0.9	735	.24	2.2(-11)	534	.28	2.1(-11)
9.2(+5)	1.0	707	.23	2.1(-11)	530	.28	2.0(-11)
	1.1	679	.22	2.2(-11)	531	.28	2.3(-11)
	1.5	600	.19	2.1(-11)	673	.35	2.2(-11)
	1.9	570	.18	2.0(-11)	1219	.64	2.2(-11)
n=16384	0	7851	8.09	1.6(-11)		13.21	
K(A) =	0.5	1229	4.71	3.1(-12)	1033	6.83	3.2(-12)
3.1(+8)	1.1	959	3.67	3.2(-12)	706	4.70	2.7(-12)
	1.5	820	3.14	2.8(-12)	839	5.55	2.7(-12)
	1.8	746	2.86	3.2(-12)	1325	8.78	3.7(-12)
	1.9	731	2.80	3.1(-12)	1676	11.11	3.7(-12)

REFERENCES

[1] Adams, L.: m-Step preconditioned conjugate gradient methods. *SIAM J. Sci. Stat. Comput.* 6, 2 (1985), 452-463.

[2] Adams, L.M., Ong E.G.: Additive polynomial preconditioners for parallel computers. *Parallel Computing* 9 (1988/89), 333-345.

[3] Eisenstat, S.C., Elman, H.C., Schultz, M.H.: Variational iterative methods for non symmetric systems of linear equations. *SIAM J. Numer. Anal.* 20, 2 (1983), 345-357.

[4] Engeli, M., Ginsburg, M., Rutishauser, H., Stifel, E.: Refined iterative methods for the computation of the solution and the eigenvalues of self-adjoint boundary value problems. *Mitt. Inst. Angew. Math.* ETH, Zurich, n. 8, Basel-Stuttgart (1959).

[5] Evans, D.J.: *Preconditioning Methods. Theory and Applications.* Gordon and Breach, New York (1983).

[6] Fletcher, R.: Conjugate gradient methods for indefinite systems. *Proc. of the Dundee Conference on Numerical Analysis* (G.A. Watson, ed.), Springer-Verlag, Berlin (1975), 73-89.

[7] Galligani, E., Ruggiero, V.: The arithmetic mean preconditioner for multivector computers (to appear).

[8] Hageman, L.A., Young, D.M.: *Applied Iterative Methods.* Academic Press, New York (1981).

[9] O'Leary, D.P., White, R.E.: Multisplittings of matrices and parallel solution of linear systems. *SIAM J. Alg. Disc. Meth.* 6, 4 (1985), 630-640.

[10] Ruggiero, V., Galligani, E.: An iterative method for large sparse linear systems on a vector computer. *Computers Math. Applic.* 20, 1 (1990), 25-28.

[11] White, R.E.: Multisplitting with different weighting schemes. *SIAM J. Matrix Anal. Appl.* 10, 4 (1989), 481-493.

[12] Young, D.M.: *Iterative Solution of Large Linear Systems.* Academic Press, New York (1971).

SCHEDULING PARALLEL FACTORIZATION ALGORITHMS ON A SHARED MEMORY COMPUTER

Michel COSNARD
Laboratoire de l'Informatique du Parallélisme - CNRS,
Ecole Normale Supérieure de Lyon
46 allée d'Italie, 69364 LYON Cedex 07
FRANCE

Abstract: We review some of the results obtained in the last decade on the design and analysis of parallel factorization algorithms for solving dense linear systems of equations. After defining a model of computation, we introduce a graph-theoretic approach which allows to separate the design of a parallel algorithm into two independent parts

- construction of a task system, determination of the precedence constraints and of the task graph associated to a sequential algorithm
- construction of scheduling algorithms compatible with the task graph.

We present some theoretical tools for analysing the complexity of the various parallel algorithms associated to a sequential algorithm. We then apply these tools to Gauss-Jordan method, several Gaussian elimination algorithms, Cholesky decomposition and Doolittle reduction and QR factorization. In a second part, we examine the parallelization of Givens factorization and review several complexity results.

Keywords: Parallel algorithms, scheduling, linear systems of equations, shared memory computers.

1. INTRODUCTION

The advances in parallel computing have been really impressive during the last 15 years. Starting with purely theoretical works or laboratory experimental researches on the design of parallel computers and algorithms, this period has revealed the power of parallelism. It is now well admitted that supercomputers are parallel machines. In the medium range of computing power are specialized machines which also use parallel processing as a way to increase computing speed. The next decade will certainly see the entrance of parallel computing in the personal computer market. This progress has been possible not only because of the advances in technology and circuit integration but also because of intensive research in the field of software and algorithms: parallel computing is by now present in all domains of sciences. However this issue has been really economically viable due to the incredible speed increases in the processing of number crunching problems especially by the use of parallel linear algebra codes.

In this work we review some of the results obtained in the last decade on the design and analysis of parallel factorization algorithms for solving dense linear systems of equations. We shall restrict ourselves to shared memory model of computation. We refer to Robert's paper in this volume [Rob90] for a review on the implementation of linear algebra algorithms on distributed memory architectures such as systolic arrays and general-purpose hypercubes. Our concern is mainly theoretical in the sense that we shall present the new tools and methodologies for the design and analysis of parallel algorithms but no experimental results on real computers.

NATO ASI Series, Vol. F 77
Computer Algorithms for Solving Linear Algebraic Equations
Edited by E. Spedicato

Parallel complexity analysis are now based on very realistic models of computation and their conclusions are highly accurate.

After defining our model of parallel architecture (a shared memory computer), we introduce a graph-theoretic approach which allows to separate the design of a parallel algorithm into two independent parts

- construction of a task system, determination of the precedence constraints and of the task graph associated to a sequential algorithm
- construction of scheduling algorithms compatible with the task graph.

We present some theoretical tools for analysing the complexity of the various parallel algorithms associated to a sequential algorithm. We then apply these tools to Gauss-Jordan method, several Gaussian elimination algorithms, Cholesky decomposition and Doolittle reduction and QR factorization. In a second part, we examine the parallelization of Givens factorization and review several complexity results.

2.MODEL FOR PARALLEL COMPUTATION

To carry out a study of complexity, we need to simplify the architecture module. In the first place, we shall concern ourselves only with the parallelization of 'compute bound' algorithms derived from linear algebra, especially matrix factorization algorithms. In a 'compute bound' algorithm, the number of elementary calculations is greater than the number of input-output data transfers (in the opposite case, the problem is 'I/O-bound'). For example the matrix product is compute bound whereas the addition of two matrices is I/O-bound.

Having made this restriction, we can base our analysis on an ideal case of execution time, by assuming that data read and write times in common memory and the cost of arithmetic operations are constant. As our unit of time, we choose the time needed to read two data items from memory, to execute one of the four elementary operations and writing of the result to memory. This model if often referred as a Parallel Random Access Memory (PRAM), but as we shall see later, our model differs in the sense that we do not assume any synchronization between the read, write or compute operations done in parallel by several processors.

To be realistic, it is now absolutely necessary to limit the number of processors p by a linear factor of the size of the problem. Saad [Saad85] has shown how data access time predominates over computation time to solve a dense linear system of size n with n^2 processors. More precisely we shall set $p = \alpha n$, with $\alpha \leq 1$.

Moreover, it is clear that the model is a better representation of architectures where communication between processors takes place by means of a shared memory accessible by an interconnection network, than by data exchanges between local memories for which the data transfer time between processors is not constant, as it depends on the location of the processors [HB84]. However it is worth noting that the gap between these two main models of parallel architectures is being reduced by the use of dedicated hardware for routing messages (iPSC of Intel) or dynamically reconfigurable networks (Transputer based computers).

Finally, we shall impose a certain form of access to data. So, for parallel processing of the elements in a matrix A, we shall assume in what follows that we can access elements either by row, or by column. The possible transfer operations are then:

(i) transmission of a row or column of A from the storage device to a processing device

(ii) transmission of a row or column of A from a processing device to a storage device.

We assume that it is possible to transfer the same data item simultaneously to several processors. In this case, no processor may modify the data. Conversely, a processor may only modify data if it is the only one with copy of it. This corresponds clearly to the CREW (Concurrent Read Exclusive Write) type of the PRAM model.

Apart from the restrictions already mentioned, we should stress that the matrix problems studied are based only on transfers of row or column vectors. Assuming that these are stored in a contiguous way in central memory, as is done in the usual approach for vectors computers and architectures using a cache memory, it is possible to pipeline the transfer of their components, and therefore to execute data transfer and processing operations on processors simultancously. We shall not be concerned with the computer mechanisms making it possible to resolve the problem of data allocation without conflict. Let us simply mention that the preceding constraints and the mechanism managing them ensure unambiguous execution of the algorithm and a certain degree of synchronization.

A single algorithm may lead to several different parallel versions depending on access to data, decomposition into tasks, assignment of tasks to processors and the synchronization of processors. Certain versions will be more suitable for particular computer structures: in particular, a SIMD machine would be unable to execute an asynchronous algorithm. For MIMD machines, it is simply necessary to specify explicitly the necessary synchronization between tasks, and the hardware and software system will handle the synchronization. We are interested here in architectures and systems for which the programmer explicitly specifies the ordering of tasks and not dependencies alone.

3. TASK GRAPH AND SCHEDULING

As mentionned by Gerasoulis and Nelken [GN89], the parallelization problem consists of three important stages:

° Identifying parallelism and finding the data dependencies

° Partitioning the algorithm into indivisible tasks and the data into corresponding data items (the size of the tasks depends on the problem as well as the target architecture)

° Scheduling the execution of these tasks and mapping the data items into a given multiprocessor.

The concept of systems and tasks, as they are used by Kumar [Kum82] and Coffman and Denning [CD73] allows the introduction of a formalism which will be the basis of all our studies on algorithm parallelization. An algorithm may be segmented into a set of tasks, associated with their input and output variables, making it possible to construct a precedence system which expresses the internal parallelism of the algorithm. In such a system, there may be precedence constraints between tasks, and in this case their execution is sequential. Otherwise tasks are non-interfering and their execution can take place in parallel. We shall start by defining these intuitive notions more precisely.

A task is an indivisible processing unit characterized exclusively by its external behaviour: input, output, function and execution time. The task can start execution once all of its inputs have been read, carries on to completion without interruption (preemption is not allowed), and writes the outputs as soon as it finishes execution. Consequently, task T will be represented by a quadruplet (I_T, O_T, f_T, t_T) such that

°I_T is the set of inputs to T

°O_T is the set of outputs from T

°f_T is an operator on I_T in O_T, i.e. an ordered sequence of elementary operations of I_T in O_T

°t_T is an integer representing the execution time for task T.

The execution time (or duration) of a task is in general the sum of two terms: the computation time, which corresponds to the number of elementary operations carried out in the task, and the communication time associating the number of transfers between shared memory and the processor executing the task. The assumption that the execution depends only of the task itself is valid for shared memory parallel computers but cannot be used for distributed memory parallel computers with message passing communication means, as in this last case the communication time is also depending on the relative place of the processor executing the task in the communication network topology. Problems of availability and integrity of data are not taken into account at this level, as they depend on the ordering of tasks for their execution.

A *sequential algorithm* is a set of tasks with a total ordering relation: A = $(T_1, \ldots, T_n, \rightarrow)$. The relation $T_i \rightarrow T_k$ (i≠k) means that the execution of task T_i must be complete before execution of T_k starts.

The preceding definition of a sequential algorithm may seem restricive. Strictly speaking, it is not restrictive since an algorithm is a task, which is a fairly general concept (still more general if we do not regard execution time as determined a priori). In fact it is not elaborate enough to allow us to study algorithms with termination tests or conditional branching, in which cases the execution time is data dependent and is known only at the end of the execution. There is not yet a methodology to solve such problems within the context considered here (explicit assignement). Dynamic assignment techniques allow them to be taken into account. This definition will, however, be sufficient for a study of parallelization of the main algorithms for the direct solution of dense linear systems.

The decomposition of an algorithm into tasks can be carried out by program or by the user. Is is the latter case that we shall study in what follows. Several different decompositions are possible for a single algorithm. The grain of a decomposition is a rough measure of the relation between the mean execution time of a task and the total time for the algorithm: it may be fine, medium or coarse. The choice of the best decomposition is a difficult problem associated with the execution time of tasks and the underlying architecture. The main factors guiding choice in this area are the number of processors, the ratio between the unit of communication time and the unit of computation time and memory access.

A *task system* $S = (T_1, \ldots, T_n, <<)$ is a set of tasks with a partial ordering relation written $<<$: $T_i << T_k$ ($i \neq k$) meaning that the execution of task T_i must be complete before execution of T_k can begin. A sequential algorithm is a task system in which all the tasks are ordered.

Let $S = (T_1, \ldots, T_n, <<)$ be a *task system*. Two tasks T_i and T_k are independent if they modify no common variable. The independence relation can be written as follows:

$$(O_{T\,i} \cap O_{T\,k}) = (O_{T\,i} \cap I_{T\,k}) = (I_{T\,i} \cap O_{T\,k}) = \emptyset$$

T_i and T_k are consecutive if $T_i << T_k$ and there is no other task T_i such that $T_i << T_j << T_k$.

A task system is a precedence system if , for all i and k, one and only one of the following three conditions is satisfied:

1. $T_i << T_k$
2. $T_k << T_i$
3. T_i and T_k are independent.

The *precedence graph* (or *task graph*) G associated with a precedence system is defined as follows:

(a) The set of vertices G is a set of tasks in S.

(b) T_i and T_k are linked by an edge if and only if T_i and T_k are consecutive.

Given a sequential algorithm $A = (T_1, \ldots, T_n, \rightarrow)$, it can be shown that there exists a single precedence system $S = (T_1, \ldots, T_n, <<)$ such that $<<$ is a subordering of $\rightarrow$. $<<$ is the following relation:

$$T_i << T_k \iff (T_i \rightarrow T_k \text{ and } (O_{T\,i} \cap O_{T\,k}) \cup (O_{T\,i} \cap I_{T\,k}) \cup (I_{T\,i} \cap O_{T\,k}) \neq \emptyset)$$

To parallelize a sequential program, we always therefore start by computing the relation $<<$ taking the relation $\rightarrow$ as starting point. In what follows we shall assume that this stage has already been carried out. Here again, it can be left to the programmer, or carried out by program (automatic construction of the task graph).

We then have to assign tasks to the available processors respecting the constraints in the task graph and those associated with the execution times. This question is part of a more general problem of scheduling a task system. We shall not consider this general problem, which is difficult since the scheduling of any task system is an NP-complete problem [CD72]. On the other hand we shall attempt to solve the problem for particular precedence systems. Recall that throughout this section the number of processors will be written p.

A compatible scheduling for a precedence system S = $(T_1, \ldots, T_n, \ll)$ is a mapping *Sch* from $\{T_1, \ldots, T_n\}$ into $\{1, \ldots, p\} \times N$, $Sch(T_k) = (proc(T_k), time(T_k))$, such that:

(i) $T_i \ll T_k$ implies $time(T_i) + t_i \leq time(T_k)$

(ii) $proc(T_i) = proc(T_k)$ implies $(time(T_i) + t_i \leq time(T_k)$ or $time(T_k) + t_k \leq time(T_i))$

$proc(T_i)$ is the processor which will execute T_i and $time(T_i)$ is the date at which execution of T_i starts. Condition (i) simply means that if T_i and T_k are not independent, the execution of T_k cannot start until completion of the execution of T_i. Condition (ii) means that a processor cannot be assigned at the same time to two different tasks.

A *parallel algorithm* associated with a sequential algorithm A = $(T_1, \ldots, T_n, \rightarrow)$ is a pair (S, *Sch*) formed by the precedence system S associated with A and by a scheduling compatible with S. The execution time of the parallel algorithm is the date of the end of execution of the last task executed. It is therefore:

$$T_{PA} = Max_{k=1,n}(time(T_k) + t_k)$$

A parallel algorithm associated with a sequential algorithm A is optimal if there is no other parallel algorithm associated with A with a lower execution time. The basic problem studied in the framework of algorithm parallelization can now be stated as follows: if A is a sequential algorithm and p is the number of processors available, determine an optimal parallel algorithm associated with A and calculate its execution time.

The basic problem studied in the framework of algorithm parallelization can now be stated as follows: *if A is a sequential algorithm and p is the number of processors available, determine an optimal parallel algorithm associated with A and calculate its execution time.*

There is no general solution to this problem. Other problems, derived from the preceding one, are therefore studied. Here are some of them:

1. Assuming that there is no limit to the number of processors available, what is the execution time t_{opt} of an optimal parallel algorithm? If $t_{opt}(p)$ is the execution time of an optimal parallel algorithm with p processors, t_{opt} is defined by

$$t_{opt} = Min_{p=1,n}(t_{opt}(p))$$

2. What is the minimum number of processors p_{opt} making it possible to implement an algorithm executing in time t_{opt}? p_{opt} is defined by

$$p_{opt} = Min_{p=1,n}(p \text{ such that } t_{opt}(p)=t_{opt})$$

Measuring the performances is classically done by comparing the ratio of the sequential computing time and the parallel execution time, defining the speed up with p processors as:

$$S_p = t_{seq} / t_{opt}(p)$$

The asymptotic speed up is obtained by letting the dimension of the problem go to infinity with fixed p. The (asymptotic) efficiency is obtained by normalizing the (asymptotic) speed up:

$$E_p = S_p / p$$

4. GENERAL BOUNDS ON SCHEDULING

In this section, we derive some tools and general results for scheduling tasks graphs. We first concentrate on determining t_{opt} and obtaining lower and upper bounds on p_{opt}. When considering a number of processors as large as necessary, we shall speak of unbounded parallelism. Remark that p is always less than the total number of tasks of the sequential algorithms.

The execution time of a path in the task graph is the sum of execution times of the tasks that make it up. A longest path through the task graph is one for which the execution time is the greatest.

Theorem 1: The execution time t_{opt} of an optimal algorithm is equal to the time of execution of the longest paths in the task graph.

The longest paths in G are called critical paths. We shall use the term *decomposition in levels* of the task graph $D(G) = \{N_1, ..., N_{t_{opt}}\}$ for a partition into t_{opt} subsets (levels) of vertices of this graph satisfying the following conditions. Level 1 is made up of tasks with no predecessor. Level k is made up of tasks all of whose predecessors are in lower levels and all of whose successors are in higher levels. Level t_{opt} is made up of tasks with no successor. Note that a single graph will allow more than one decomposition into levels.To any decomposition, we can associate a parallel algorithm by simply starting the execution of the tasks in level k at time k. Two particular decompositions are decomposition by predecessors, where each level contains all the possible predecessors, and decomposition by successors. The first decomposition is also called earliest decomposition, and the second decomposition latest decomposition, as scheduling by levels leads to a minimum execution start date in the first place, and a maximum date in the second.

In case of a unit execution time (UET) graph, each level is non empty and the cardinality L(D) of the largest level is the minimum number of processors in order to execute the corresponding scheduling scheme. Hence p_{opt} is given by the minimum of L(D) over all the possible decompositions in levels of G. This minimum will be called the *width of G*. These notions can be extended to general tasks graphs by simply considering that a task with a duration t greater than 1 belongs to t consecutive levels.

Theorem 2: The minimum number of processors p_{opt} in order to execute the algorithm in optimal time t_{opt} is equal to the width of the task graph.

However there is no direct method for determining the width of a general task graph. The cardinalities of the largest levels in the decompositions by predecessors and by successors give easily computable upper bounds. Lord, Kowalik and Kumar [LKK84] have developed a general method for obtaining a lower bound. Let us call total amount of work tw the sequential execution time (the sum of the durations of all the tasks). Let t_p be the execution time of an algorithm with p processors. The *computational area ca* of the parallel algorithm is the product of the number of processors and the execution time less the *lost area la*, where not all the processors can be doing work (idle processors):

$$ca := p.t_p - la$$

Clearly ca must be greater than or equal to tw. An upper bound on ca is obtained as follows:

- in the decomposition by predecessors, assume that the first q levels have less than p tasks and call $p_1, \ldots, p_q$ the number of used processors in these levels
- in the decomposition by successors, assume that the last r levels have less than p tasks and call $p_{t_{p-r}}, \ldots, p_{t_p}$ the number of used processors in these levels.

Without loss of generality we can assume that $q \leq t_{opt}$-r (else p is greater than p_{opt} and can be reduced). Since any decomposition has less tasks in the first q levels than the decomposition by predecessors and more tasks in the last r levels than the decomposition by successors, the computational area is bounded above as follows:

$$ca \leq p.t_p - p.(q+r) + p_1 + \ldots + p_q + p_{t_{p-r}} + \ldots + p_{t_p}$$

Thus a lower bound on the number of processors required for a schedule of length t_p is the smallest p for which ca $\geq$ tw.

Theorem 3: The minimum number of processors p_{opt} in order to execute the algorithm in optimal time t_{opt} is equal to the solution of the following equation:

$$p.t_p - la = tw$$

It is greater than or equal to the solution of:

$$p.t_{opt} - p.(q+r) + p_1 + \ldots + p_q + p_{t_{opt-r}} + \ldots + p_{t_{opt}} = t_{seq}$$

The preceding method can be applied to the computation of a lower bound of $t_{opt}(p)$, for the general problem of determining the best parallel algorithm with p processors.

Notations: Let
- ac(t) be the number of active processors at step t
- id(t) be the number of idle processors at step t
- T(p) be the execution time with p processors

The following relation holds p = id(t) + ac(t). Define:

$$P_{ac} = \sum_{t=1}^{T(p)} ac(t) \qquad P_{id} \sum_{=t=1}^{T(p)} id(t)$$

Using theorem 3, Lord et al. derive the following result [LKK84]:

Corollary 1: $T(p) \geq \frac{P_{id} + P_{ac}}{p}$

There is no theoretical method for determining exactly $t_{opt}(p)$. Moreover computing $t_{opt}(p)$ and scheduling the tasks are NP-hard problems. Hence static scheduling heuristics are used. The critical path heuristic is one of the most popular. It uses the function *cp(T)* which is defined as the length of the longest path from task T to the final tasks. A practical and efficient implementation of the critical path heuristic can be described as follows [GN89]:

1. Compute *cp(T)* for each task of the graph
2. Construct a priority list in descending order of *cp(T)*. If *cp(T)* is equal for two distinct tasks, then the task with most successors has a higher priority
3. Perform a list scheduling on the priority list.

5. PARALLEL GAUSS-JORDAN METHODS

Let us first study the parallelization of Gauus-Jordan method to solve a dense linear system Ax=b. We shall use a very simple form of the algorithms in order to concentrate on the scheduling problem. Hence we shall not consider pivoting strategies. The sequential algorithm is as follows:

{sequential JORDAN}
For k := 1 **to** n **do**
    For j := k + 1 **to** n + 1 **do**
        $a_{kj} := a_{kj} / a_{kk}$
    For i := 1 **to** n, $i \neq k$ **do**
        For j := k + 1 **to** n + 1 **do**
            $a_{ij} := a_{ij} - a_{ik} * a_{kj}$

The mode of access to data determines the segmentation of the sequential program into tasks. We consider accesses by rows or by columns:

{JORDAN by rows}
For k := 1 **to** n **do**
    execute T_{kk} : < **For** j := k + 1 **to** n + 1 **do** $a_{kj} := a_{kj} / a_{kk}$ >
    For i := 1 **to** n, $i \neq k$ **do**
        execute T_{ki} : < **For** j := k + 1 **to** n + 1 **do** $a_{ij} := a_{ij} - a_{ik} * a_{kj}$ >

{JORDAN by columns}
For k := 1 **to** n **do**
    For j := k + 1 **to** n + 1 **do**
        execute T_{kj} : < $a_{kj} := a_{kj} / a_{kk}$
            For i := 1 **to** n, $i \neq k$ **do** $a_{ij} := a_{ij} - a_{ik} * a_{kj}$ >

Jordan by rows

Task T_{kk} executes in n+1-k units of time, while T_{ki}, $i \neq k$, requires 2(n+1-k) units. The order of constraints is as follows:

$$T_{kk} << T_{ki}, \text{ for } i \neq k \qquad T_{ki} << T_{k+1,i}, \text{ for } i \neq k$$

There exist n-1 longest paths in this graph which only differ in the last task and are made up of the following tasks:

$$T_{11}, T_{12}, T_{22}, \ldots, T_{kk}, T_{kk+1}, T_{k+1k+1}, \ldots, T_{nn}, T_{ni} \qquad \text{for } i = 1, \ldots, n-1$$

Then we have

$$t_{opt} = 3\,n\,(n+1) / 2$$

The decomposition by predecessors corresponds to assigning the tasks with the same k index to the same level. The width of this decomposition is equal to n-1. The same is true for decomposition by successors. We therefore deduce that $p_{opt} \leq n-1$. The structure of the n-1 different paths of length t_{opt} shows that it is impossible to execute all the tasks making up these paths with less than n-1 processors. We therefore deduce that

$$p_{opt} = n-1$$

The value of p_{opt} is fairly disappointing, since n-2 processors are free during the execution of T_{kk}. The question is then posed of discovering whether a slight increase in execution time would not produce a substantial reduction in the number of processors. A parallel algorithm is *asymptotically optimal* (or *asymptotically optimal for p propcessors*) if the ratio between its execution time t_{alg} and t_{opt} or $t_{alg}(p)$ and $t_{opt}(p)$ tends towards 1 as n goes to infinity.

An asymptotically optimal algorithm as been constructed in [CR87]. It is based on the remark that T_{kk} has an execution time equal to half the time of $T_{k-1,i}$. Hence by dividing each of the $T_{k-1,i}$ into two subtasks, we obtain 2n-2 tasks with execution times very nearly equal to the time of T_{kk}. Using $p = 2n / 3 + O(1)$ processors, we then obtain an asymptotically optimal algorithm in time $t_{opt} + 2n$. Following the same idea, it has been shown that

$$t_{opt}(p) = n^3 / p + O(n)$$

This implies that for $p \leq 2n / 3$, the algorithms are of asymptotic efficiency equal to 1.

Jordan by columns

All tasks T_{kj} execute in the same execution time t=n+1-k time units. The order of constraints is as follows:

$$T_{kk+1} << T_{kj}, \text{ for } j > k+1 \qquad T_{kj} << T_{k+1,j}, \text{ for } j > k$$

If we call the set $\{T_{kk+1}, T_{kk+2}, \ldots, T_{kn+1}\}$ level k, we now have a graph where the number of tasks per level decreases, but where all the tasks of a level have the same execution time.

All paths between the tasks T_{ij} and T_{nn+1} are the longest paths in the graph. As there are n of them, we obtain

$$t_{opt} = 2n^2 - n \qquad\qquad p_{opt} = n$$

Let us now consider a fixed number of processors p and attempt to evaluate $t_{opt}(p)$. To do so, let us introduce the greedy scheduling algorithm, defined as follows. Up to level n - p + 1 of the task graph, there are more than p tasks per level. Let us divide all the tasks into groups of p tasks, scanning the levels of the precedence graph from left to right.The last of these groups is therefore made up of the remaining tasks in level n - p +1. The following groups are formed, respectively, of levels n - p + 2, ..., n of the graph. The greedy scheduling algorithm assigns each of these groups to the p processors, with the execution of tasks in group i beginning at time (i-1) (2n-1). It has been shown in [CR87] that this algorithm is optimal.

Jordan by rows versus Jordan by columns

The preceding results can be summarized in the following theorem.

Theorem 4:

1. For thc Jordan method by rows:

$$t_{opt} = 3\,n\,(n+1)\,/\,2 \qquad p_{opt} = n-1 \qquad t_{opt}(p) = n^3\,/\,p + O(n)$$

For $p \leq 2n / 3$, the algorithms are of asymptotic efficiency equal to 1

2. For the Jordan method by columns:

$$t_{opt} = 2n^2 - n \qquad\qquad p_{opt} = n$$

For $p \leq n$, the greedy scheduling algorithm produces an optimal parallel algorithm with an execution equal to

$$t_{opt}(p) = (\,(n - p)\,(n - p + 1)\,/\,2p)\,(2n - 1)$$

An immediate consequence of this theorem 4 is that the method by rows is always preferable to the method by columns, whatever the number of processors.

6. GAUSSIAN ELIMINATION

To implement Gaussian elimination on a conventional sequential computer we can use various algorithms. Generally speaking, these algorithms are all equivalent in term of performances. This is no longer the case if we consider less conventional computers. Six different versions for a vector pipeline machine have been considered in [DGK84]. Three implementations where data are accessed columnwise have been discussed in detail, each of them corresponding to a given permutation of the loop indices i, j, k of the sequential algorithm: namely the SAXPY version (form kji), the GAXPY version (form jki) and the DOT version (form ijk).We shall see in this section that the number of versions is increased if we consider parallel computers, even in the case of shared memory machines.

Parallelization of the standard version

The standard column-oriented KJI version with partial pivoting can be split into tasks as follows:

{KJI version with partial pivoting}
For k := 1 **to** n-1 **do**
  execute T_{kk} : < **find** pivot(k) st $|a_{pivot(k)k}|$ = max { $|a_{kk}|$, ..., $|a_{nk}|$ };
    swap($a_{pivot(k)k}$, a_{kk}) ;
    For i := k + 1 **to** n **do** $a_{ik} := a_{ik} / a_{kk}$ >
  For j := k + 1 **to** n **do**
    execute T_{kj} : < **swap**($a_{pivot(k)k}$, a_{kj}) ;
      For i := k + 1 **to** n **do** $a_{ij} := a_{ij} - a_{ik} * a_{kj}$ >

Task T_{kk} executes in 2(n+1-k) units of time, while T_{kj}, k<j, requires 2(n-k) units. The order of constraints is as follows:

(A) $T_{kk} << T_{kj}$, $k+1 \le j \le n,\ 1 \le k \le n-1$

(B) $T_{kj} << T_{k+1,j}$ $k+1 \le j \le n,\ 1 \le k \le n-1$.

which produces the task graph in figure 1. Constraint (A) means that preparation of column k at step k must be completed before updating any column j>k. Constraint (B) means that updating column j at step k, j>k, must be completed before updating this column for step k+1

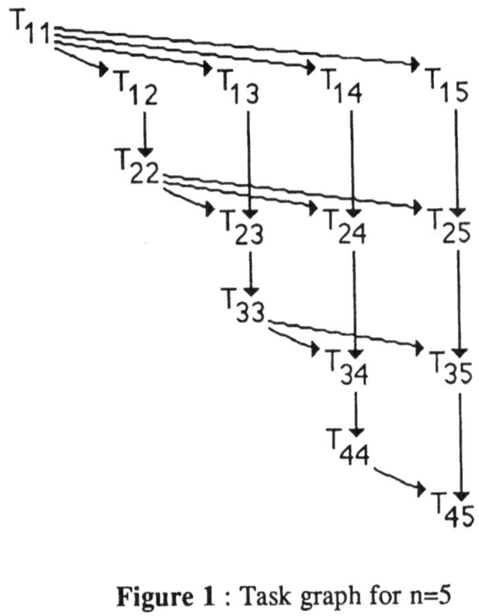

Figure 1 : Task graph for n=5

The sequential execution time is $t_1 = 2n^3/3 + O(n^2)$. The longest path s_1 of the graph is composed of the tasks T_{11}, T_{12}, T_{22}, ..., T_{kk}, $T_{k,k+1}$, ..., $T_{n-1,n-1}$, $T_{n-1,n}$. Its length $L(s_1)$ is $L(s_1) = 2n^2-2$. Hence we deduce that

$$t_{opt} = L(s_1) = 2n^2-2.$$

It is obvious to design a parallel algorithm with $p = n-1$ processors whose execution time is t_{opt}. Processor 1 executes all the tasks in s_1. While processor 1 executes a task T_{kk}, the p-1 other processors remain idle. While processor 1 executes a task T_{kk+1}, we let n - k processors execute in parallel the n - k independent tasks T_{kk+2}, T_{kk+3}, ..., T_{kn}, which all have the same execution time as T_{kk+1}. It follows that $t_{opt} = L(s_1)$ and also that $p_{opt} \le n - 1$.

Define α as the ratio between p and n and let e_α denote the asymptotic efficiency of an algorithm using $p = \alpha\, n$ processors. The asymptotic efficiency of the preceding algorithm is $e_1 = 1/3$. Lord, Kowalik and Kumar [LKK84] propose a parallel algorithm with $p = n/2$ processors ($\alpha = 1/2$) which achieves the same execution time t_{opt}. Its asymptotic efficiency is thus $e_{1/2} = 2/3$. Their algorithm is briefly the following. Let level k be the set of tasks T_{kk+1}, T_{kk+2}, ..., T_{kn}, plus T_{k+1k+1}. As before Processor 1 executes all the tasks in s_1. Hence it executes two tasks per level, namely T_{kk+1} and T_{k+1k+1} at level k. During the same time, the other processors can execute two tasks of the same level. Since the maximum number of tasks per level is n, n/2 processors are sufficient. The asymptotic efficiency of this algorithm is $e_{1/2} = 2/3$.

Lord, Kowalik and Kumar raised the question as to whether a schedule of length t_{opt} exists with $p < n/2$: is $p_{opt} \le n/2$? For computing a lower bound for this particular problem they introduced the method that we have presented in theorem 3 and corollary 1. The minimum number of processors p_{opt} in order to execute the algorithm in optimal time t_{opt} is equal to the solution of the following equation:

$$p.t_{opt} - la = t_1$$

Using the same method as in theorem 3, we can show that $la \ge p^3/3 + O(n^2)$. Since $t_1 = 2n^3/3 + O(n^2)$ and $t_{opt} = 2n^2-2$. We deduce [LKK84] that $\alpha_{opt} \ge \alpha_0$, where $\alpha_0 \approx 0.347$ is a solution to the equation

$$3\alpha - \alpha^3 = 1$$

Hence $$p_{opt} \ge \alpha_0\, n + O(1) \approx 0.347\, n$$

The question of computing p_{opt} and finding algorithms for smaller values of α has been addressed by several authors and solved recently. Veldhorst [Vel89] designed an algorithm in time t_{opt} using only $p = \sqrt{2}/4\, n$ processors. For $\alpha \ge \sqrt{2}/4 \approx 0.354$, we deduce that there exist algorithms in time t_{opt}, of efficiency $e_\alpha = 1/(3\alpha)$, and this value is optimal. The value of $\sqrt{2}/4$ is close to α_0. Cosnard et al. [CMRT88], Gerasoulis and Nelken [GN88] and Robert and Trystram [TR88] have successively developed new algorithms.

Marrakchi and Robert [MR89] solved the problem. They have constructed an algorithm with $\alpha_0 n$ processors which executes asymptotically in time t_{opt}, thereby achieving the preceding lower bound:

$$p_{opt} = \alpha_0 n + O(1) \approx 0.347 n$$

For $\alpha = \alpha_0$, the asymptotic efficiency is excellent: we have $e_{\alpha_0} \geq 0.959$, which evidences the high degree of parallelism that can be achieved.

They also proposed asymptotically optimal algorithms for $\alpha \leq \alpha_0$, leading to an asymptotic effciency equal to $e_\alpha = 1/(1+\alpha^3)$.

Parallelization of the other versions

In this part we briefly described the results presented by Cosnard et al. in [CMRT88]. Together with the SAXPY, GAXPY and DOT versions, they have analysed several Gaussian-like elimination algorithms: LDM^t, LDL^t, Doolittle and Cholesky algorithms. The decomposition in tasks of the algorithms is the following:

(A) { Generic Gaussian elimination - KJI - SAXPY Form }

For k=1 **to** n-1 **do**

  execute T_{kk} : < **For** i := k + 1 **to** n **do** $a_{ik} := a_{ik} / a_{kk}$ >

  For j := k + 1 **to** n **do**

    execute T_{kj} : < **For** i := k + 1 **to** n **do** $a_{ij} := a_{ij} - a_{ik} * a_{kj}$ >

T_{kk} :	n-k	arithmetic operations	$1 \leq k \leq n-1$
T_{kj} :	2(n-k)	arithmetic operations	$k+1 \leq j \leq n$, $1 \leq k \leq n-1$

Sequential execution time: $2n^3/3 + O(n^2)$.

Precedence relations :	$T_{kk} \ll T_{kj}$	$k+1 \leq j \leq n$, $1 \leq k \leq n-1$
	$T_{kj} \ll T_{k+1,j}$	$k+1 \leq j \leq n$, $1 \leq k \leq n-1$

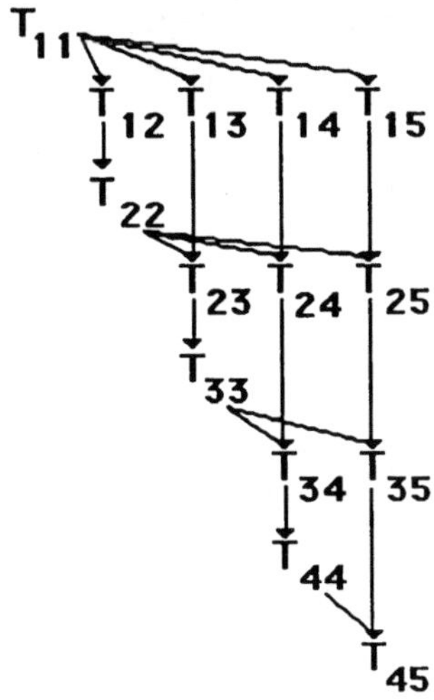

Figure 2: Task graph of Generic Gaussian elimination - KJI - SAXPY Form (n=5)

(B) { Generic Gaussian elimination - JKI - GAXPY }

For j=1 **to** n **do**

  For k=1 **to** j-1 **do**

    execute T_{kj} : < **For** i := k + 1 **to** n **do** $a_{ij} := a_{ij} - a_{ik} * a_{kj}$ >

  execute T_{jj} : <**For** i := j + 1 **to** n **do** $a_{ij} := a_{ij} / a_{jj}$ >

T_{kj}	: 2(n-k)	arithmetic operations	$1 \le k \le j-1,\ 1 \le j \le n$
T_{jj}	: n-j	arithmetic operations	$1 \le j \le n$
Sequential execution time:		$2n^3/3 + O(n^2)$.	
Precedence relations :		$T_{kj} << T_{k+1,j}$	$1 \le k \le j-1,\ 1 \le j \le n$
		$T_{jj} << T_{jk}$	$j+1 \le k \le n-1,\ 1 \le j \le n$

The tasks graphs of (A) and (B) are identical. Two different sequential algorithms can lead to the same parallel implementation.

(C) { Generic Gaussian elimination - IJK - DOT }

For i := 2 **to** n **do**

  For j := 2 **to** i **do**

    execute T_{ij} : < $a_{ij-1} := a_{ij-1} / a_{j-1j-1}$

      For k := 1 **to** j-1 **do** $a_{ij} := a_{ij} - a_{ik} * a_{kj}$ >

  For j := i+1 **to** n **do**

    execute U_{ij} : < **For** k := 1 **to** i-1 **do** $a_{ij} := a_{ij} - a_{ik} * a_{kj}$ >

T_{ij} : 2j-1	arithmetic operations	$2 \le j \le i,\ 3 \le i \le n$
U_{ij} : 2(k-1)	arithmetic operations	$i+1 \le j \le n,\ 2 \le i \le n$

Sequential execution time: $2n^3/3 + O(n^2)$.

Precedence relations :

$T_{ij} << T_{i,j+1}$	$2 \le j \le i-1,\ 3 \le i \le n$
$T_{ii} << U_{ij}$	$i+1 \le j \le n,\ 2 \le i \le n$
$U_{ij} << U_{i+1j}$	$i+2 \le j \le n,\ 2 \le i \le n$
$U_{jj+1} << T_{ij+1}$	$i \le j \le n,\ \ 2 \le i \le n$

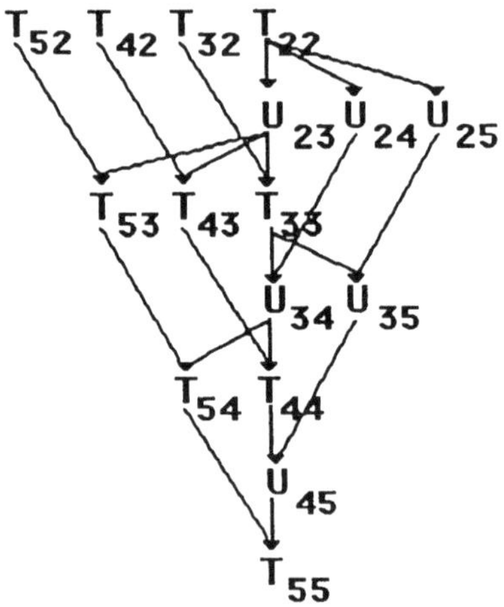

Figure 3: Task graph of Generic Gaussian elimination - IJK - DOT Form (n=5)

(D) { Generic Gaussian elimination - KJI - SAXPY modified Form }

For k=1 **to** n-1 **do**

  For j := k + 1 **to** n **do**

    execute T_{kj} : < $a_{kj} := a_{kj} / a_{kk}$

      For i := k + 1 **to** n **do** $a_{ij} := a_{ij} - a_{ik} * a_{kj}$ >

T_{kj} : 2(n-k)+1	arithmetic operations	$k+1 \le j \le n,\ 1 \le k \le n$

Sequential execution time: $2n^3/3 + O(n^2)$.

Precedence relations :

$T_{k,k+1} << T_{k+1,j}$	$k+2 \le j \le n,\ 1 \le k \le n-1$
$T_{kj} << T_{k+1,j}$	$k+2 \le j \le n,\ 1 \le k \le n-1$

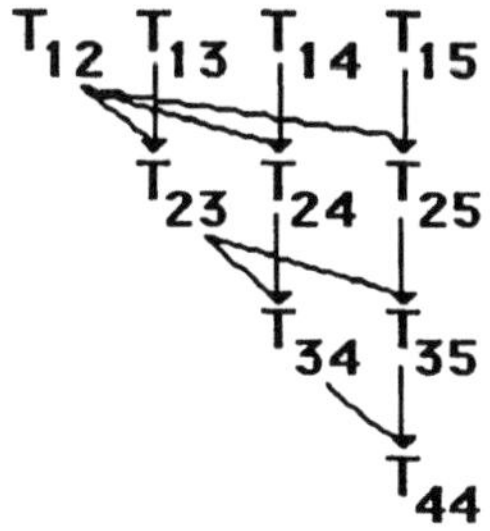

Figure 4: Task graph of Generic Gaussian elimination - KJI - SAXPY modified Form

(E) { Doolittle reduction [GVL83] }

For k=1 **to** n **do**

  For j := k **to** n **do**

    execute T_{kj} : < **For** p := 1 **to** k-1 **do** $a_{kj} := a_{kj} - a_{kp} * a_{pj}$ >

  For i := k + 1 **to** n **do**

    execute U_{ki} : < **For** p := 1 **to** k-1 **do** $a_{ik} := a_{ik} - a_{ip} * a_{pk}$

      $a_{ik} := a_{ik} / a_{kk}$ >

T_{kj} : 2(k-1) arithmetic operations $k \le j \le n,\ 1 \le k \le n$

U_{ki} : 2k-1 arithmetic operations $k+1 \le i \le n, 1 \le k \le n$

Sequential execution time: $2n^3/3 + O(n^2)$.

Precedence relations :

$T_{kj} << T_{k+1,j}$ $k+1 \le j \le n,\ 2 \le k \le n$

$T_{kk} << U_{k,j}$ $k+1 \le j \le n,\ 2 \le k \le n$

$U_{ki} << U_{k+1,i}$ $k+2 \le i \le n,\ 2 \le k \le n$

$U_{kk+1} << T_{k+1,j}$ $k+1 \le j \le n,\ 2 \le k \le n$

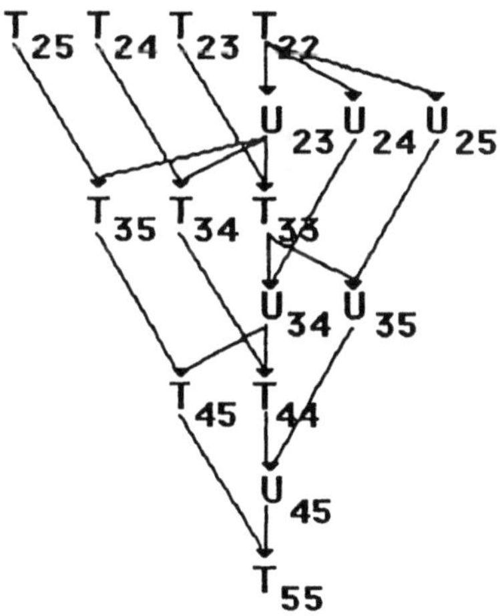

Figure 5: Task graph of Doolittle reduction (n = 5)

(F) { LDM^t decomposition - non optimized version [GVL83] }

For k=1 **to** n **do**

  execute T_{kk} : < **For** p := 1 **to** k+1 **do** $a_{kk} := a_{kk} - a_{kp} * a_{pp} * a_{pk}$ >

  For i := k + 1 **to** n **do**

    execute T_{ik} : < **For** p := 1 **to** k-1 **do** $a_{ik} := a_{ik} - a_{ip} * a_{pp} * a_{pk}$

      $a_{ik} := a_{ik} / a_{kk}$ >

  For j := k+1 **to** n **do**

    execute T_{kj} : < **For** p := 1 **to** k-1 **do** $a_{kj} := a_{kj} - a_{kp} * a_{pp} * a_{pj}$

      $a_{kj} := a_{kj} / a_{kk}$ >

T_{kk}	: 3(k-1)	arithmetic operations	$1 \le k \le n$
T_{ik}	: 3k-2	arithmetic operations	$k+1 \le i \le n,\ 1 \le k \le n$
T_{kj}	: 3k-2	arithmetic operations	$k+1 \le j \le n,\ 1 \le k \le n$

Sequential execution time: $n^3 + O(n^2)$.

Precedence relations :

	$T_{kk} << T_{kj}$	$k+1 \le j \le n,\ 2 \le k \le n$
	$T_{kk} << T_{ik}$	$k+1 \le i \le n,\ 2 \le k \le n$
	$T_{ik} << T_{i,k+1}$	$k+1 \le i \le n,\ 1 \le k \le n-1$
	$T_{kj} << T_{k+1,j}$	$k+1 \le j \le n,\ 1 \le k \le n-1$

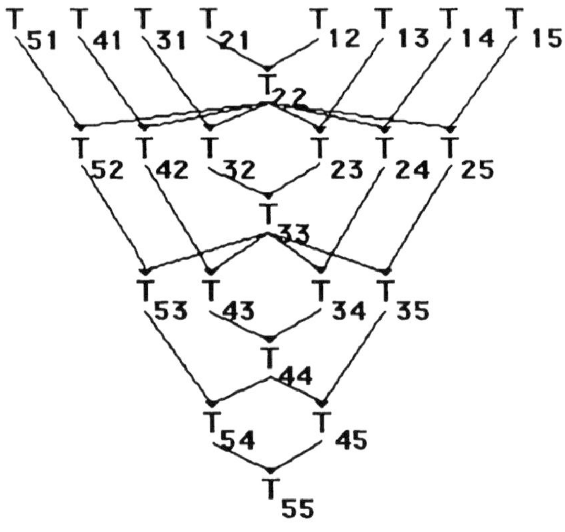

Figure 6: Task graph of LDM^t decomposition - non optimized version (n = 5)

(G) { LDM^t decomposition - optimized version [GVL83] }

For k=1 **to** n **do**

  execute T_{kk} : $<$ **For** p := 1 **to** k-1 **do**

    $r_p := a_{pk} * d_p$

    $w_p := a_{kp} * d_p$

    $a_{kk} := a_{kk} - a_{kp} * r_p >$

  For i := k + 1 **to** n **do**

    execute T_{ik} : $<$ **For** p := 1 **to** k-1 **do** $a_{ik} := a_{ik} - a_{ik} * r_p$

      $a_{ik} := a_{ik} / a_{kk} >$

  For j := k+1 **to** n **do**

    execute T_{kj} : $<$ **For** p := 1 **to** k-1 **do** $a_{kj} := a_{kj} - w_p * a_{pj}$

      $a_{kj} := a_{kj} / a_{kk} >$

T_{kk}	: 4(k-1)	arithmetic operations	$1 \le k \le n$
T_{ik}	: 2k-1	arithmetic operations	$k+1 \le i \le n,\ 1 \le k \le n$
T_{kj}	: 2k-1	arithmetic operations	$k+1 \le j \le n,\ 1 \le k \le n$

Sequential execution time: $2n^3/3 + O(n^2)$.

The precedence constraints and the precedence graph are the same as for the non-optimized LDM^t decomposition.

(H) { LDL^t decomposition - non optimized version [GVL83] et [KK84] }

For k=1 **to** n **do**

  execute T_{kk} : $<$ **For** p := 1 **to** k-1 **do** $a_{kk} := a_{kk} - a_{kp} * a_{pp} * a_{kp} >$

  For i := k + 1 **to** n **do**

    execute T_{ik} : $<$ **For** p := 1 **to** k-1 **do** $a_{ik} := a_{ik} - a_{ip} * a_{pp} * a_{kp}$

      $a_{ik} := a_{ik} / a_{kk} >$

T_{kk}	: 3(k-1)	arithmetic operations	$1 \le k \le n$
T_{ik}	: 3k-2	arithmetic operations	$k+1 \le i \le n,\ 1 \le k \le n$

Sequential execution time: $n^3/2 + O(n^2)$.

Precedence relations : $T_{kk} << T_{ik}$ $k+1 \le i \le n,\ 2 \le k \le n$

$T_{ik} << T_{i,k+1}$ $k+1 \le i \le n,\ 2 \le k \le n$

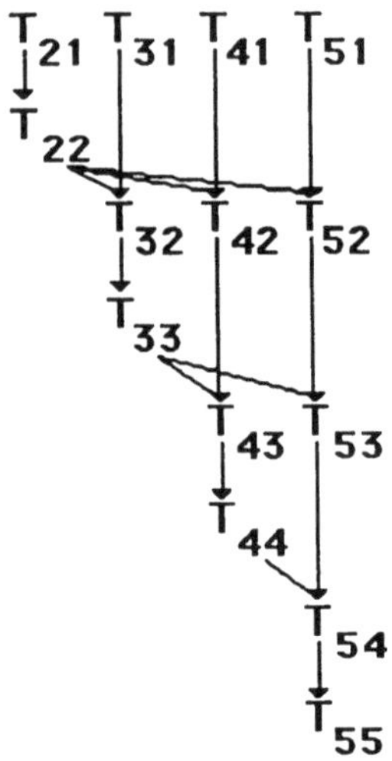

Figure 7: Task graph of LDL^t decomposition - non optimized version (n = 5)

(I) { LDL^t decomposition - optimized version [GVL83] }

For k=1 **to** n **do**

execute T_{kk} : < **For** p := 1 **to** k-1 **do**

$r_p := a_{pk} * d_p$

$a_{kk} := a_{kk} - a_{kp} * r_p$ >

For i := k + 1 **to** n **do**

execute T_{ik} : < **For** p := 1 **to** k-1 **do** $a_{ik} := a_{ik} - a_{ip} * r_p$

$a_{ik} := a_{ik} / a_{kk}$ >

T_{kk} : 3(k-1) arithmetic operations $1 \leq k \leq n$

T_{ik} : 2k-1 arithmetic operations $k+1 \leq i \leq n,\ 1 \leq k \leq n$

Sequential execution time: $n^3/3 + O(n^2)$.

The precedence constraints and the precedence graph are the same as for the non-optimized LDL^t decomposition.

(J) {Cholesky decomposition GG^t [GVL83] }

For k=1 **to** n **do**

execute T_{kk} : < **For** p := 1 **to** k-1 **do** $a_{kk} := a_{kk} - a_{kp} * a_{kp}$

$a_{kk} := \mathrm{sqrt}(a_{kk})$ >

For i := k + 1 **to** n **do**

execute T_{ik} : < **For** p := 1 **to** k-1 **do** $a_{ik} := a_{ik} - a_{ip} * a_{kp}$

$a_{ik} := a_{ik} / a_{kk}$ >

T_{kk} : 2k-1 arithmetic operations $1 \le k \le n$

T_{ik} : 2k-1 arithmetic operations $k+1 \le i \le n$, $1 \le k \le n$

Sequential execution time: $n^3/3 + O(n^2)$, assuming that a square root takes a unit time.

The precedence constraints and the precedence graph are the same as for the non-optimized LDL^t decomposition.

These ten algorithms lead to four tasks graphs, with differences in the execution times of the tasks. We refer to [CMRT88] for a detailed analysis of these graphs. We shall only in this paragraph recall the main properties of two of them:

- 'greedy graph': the task graph of algorithm (D)
- '2-steps graph': the task graphs of algorithms (A), (B, (H), (I), (J).

Note that we have already encountered the greedy graph as the task graph of the Jordan method by columns and the 2-steps graph as the task graph of the standard Gaussian elimination algorithm. However, the differences on the execution times of the elementary tasks make difficult the use of the results obtained in these two cases.

Greedy graph

The general form of the greedy graph is the following:

For k=1 **to** n-1 **do**

  For j := k + 1 **to** n **do** execute T_{kj}

We assume that the elementary task T_{kj}, $k+1 \le j \le n$, $1 \le k \le n$, can be processed in bt_k time units, where b is an integer and $t_k = k$, for all k, or $t_k = n - k$, for all k. The precedence relations are

(A) $T_{k,k+1} << T_{k+1,j}$ $k+2 \le j \le n$, $1 \le k \le n-1$

(B) $T_{kj} << T_{k+1,j}$ $k+2 \le j \le n$, $1 \le k \le n-1$

For a given value of k, there are n - k tasks to be executed. Call level k the set of these tasks.

The *Greedy Scheduling Algorithm GSA* executes the tasks from one level of the task graph to another, from left to right in each level, and it starts the maximum of tasks at each time (hence the name of greedy). More precisely the algorithm *GSA* executes the tasks in the following order:

$$T_{12} << T_{13} << \ldots << T_{1n} << T_{23} << T_{24} << \ldots << T_{2n} << T_{34} << \ldots << T_{n-1,n}$$

Theorem 5 : The *Greedy Scheduling Algorithm GSA* is asymptotically optimal. Its execution time is equal to:

$$T_{GSA} = (b/p) \sum_{k=1,n-p} (n - k)\, t_k + b \sum_{k=n-p+1,n-1} t_k + O(n)$$

With $p = \alpha\, n$ processors, the execution times are respectively:

- $t_k = k$, then $T_{GSA} = b\, n^2\, [(1 - \alpha)/(6\alpha) + 1/2] + O(n)$
- $t_k = n - k$, then $T_{GSA} = b\, n^2\, [(2 + \alpha^3)/(6\alpha)] + O(n)$

2-steps graph

The general form of the 2-step graph is the following:

For k=1 **to** n-1 **do**

  For i := k + 1 **to** n **do** execute T_{ki}

  execute $T_{k+1,k+1}$

We assume that the elementary task T_{ki}, $k+1 \le i \le n$, $1 \le k \le n$, can be processed in bt_k time units and $T_{k+1,k+1}$ in at_k time units, where a and b are integers and $t_k = k$, for all k, or $t_k = n - k$, for all k. The precedence relations are

(A) $T_{kk} << T_{kj}$ $k+1 \le i \le n$, $1 \le k \le n-1$

(B) $T_{ki} << T_{k+1,i}$ $k+1 \le i \le n$, $1 \le k \le n-1$

We have already encountered the 2-steps graph in the study of the standard gaussian elimination, which corresponds to $a = b = 2$ and $t_k = n - k$.

Theorem 6: The *Greedy Scheduling Algorithm GSA* is asymptotically optimal. Its execution time is equal to:

$$T_{GSA} = (b/p) \sum_{k=1,n-p} (n - k)\, t_k + b \sum_{k=n-p+1,n-1} t_k + O(n)$$

With $p = \alpha n$ processors, the execution times are respectively:

- $t_k = k$, then $T_{GSA} = b\, n^2 [(1 - \alpha)/(6\alpha) + 1/2] + O(n)$
- $t_k = n - k$, then $T_{GSA} = b\, n^2 [(2 + \alpha^3)/(6\alpha)] + O(n)$

7. COMMUNICATION COMPLEXITY

In this section, we analyse the influence of communication costs on the computational complexity of parallel Gaussian elimination algorithms. In earlier publications on parallel algorithms, the communications costs were often not taken into account. However, in some cases, the time for transmitting one data item is not negligible relative to the time for performing a floating point operation. In [Saa85], the communication complexity of the Gaussian elimination algorithm is studied on three parallel architectures: processors connected by a bus, a ring and a two dimensional grid array. It is shown that on the bus and the ring topologies the algorithm requires a communication time of at least $O(n^2)$ no matter what the number of processors is. On the grid, the total time including arithmetic can be reduced to $O(n^{5/3})$, and the communication time is at least $O(n^2/\sqrt{p})$. In [ISS85] a similar analysis has been presented on a ring consisting of a small number of processors relative to the size of the problem. In this case it was shown that the communication time is a low order term as compared with the computation time.

The target architecture is assumed to be a shared memory MIMD computer, as defined previously. For the sake of simplicity we assume that a processor can begin a task only when all its data are stored in its local memory, but the results of the computation can be stored in the central memory as soon as they are available. Hence the reads precede the processes but the writes overlap the processes. These are the only forms of hardware synchronization that we shall assume in the remaining. The cost analysis of the algorithm is based on the assumptions that each processor can perform any of the four arithmetic operations in one computation unit, denoted t_{exec}, and that there are no memory conflicts when accessing the shared memory. The time for transfering a data from the central memory to the individual memory of a processor will be called t_{com}. Without loss of generality, we can assume that t_{exec} is constant for the four arithmetic operations. On the contrary t_{com} is machine dependent: in the remaining we shall assume that it is an increasing function of the size of the problem and the number of processors.

We shall concentrate our study on the form KJI-SAXPY modified of the generic Gaussian elimination algorithm (KJI-algorithm for short) since, among the ten parallel versions analysed in the preceding paragraph, this form is the most suitable for parallelization on an MIMD computer with zero communication cost. Each elementary task T_{kj} in KJI is a linear combination of two columns. The computation cost and communication cost of T_{kj} are respectively $(2(n-k)+1)t_{exec}$ and $(3(n-k)+3)t_{com}$.

Note that these new assumptions imply no modification on the precedence constraints and the precedence graph. The only difference is the increase in the total execution time of the elementary tasks. Hence we could apply theorem 5 which states that the greedy scheduling algorithm is asymptotically optimal.

Theorem 7: Including communication costs, the greedy scheduling algorithm is asymptotically optimal for the pointwise KJI-algorithm. The optimal execution time is equal to:

$$t_{point} = t_{exec} [2n^3/(3p) + p^2/3] + t_{com} [n^3/p + p^2/2]$$

Hence taking into account the communication costs, the complexity is increased by t_{com} $[n^3/p + p^2/2]$. If we consider that t_{com} is equal to t_{exec}, then the complexity is more than twice the zero communication case. Hence, communications costs can no longer be neglected on such a model. In order that arithmetic could dominate communication, we must increase the granularity of the elementary task. This is done by considering block methods.

Block KJI-Algorithm

Let n=qr. The matrix A is decomposed in q^2 square blocks of size r. The form KJI - SAXPY modified of the block generic Gaussian elimination algorithm is the following:

{Block KJI - algorithm}

For k := 1 to q **do**

  For j := k+1 **to** q **do**

    execute $\mathbf{TB}_{kj}$: < $A_{kj} = A_{kk}^{-1} * A_{kj}$

      For i := k+1 **to** q **do** $A_{ij} := A_{ij} - A_{ik} * A_{kj}$ >

Of course here, the inverse of the diagonal blocks A_{kk} are not directly computed. We solve r linear systems using pointwise Gaussian elimination. In order to derive the computation cost and communication cost of TB_{kj} we must make some new assumptions. Since we are interested in asymptotic analysis and we want to increase the granularity of TB_{kj} we assume that r and q go to infinity with n (recall that rq=n). Processing block methods implies some modifications on the underlying parallel architecture. In particular the local memory of each processor must be big enough in order to store two-dimensional arrays instead of vectors. Moreover the number of processors will be smaller than in the case of the pointwise algorithm. Clearly the computation time unit t_{exec} remains the same. In the contrary since the communication time unit depends on the number of processors, the communication time will be smaller, but still denoted by t_{com}.

The computation of $A_{kj} = A_{kk}^{-1} * A_{kj}$ is equivalent to solve r systems of size r with the same matrix. Neglecting low order terms it requires $(8r^3/3)t_{exec}$ units (see [GVL83]). The computation of $A_{ij} := A_{ij} - A_{ik} * A_{kj}$ corresponds to (q-k) matrix products and matrix additions, hence leading to $(2r^3(q-k))t_{exec}$ units. Then performing TB_{kj} requires $2r^3[(q-k)+4/3]\ t_{exec}$ units. These computations require the knowledge of A_{kk}, A_{kj}, A_{ij} and A_{ik} for i=k+1 to q. Hence the communications costs are $3r^2[(q-k)+1]t_{com}$ units.

Let us call t_{block} the optimal execution time of the block KJI-algorithm. Since the precedence constraints for TB_{kj} are the same as for the pointwise KJI-algorithm, the preceding analysis applies.

Theorem 8: If $n = qr$ and $p \leq q$, and if q and r go to infinity with n, then the greedy scheduling algorithm is asymptotically optimal for the block KJI-algorithm. The optimal time is given by:

$$t_{block} = t_{exec}\,[\,2n^3/(3p) + p^2r^3/3\,] + t_{com}\,[\,n^3/(pr) + p^2r^2/2\,]$$

The adaptive KJI-algorithm

In order to take full advantages of both algorithms, we must construct an algorithm which keeps efficiency one as long as the pointwise algorithm but preserves a level of granularity large enough to maintain low communications costs. This can be done by reducing gradually the blocks size, hence adapting the size of the blocks to the number of processors and the size of the remaining matrix. Such adapting stategies have been advocated recently for matrix factorizations [BL90], [Bis89].

Let p and n be given and set $q=p+1$. At the first step, we set $s_1=n$ and $r_1= s_1/q$ (take the ceiling function if necessary). We use the block-KJI algorithm to a matrix of size s_1 decomposed in q^2 blocks. Hence only one processor could have a task of length smaller than the others, which is negligible since p goes to infinity with n. At the second step, we set $s_2=s_1-r_1$ and $r_2= s_2/q$ and apply the second step of the block-KJI algorithm to a matrix of size s_2 decomposed in q blocks. We repeat this process as long as the size s_k of the remaining matrix is greater than q. Once $s_k<q$, we use the pointwise KJI-algorithm. The complete description of the algorithm and a detailed complexity analysis can be found in [CT89]. The main result is the following:

Theorem 9: If p and n/p go to infinity with n, then the asymptotic execution time of the adaptive KJI-algorithm is equal to:

$$t_{adap} = t_{exec}\, 2n^3/(3p) + t_{com}\, 3n^2/2$$

Comparisons - Influence of communications

Recall that:

$$t_{point} = t_{exec}\, [\, 2n^3/(3p) + p^2/3\,] + t_{com}\, [\, n^3/p + p^2/2\,]$$
$$t_{block} = t_{exec}\, [\, 2n^3/(3p) + p^2r^3/3\,] + t_{com}\, [\, n^3/(pr) + p^2r^2/2\,]$$
$$t_{adap} = t_{exec}\, 2n^3/(3p) + t_{com}\, 3n^2/2$$

We deduce that:

- the adaptive algorithm is always more efficient than the block algorithm: the former has low communication cost due to the use of block decompositions without increasing the computation cost, by adapting dynamically the size of the decompositions

- the adaptive algorithm is always more efficient than the pointwise algorithm: the computation is of the same order while the use of block decomposition allows lower communication cost.

For purpose of illustration, let us consider $r = q = p = n^{1/2}$. We obtain:

$$t_{point} = t_{exec}\, [\, (2/3)n^{5/2}\,] + t_{com}\, [\, n^{5/2}\,]$$
$$t_{block} = t_{exec}\, [n^{5/2}\,] + t_{com}\, [\, (3/2)n^2\,]$$
$$t_{adap} = t_{exec}\, [\, (2/3)n^{5/2}\,] + t_{com}\, [\, (3/2)n^2\,]$$

We now study briefly the influence of the architecture on the execution cost. Without loss of generality and in order to simplify the presentation, we take $t_{exec} = 1$. Determining the possible values of t_{com} is difficult since it depends strongly on the architecture, more precisely on the interconnection network betweeen the shared memory divided in several banks and the processors. Hence t_{com} is an increasing function of the size of the memory and of the number of processors. A complete network is composed of p.b switches, where b is the number of banks, a Benes network is composed of 2p.Logp (b=p in this case) and an Omega network possesses half this number but may require two passes through the network for some permutations [HB84]. Assuming a constant time for a single switch, we deduce that t_{com} can be of the generic following forms:

$$t_{com} = v \qquad t_{com} = v.f(n) \qquad t_{com} = v.g(n,p)$$

where v stands for a constant independent of n and p and f and g are increasing functions of their variables.

Assuming first a constant communication unit, for the pointwise algorithm communication and computation costs are of the same order, whereas for that block and the adaptive algorithms the communication cost is negligible relative to the computation cost. Taking for instance $r = q = p = n^{1/2}$, we obtain:

$$t_{point} = ((2/3) + v)\, n^{5/2} \qquad t_{block} = n^{5/2} \qquad t_{adap} = (2/3)n^{5/2}$$

which clearly demonstrates the advantage of the adaptive algorithm on the classical one.

For $t_{com} = v.f(n)$ or $t_{com} = v.g(n,p)$, the communication overcome the computation in the pointwise algorithm. Since the adaptive algorithm is always more efficient than the block algorithm, we only study the former: if $t_{com} = v.g(n,p)$, then $t_{adap} = 2n^3/(3p) + 3vn^2/(2g(n,p))$. Hence:

- if $p = O(p/g(n,p))$, communication and computation times are of the same order
- if $p > O(p/g(n,p))$, the communication time overcomes the computation time
- if $p < O(p/g(n,p))$, the computation time overcomes the communication time.

Hence the execution time is not stricly decreasing as p increases: there exist optimal values of p. Taking for instance $g(n,p) = (np)^{1/2}$, we obtain:

- for the pointwise algorithm, $popt = n / (5)^{1/3}$
- for the adaptive algorithm, $popt = (64n / (81v^2))^{1/3}$

8. COMPLEXITY OF GIVENS METHOD

Computing the orthogonal decomposition of a mxn matrix is a classical problem in scientific computation. Two well known methods are available for solving such a problem: Householder reduction and Givens rotations. The parallelization of Givens' method gives rise to very interesting algorithmic problems. Such a factorization requires mn - n(n+1)/2 steps on a sequential machine, each step being the time necessary to achieve a Givens rotation. This O(m*n) number of steps to obtain the factorization motivates its parallelization.

Throughout the paper, A is an m by n rectangular matrix. We let R(i,j,k), $i \neq j$, $1 \leq i,j \leq m$ and $1 \leq k \leq n$, denote the rotation in plane (i,j) which annihilates the element A(i,k). The sequential method of orthogonal factorization of A eliminates the subdiagonal elements of A one at a time. The elimination process is performed sequentially by applying Givens plane rotation to A in such a way that the previously introduced zeros are not destroyed. For each column j of A, n - j rotations are required. This can be accomplished by the following algorithm:

{Sequential GIVENS}

For j := 1 **to** n - 1 **do**

  For i := j + 1 **to** n **do** execute R(i, j, j)

In a parallel implementation of the fast Givens method more than one plane rotation could be applied concurrently. Sameh and Kuck [SK78] described details of such schemes which assume that $p = O(n^2)$. They constructed an algorithm based on square root free Givens rotations with time (8n - 7) using $p = n^2$ processors. For the standard method they proposed another parallel algorithm with time (10n - 15) using $3n^2/2 + O(n)$ processors. Daoudi [Dao89] improved this result by designing a parallel algorithm with time (8n - 12) using $p = 3n^2/4 + O(n)$ processors. He also presented results on various parallel architectures.

We shall present in detail the case p = O(n) and concentrate on synchronous parallel algorithms: we call time step the duration of a Givens rotation and assume that such a transformation takes a time independent of the size of the vectors.

Determination of the optimal algorithm

Clearly at each step no more than [m/2] elements can be simultaneouly annihilated, where [.] stands for the floor function. A rather complete analysis of the case p = [m/2] can be found in the literature. In this section we shall make a review of the existing results. An algorithm M requiring T steps to achieve the Givens factorization of A is represented by M = (M(1),...,M(T)) where for $1 \le t \le T$, M(t) is a group of r(t) disjoint rotations ($r(t) \le p$), which can be performed simultaneously. M is used in order to construct a sequence A(t) such that : A(0)=A and for $1 \le t \le T$, A(t) is obtained by applying in parallel the rotations in M(t) to A(t-1). For short we use (M,T,R) where R = r(1)+...+r(T).

To describe informally the parallel algorithms, let us fix m = 13 and n = 6. The table of figure 8 illustrates Sameh and Kuck's annihilation scheme [SK78]. An integer r is entered when zeros are created at the r-th step. We do not specify completely each rotation. A zero can be created in a row using any other row with the same number of annihilated elements. For instance at step 3, it can perform simultaneously the rotations R(11,10,1) and R(13,12,2). R(13,12,2) is the only possible choice, but R(11,10,1) can be replaced by R(11,x,1) for any $x \le 10$. Sameh and Kuck's scheme is easy to program and to analyze: a possible choice is to use R(i,i-1,k) in order to annihilate the element in position (i,k). Clearly, the total number of steps is m+n-2 if m>n (17 in the example) and 2n-3 otherwise.

Cosnard and Robert [CR86] and independently Modi and Clarke [CR84] have introduced a greedy algorithm which performs at each step as many rotations as possible, annihilating the elements in each column from bottom to top and in each row from left to right. This scheme is depicted in the second table of figure 1: it begins very fast since it performs 6 rotations from step 1 to step 2, 5 rotations from step 4 to step 8, 4 rotations at the steps 9 and 10, but it terminates slowly with only 2 rotations at the steps 11 and 12 and one at the last two steps.

*					
12	*				
11	13	*			
10	12	14	*		
9	11	13	15	*	
8	10	12	14	16	*
7	9	11	13	15	17
6	8	10	12	14	16
5	7	9	11	13	15
4	6	8	10	12	14
3	5	7	9	11	13
2	4	6	8	10	12
1	3	5	7	9	11

Sameh and Kuck's

*					
4	*				
3	6	*			
3	5	8	*		
2	5	7	10	*	
2	4	7	9	12	*
2	4	6	8	11	14
1	3	6	8	10	13
1	3	5	7	9	12
1	3	5	7	9	11
1	2	4	6	8	10
1	2	4	6	8	10
1	2	3	5	7	9

Greedy

Figure 8: Annihilation schemes

Numerical experiments show that the greedy algorithm takes appreciably fewer stages than Sameh and Kuck's scheme, which is confirmed by an asymptotic theoretical analysis: Modi and Clarke obtain the approximation

$$\log m + (n-1)\log\log m$$

which is valid if m goes to infinity, n fixed (when not stated explicitly the basis of the logarithms will be 2). In their paper, Modi and Clarke introduce another class of algorithms, namely the Fibonacci schemes, and discuss in detail their performances. However, they observe that these schemes seem to be less efficient than the greedy algorithm.

The Fibonnacci annihilation scheme of order 1 is derived as indicated in figure 9. We fill up the first column from bottom to top. There is a single zero, below are $u_2=2$ successive copies of -1, then $u_3=3$ copies of -2, and so on (there are $u_k=u_{k-1}+1$ values of $-(k-1)$, with $u_1=1$). The second column is like the first except that all the entries are increased by 2, and the whole column moved down one place. Other columns are filled up using the same rule. Now, to get an annihilation scheme, just add u+1 to all integers in the lower part of the matrix, where u is the integer in the left hand bottom corner. It is not straightforward to ensure that this scheme is actually a Givens annihilation scheme: we would need to check carefully whether it is possible to annihilate the elements of the matrix in such a way. However, this can be done owing to the very systematic construction of the scheme.

*							*					
0	*						5	*				
-1	2	*					4	7	*			
-1	1	4	*				4	6	9	*		
-2	1	3	6	*			3	6	8	11	*	
-2	0	3	5	8	*		3	5	8	10	13	*
-2	0	2	5	7	10		3	5	7	10	12	15
-3	0	2	4	7	9		2	5	7	9	12	14
-3	-1	2	4	6	9		2	4	7	9	11	14
-3	-1	1	4	6	8		2	4	6	9	11	13
-3	-1	1	3	6	8		2	4	6	8	11	13
-4	-1	1	3	5	8		1	4	6	8	10	13
-4	-2	1	3	5	7		1	3	6	8	10	12

Figure 9: Fibonnacci scheme of order 1 (Add 5 to get the annihilation scheme)

The Fibonacci scheme of order 2 is obtained in a similar way, replacing the relation $u_k=u_{k-1}+1$ by $u_k=u_{k-1}+u_{k-2}+1$ (with $u_1=1$, $u_2=2$), and adding 3 instead of 2 to the entries of column j to get the entries of column j+1. The number of steps for the Fibonnacci scheme of order 1 is equal to k+1+2(n-1), where k is the least integer such that $k(k+1)/2 \geq m-1$ (leading to 15 in our example). The number of steps for the scheme of order 2 is 19 in our example. Rather than providing the exact expression (there is no simple formula), we recall the asymptotic value $\log_\alpha m + 3n$, where $\alpha = (1+\sqrt{5})/2$ given in [MC84].

All the algorithms presented by Sameh and Kuck [SK78] and Modi and Clarke [MC84] are Givens sequences, that is sequences of Givens rotations in which zeros once created are preserved, i.e. for any couple of distinct rotations R(i,j,k) and R(i',j',k') in M we have $(i,k) \neq (i',k')$.. The question whether temporarily annihilating elements and introducing zeros that will be destroyed later on can lead to any additional parallelism is a nice example of specific questions that parallelism can rise.

In the remaining of this section, we describe the results obtained by Cosnard and Robert [CR86] and Cosnard, Muller and Robert [CMR86]. (M,T,R) is called a Standard Parallel Givens Sequence (SPGS for short) if it is a Givens sequence which reduces A to upper triangular form and annihilates elements from left to right and from bottom to top, i.e. if R(i,j,k) â M(t) and R(i',j',k') â M(t') where $t \leq t'$, then $k<k'$ or ($k=k'$ and $i \geq i'$). Clearly, all the algorithms introduced so far are SPGS, in particular the Greedy algorithm.

After showing that in the set of optimal algorithms there exists a SPGS, Cosnard and Robert [CR86] prove that, for any value of m and n (with $m \geq n$), and for p =[m/2], the greedy SPGS is an optimal algorithm. Hence the complexity of the QR decomposition using plane rotations is the number of steps OPT(m,n) of the greedy algorithm and the asymptotic efficiency ASE(m,n) could be easily deduced.

Theorem 10: Let p = [m/2].

1. If n = m (A is a square matrix), then OPT(m,n) = 2n + o(n) and ASE(m,n) = 1/2.
2. If n is fixed, then $OPT(m,n) = \log_2 m + o(\log_2 m)$ and ASE(m,n) = 2n/logm.

3. If $m = o(n^2)$, then $OPT(m,n) = 2n + o(n)$ and $ASE(m,n) = 1$.
4. If $m = o(n^k)$, with $k \geq 3$, then $2n + o(n) \leq OPT(m,n) \leq 3n + o(n)$ and $ASE(m,n) \geq 2/3$.

An interesting consequence [CR86], [CMR86] of this asymptotic analysis is that it allows to select among various parallel algorithms those which are aymptotically optimal. This is important since the computation of the indices of the rows involved in plane rotations at a given step of the greedy algorithm is somewhat difficult. Hence it must be avoided. This can be done by replacing the greedy SPGS by another algorithm which, although not optimal, will have good performances and could be implemented in a simple way.

Corollary 2: Let $p = [m/2]$.
1. If $m = n$, then Sameh and Kuck's scheme is asymptotically optimal.
2. If $m = o(n^2)$, then the Fibonacci scheme of order 1 is asymptotically optimal.
3. If $m = o(n^k)$ and $m \geq n2$, then the Fibonacci scheme of order 2 is of asymptotic efficiency 2/3.

When n is fixed and m goes to infinity, none of the previous algorithms but the greedy is asymptotically optimal.

Reduction of the number of processors

In this section, we shall limit the number of plane rotations which can be performed simultaneously to a given integer $p \leq [m/2]$. A detailed presentation can be found in [CD89]. The first following results are straightforward extensions of results presented above. For any algorithm (M,T,R), call r(i,t) the number of zeros in column i introduced at time t and s(i,t) the total number of zeros in column i after step t. Hence for Givens Sequences we have $s(i,t) = r(i,1) + \ldots + r(i,t)$. A simple proof enables us to characterize the SPGS among the Givens Sequences and to show that there exists an optimal SPGS:

Lemma 1: Let $p \leq (m/2)$ and M be a Givens Sequence. M is a SPGS if and only if :

* $s(0,t) = m$ for $t \geq 0$ * $s(i,0) = 0$ for $1 \leq i \leq m$

* $r(i,t) \leq \min \{ p - \sum_{j=1,\ldots,i-1} r(j,t) \ ,\ [(s(i-1,t-1) - s(i,t-1))/2] \}$ for $i,t \geq 1$.

Lemma 3 means that the total number of zeros introduced at time t cannot exceed p and that, just as before, no more zeros than half the difference between the number of zeros in column i and i-1 at step t-1 can be created in column i at step t. We shall see later on that these constraints make the analysis of the problem much more difficult.

Lemma 2: In the set of optimal algorithms there exists a SPGS.

A great difference with the case p=[m/2] is the fact that there is not a unique greedy SPGS. The greedy algorithms can be unformally described in the following manner: at each step do the maximum number of rotations.

Definition: Let M be a SPGS. M is a greedy algorithm if and only if:

$$\sum_{j=1,\ldots,n} r(j,t) = \min \{ p , \sum_{j=1,\ldots,n} [(s(j-1,t-1) - s(j,t-1))/2] \}$$

Clearly this can be achieved with different strategies. To illustrate this fact we present two algorithms: the vertical greedy, which fills up the columns, and the horizontal greedy, which fills up the rows.

Vertical greedy:

- $svg(0,t) = m$ for $t \geq 0$
- $svg(i,0) = 0$ for $1 \leq i \leq m$
- $rvg(i,t) = \min \{ p - \sum_{j=1}^{i-1} rvg(j,t) , ((svg(i-1,t-1) - svg(i,t-1))/2) \}$ for any $i,t \geq 1$.

Horizontal greedy:

- $shg(0,t) = m$ for $t \geq 0$
- $shg(i,0) = 0$ for $1 \leq i \leq m$
- $rhg(i,t) \leq \min \{ p - \sum_{j=i+1}^{n} rhg(j,t) + ((shg(i-1,t-1) - shg(i,t-1))/2) \}$ for any $i,t \geq 1$.

The problem of optimality will be solved in the following sections. Remark that none of the two preceding greedy algorithms is optimal. However for particular choices of m, n or p both can be optimal.

Theorem 11: Given m, n and p, there exists an optimal algorithm in the set of greedy SPGS.

From the theorem, it can be easily seen that, although there exists an optimal greedy algorithm for any m, n and p, we do not know how to construct it. Figure 10 shows that this optimal greedy SPGS can be different from the vertical or the horizontal greedy. We choose m=n=12 and p=3.

*											
5	*										
4	8	*									
3	7	11	*								
3	7	10	14	*							
3	6	10	13	16	*						
2	6	9	12	15	18	*					
2	6	9	12	15	17	20	*				
2	5	9	12	14	17	19	22	*			
1	5	8	11	14	16	18	21	23	*		
1	4	8	11	13	16	18	20	22	24	*	
1	4	7	10	13	15	17	19	21	23	25	*
	*										

Vertical greedy

*											
15	*										
13	16	*									
11	14	17	*								
5	12	15	18	*							
4	7	13	16	19	*						
3	6	9	14	17	20	*					
2	5	8	10	15	18	21	*				
2	4	7	9	11	16	19	22	*			
1	3	6	8	10	12	17	20	23	*		
1	3	5	7	9	11	13	18	21	24	*	
1	2	4	6	8	10	12	14	19	22	25	

Horizontal greedy

*											
7	*										
6	10	*									
5	9	13	*								
4	8	12	15	*							
4	7	11	14	17	*						
2	6	10	13	16	19	*					
2	5	9	12	15	18	20	*				
2	4	8	11	14	17	19	21	*			
1	3	7	10	13	16	18	20	22	*		
1	3	6	9	12	15	17	19	21	23	*	
1	3	5	8	11	14	16	18	20	22	24	*

Better greedy (optimal, see later)

Figure 10: Various greedy algorithms

The remaining of this paper is devoted to the construction an asymptotically optimal algorithm with p processors for square matrices. First of all, we shall derive a lower bound for the complexity of the parallel Givens factorization in the case m=n. Recall that $T_{opt}(p)$ is the optimal time for computing the Givens factorization with p processors and T_{opt} the optimal time when an unlimited number of processors is available (in fact we deduce from the preceding section that $T_{opt}=T_{opt}(p)$ for $p \geq (n/2)$). The following relations are straightforward:

$$* \; T_{opt} = T_{opt}\left(\left(\frac{n}{2}\right)\right) \leq T_{opt}(p) \text{ for all } p \qquad * \; T_{opt} = 2n - o(n)$$

Theorem 12: For $1 \leq p \leq n/2$, we have: $B_{inf}(p) = \frac{n(n-1)}{2p} + p - 1 \leq T_{opt}(p)$.

The proof is based on the use of Corollary 1. It can be found in [CDR89]. A direct consequence of the previous theorem is that the optimal number of processors in order to obtain the optimal time is bounded below by the value of p for which $B_{inf}(p)=2n$.

Corollary 3: Let p_{opt} be the minimum number of processors for computing the Givens factorization of a dense square matrix of size n in T_{opt}. Then asymptotically we have:

$$\frac{n}{2+\sqrt{2}} \leq p_{opt} \leq \frac{n}{2}$$

In fact, it can be proved that these bound are tight [CDR89].

Theorem 13: 1. For $p=\frac{n}{2+\sqrt{2}}$ we have: $T_{opt}(\frac{n}{2+\sqrt{2}}) = 2n + o(n)$.

2. The minimum number of processors in order to obtain the Givens decomposition of a square matrix of size n in asymptotically optimal time, $T_{opt}=2n-o(n)$, is equal to:

$$p_{opt} = \frac{n}{2+\sqrt{2}} + o(n)$$

3. For $1 \leq p \leq \frac{n}{2+\sqrt{2}}$, $T_{opt}(p) = \frac{n^2}{2p} + p + o(n)$.

The proof is long and tedious. We shall only illustrate it by describing briefly the optimal algorithm $p=\frac{n}{4}$.

Corollary 4: For $p=\frac{n}{4}$ we have: $T_{opt}(\frac{n}{4}) = \frac{9n}{4} + o(n)$.

Proof: We partition the matrix into five regions as depicted on figure 11. The algorithm which we construct proceeds in four sequential phases.

First phase: The elements in regions I and I' are annihilated in full asymptotic efficiency. The execution time is $T_1(\frac{n}{4})=\frac{n}{2} + o(n)$.

Second phase: The elements in region II are annihilated in full efficiency. The execution time is equal to: $T_2(\frac{n}{4}) = \frac{n}{2}$.

Third phase: We use a cyclic elimination to annihilate the elements in region III in full efficiency. The execution time is equal to: $T_3(\frac{n}{4})=\frac{3n}{4}$.

Fourth phase: The elements in region IV are annihilated in optimal time. The execution time is equal to: $T_4(\frac{n}{4})=\frac{n}{2} -1$.

The execution time of the algorithm is the sum of the execution times of the four phases:

$$T_1(\frac{n}{4})+T_2(\frac{n}{4})+T_3(\frac{n}{4})+T_4(\frac{n}{4}) = \frac{9n}{4} + o(n) = B_{inf}(\frac{n}{4})+o(n).$$

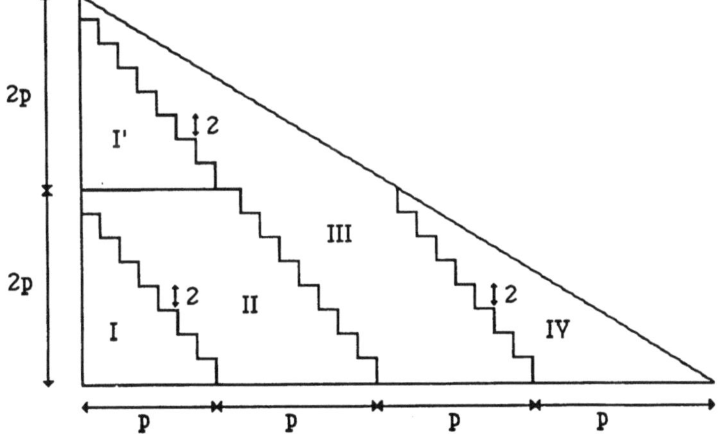

Figure 11: Decomposition of the matrix into five regions

9. CONCLUDING REMARKS

We have surveyed some of the results obtained in the last decade on the design and analysis of parallel factorization algorithms for solving dense linear systems of equations. After defining a model of computation, we have introduced a graph-theoretic approach which allows to separate the design of a parallel algorithm into two independent parts

- construction of a task system, determination of the precedence constraints and of the task graph associated to a sequential algorithm
- construction of scheduling algorithms compatible with the task graph.

We presented some theoretical tools for analysing the complexity of the various parallel algorithms associated to a sequential algorithm. We then applied these tools to Gauss-Jordan method, several Gaussian elimination algorithms, Cholesky decomposition and Doolittle reduction and QR factorization. The methodology described in this paper is clearly more general and has been applied successfully to various algorithms in numerous domains.

Acknowledgments

This work has been supported by the project C3 of the French Council for Research CNRS and of the French Ministry of Research, and by the Direction des Recherches et Etudes Techniques of the Ministry of Defense.

The author would like to thank the co-authors of the published material that has been used in this paper, namely E.M. Daoudi, M. Marrakchi, Y. Robert and D. Trystram.

REFERENCES

[BL90] BISCHOF, C.H. and LACROUTE P.G., An adaptive blocking stategy for matrix factorizations, Argonne National Laboratory MCS-P151-0490 (1990)

[Bis89] BISCHOF, C.H., Adaptive blocking in the QR factorization, The journal of Supercomputing, 3 (3) (1989), 193-208

[Bis90] BISCHOF, C.H., Fundamental linear algebra computations on high-performance computers, Argonne National Laboratory MCS-P150-0490 (1990)

[CG89] CHU, E. and GEORGE, A., QR factorization of a dense matrix on a shared-memory multiprocessor, Parallel Computing 11 (1989), 55-71

[CD72] COFFMANN E.G., DENNING P.J., Operating systems theory, Prentice Hall (1972)

[CDR89] COSNARD M., DAOUDI M. and ROBERT Y., Complexity of the parallel Givens factorization on shared memory architectures, LNCS 401, Springer (1989) 86-105

[CMR86] COSNARD M., MULLER J.M., ROBERT Y., Parallel QR decomposition of a rectangular matrix, Numerische Mathematik 48, 239-249 (1986)

[CMRT88] COSNARD, M., MARRAKCHI, M., ROBERT, Y., TRYSTRAM, D., Parallel Gaussian elimination on an MIMD Computer, Parallel Computing 6 (1988), pp. 275-296

[CR86] COSNARD M., ROBERT Y., Complexity of parallel QR decomposition, JACM 33 (4), 712-723 (1986)

[CR87] COSNARD M., ROBERT Y., Parallel algorithms : a study of complexity, Technique et Science Informatiques 6, 5 (1987), 337-348

[CT89] COSNARD, M. and TRYSTRAM, D., Communication Complexity of Gaussian elimination on MIMD Shared Memory Computers, Rev. Mat. Apl. 10 (1989), 27-48

[Dao89] DAOUDI M., Etude de la complexité de la décomposition orthogonale d'une matrice sur plusieurs modèles d'architectures parallèles, Thèse de l'Institut National Polytechnique de Grenoble (1989)

[DGK84] DONGARRA J.J., GUSTAVSON F.G. and KARP A., Implementing linear algebra algorithms for dense matrices on a vector pipeline machine, SIAM Rev. 26 (1), (1984), 91-112

[DSS86] DONGARRA J.J., SAMEH A. and SORENSEN D., Implementation of some concurrent algorithms for matrix factorization, Parallel Computing, 3(1) (1986), 25-34

[Fly72] FLYNN M.J., Some Computer Organizations and Their Effectiveness, IEEE TC 21, (1972), 948-960

[GHL] GEORGE, A., HEATH, M.T. and LIU, J., Parallel Cholesky factorization on a shared memory processor, Lin. Algebra Appl., 77 (1986), 165-187

[Gen73] GENTLEMAN W.M., Least squares computations by Givens transformations without square roots, J. Inst. Maths. Applics. 12, 329-336 (1973)

[GN89] GERASOULIS, A. and NELKEN, I., Static scheduling for linear algebra DAGs, HCCA4 (1989)

[GVL83] GOLUB G.H. ,VAN LOAN C.F. , Matrix Computations, The John Hopkins Univ. Press, 1983

[Gon77] GONZALES M.Y., Deterministic processor scheduling, ACM Comp. Surveys, 9, 3, (1977), 173-204

[Hel78] HELLER D., A survey of parallel algorithms in numerical linear algebra, SIAM Review 20, 740-777 (1978)

[HJ81] HOCKNEY R.W., JESSHOPE C.R., Parallel computers: architectures, programming and algorithms, Adam Hilger Ldt., Bristol (1981)

[HB84] HWANG K. and BRIGGS F., Parallel processing and computer architecture, Mac Graw Hill (1984)

[ISS85] IPSEN, I.C.F., SAAD, Y.,SCHULTZ, M.H., Complexity of dense linear system solution on a multiprocessor ring, Report DCS-349 Yale University, (1985)

[KK84] KOWALIK J.S. and KUMAR S.P., Parallel factorization of a positive definite matrix on an MIMD computer, Proc. ICCD 84 (1984), 410-416

[Kum82] KUMAR S.P., Parallel algorithms for solving linear equations on MIMD computers, PhD. Thesis, Washington State University (1982)

[Liu86] LIU, J.W.H., Computational models and task scheduling for parallel sparse Choleski factorization, Parallel Computing 3 (1986), 327-342

[LKK83] LORD R.E., KOWALIK J.S., S.P. KUMAR, Solving linear algebraic equations on an MIMD computer, JACM 30, (1983), 103-117

[MR89] MARRAKCHI, M. and ROBERT, Y., Optimal algorithms for Gaussian elimination on an MIMD Computer, Parallel Computing 12, 2 (1989), 183-194

[MC84] MODI J.J., CLARKE M.R.B., An alternative Givens ordering, Numerische Mathematik 43, (1984), 83-90

[OV85] ORTEGA, J.M. and ROMINE, C.H., The ijk forms of factorization methods II, parallel systems, Parallel Computing 7 (1988) 149-162

[OV85] ORTEGA, J.M., VOIGT, R.G., Solution of partial differential equations on vector and parallel computers, Siam Review 27, 2, (1985), 149-240

[Rob90] ROBERT, Y., Gaussian elimination on distributed memory architectures, this volume.

[RT89] ROBERT, Y. and TRYSTRAM, D., Optimal scheduling for parallel Gaussian elimination, A COMPLETER

[Saa85] SAAD, Y., Communication complexity of the Gaussian elimination algorithm on multiprocessors, Report DCS/348, Yale University, (1985), to appear in L.A.A. special issue on parallel computation.

[SK78] SAMEH A., KUCK D., On stable parallel linear system solvers, J. ACM 25 (1), 81-91 (1978)

[Sar89] SARKAR, V., Partitionning and scheduling parallel programs for execution on multiprocessors, The MIT Press (1989)

[Sch84] SCHENDEL, U., Introduction to Numerical Methods for Parallel Computers, Ellis Horwood Series, J. Wiley & Sons, New York (1984)

[Vel89] VELDHORST M., Gaussian elimination with partial pivoting on an MIMD computer, J. Parallel and Distributed Computing 6 (1989), 62-68

[WH80] WING, O. and HUANG, J.W., A computational model of parallel solution of linear equations, IEEE Trans. Comp. C29 (1980) 632-638

Gaussian elimination on distributed memory architectures

Yves Robert
Laboratoire de l'Informatique du Parallélisme LIP-IMAG
Ecole Normale Supérieure de Lyon
69364 Lyon Cedex 07, France
e-mail: yrobert@ensl.ens-lyon.fr, yrobert@frensl61.bitnet

1. Introduction

We deal with the implementation of linear algebra algorithms on distributed memory architectures such as systolic arrays and general-purpose hypercubes. Throughout, we use Gaussian elimination as a target example.

In the first part, we consider the implementation of Gaussian elimination on systolic arrays. We consider two-dimensional (2D) networks. We explain how to retrieve the well-known Ahmed-Delosme-Morf (ADM) array. Then we derive a spacetime-minimal systolic array, which requires only half the number of processors of the ADM array.

In the second part, we focus on the hypercube as a target machine. We discuss the performances of various allocation functions for different subtopologies of the hypercube (ring and 2D-torus) and different communication protocols (centralized versus local). We first give experimental results, and then move to a complexity analysis.

2. 2D systolic arrays for Gaussian elimination

In this section we survey systolic arrays for the solution of dense systems of linear equations, using Gaussian elimination. We refer to Kung and Leiserson [1980], Kung [1982] and Kung [1988] for a description of the systolic model of computation. In a word, systolic arrays are regular networks of tightly coupled simple processors with limited storage which have provided cost-effective high-throughput implementations of important algorithms in a variety of areas (signal and image processing, speech and pattern recognition, matrix computations). See Kung et al. [1985], Moore et al. [1987], Bromley et al. [1988], McCanny et al. [1989] among others.

2.1. The Ahmed-Delosme-Morf array

We first deal in this section with the implementation of Gaussian elimination on two-dimensional (2D) systolic arrays *without pivoting* . See section 2.2 for a discussion on pivoting issues. Let A x = b (1) be a linear system of equations, where A is a dense square matrix of order n, and b a vector with n components. We consider the matrix A augmented with the right-hand side b, that is, we let A := (A,b) be an n by (n+1) matrix. We recall the generic version of the Gaussian elimination algorithm:

NATO ASI Series, Vol. F 77
Computer Algorithms for Solving Linear Algebraic Equations
Edited by E. Spedicato

{ column-oriented version: KJI-outside }
for k = 1 **to** n-1 **do**
  execute task T_{kk} : < **for** i = k+1 **to** n **do** $a_{ik} := - a_{ik} / a_{kk}$ >
  for j = k+1 **to** n+1 **do**
    execute task T_{kj} :   < **for** i = k+1 **to** n **do** $a_{ij} := a_{ij} + a_{ik} * a_{kj}$ >

Assume that a task can be processed within a time unit. We can implement the first version on a linear array of n general-purpose processors, as illustrated in figure 1. Column j is input to the top processor at step j. Tasks are assigned to processors as shown by table 1. The program of the processors can be detailed as follows:

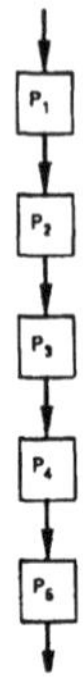

Figure 1 : The linear array of macroscopic processors (n=5)

Step	P_1	P_2	P_3	P_4	P_5
1	T_{11}				
2	T_{12}				
3	T_{13}	T_{22}			
4	T_{14}	T_{23}			
5	T_{15}	T_{24}	T_{33}		
6	T_{16}	T_{25}	T_{34}		
7		T_{26}	T_{35}	T_{44}	
8			T_{36}	T_{45}	
9				T_{46}	T_{55}
10					T_{56}

Table 1 : Task assignment for the column-oriented version (n=5)

Processor P_i
if step = 2i-1 **then**
  begin
    receive column i **from** P_{i-1}
    execute task T_{ii}
  end
for step := 2i **to** i+n **do**
  begin
    receive column j = step - i + 1 **from** P_{i-1}
    execute task T_{ij}
    send column j **to** P_{i+1}
  end

For simplicity, we have assumed that the top processor P_1 receives messages from its predecessor (i.e. the host) and that the bottom processor P_{n-1} sends messages to its successor.

To move to a systolic design, we need to replace each general-purpose processor, capable of operating on column vectors whose length is problem-size dependent, by more simple cells, capable of processing a single coefficient at a time. In other words, we move from a macroscopic time scale, where a unit is a vector-vector operation, to a microscopic time scale, where a unit is an elementary arithmetic operation such as a multiply-add. We see that P_k operates on vectors of length n-k+1. Thus, we can replace P_k by an array of n-k+1 elementary cells C_{kk}, $C_{k+1,k}$, ..., C_{nk} (see figure 2), and pipeline the execution of each task: rather than executing each task serially, we pipeline its execution on the array of cells.

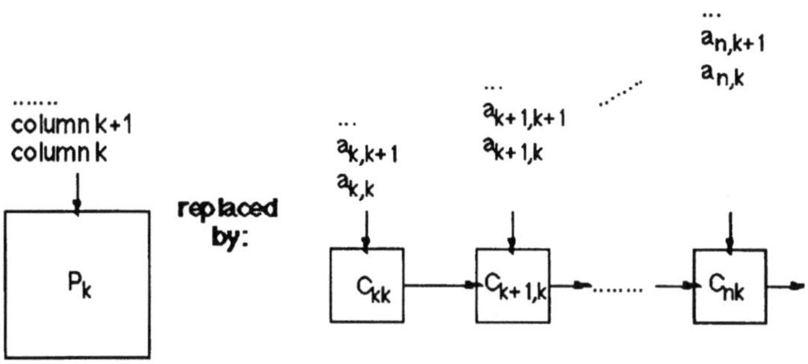

Figure 2 : Replacing each processor by an array of elementary cells

The resulting systolic array is depicted in figure 3. It is due to Ahmed, Delosme and Morf [1982]. We make explicit the operation of the cells in figure 4. The operation of a given cell depends on whether this is the first time that it operates. In the description of figure 4, we assume for simplicity that there is a boolean called *init* (for "initialization") stored in every processor which is set to "true" at the beginning of the computation. This boolean controls the operation of the processors. In an actual implementation, the instruction to start the computation would be input to the leftmost circular processor together with the first coefficient a_{11} and systolically propagated through the whole array. To triangularize a matrix of size n x (n+1) we need 3n-1 time steps and $\frac{n(n+1)}{2}$ cells.

2.2. Pivoting issues

For real matrices, it does not seem possible to extend the previous systolic array so as to include partial pivoting techniques. The global search for the maximum element (in absolute value) in an entire column of the matrix would break down the regularity of the flow in the array. However, pairwise pivoting techniques can be easily introduced: assume that we want to combine row k and row i in order to create a zero in position (k,i), $k+1 \le i \le n$. If the combination factor $|a_{ik}/a_{kk}|$ is smaller than 1, we perform the usual combination:

$$\begin{pmatrix} \text{row } k \\ \text{row } i \end{pmatrix} := M_{ik} \cdot \begin{pmatrix} \text{row } k \\ \text{row } i \end{pmatrix}, \quad M_{ik} := \begin{pmatrix} 1 & 0 \\ -\frac{a_{ik}}{a_{kk}} & 1 \end{pmatrix}$$

Otherwise we exchange row k and row i before performing the combination:

$$\begin{pmatrix} \text{row } k \\ \text{row } i \end{pmatrix} := M_{ik} \cdot \begin{pmatrix} \text{row } k \\ \text{row } i \end{pmatrix}, \quad M_{ik} := \begin{pmatrix} 1 & -\frac{a_{kk}}{a_{ik}} \\ 1 & 0 \end{pmatrix}$$

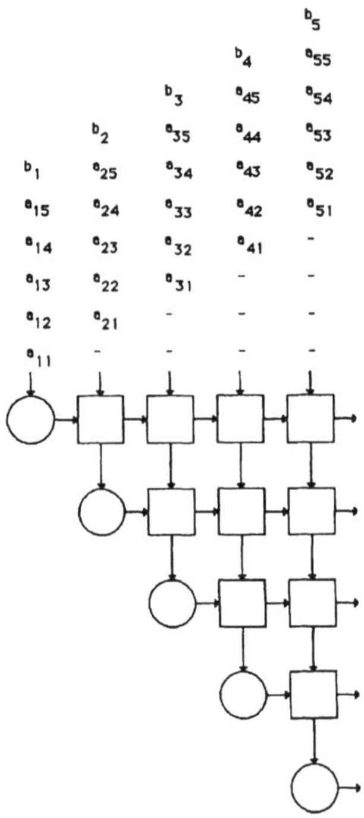

Figure 3 : The Ahmed-Delosme-Morf (ADM) array, n=5

Step t | Step t+1

{ transfer data } $a_{out} := a_{in}$;

Step t | Step t+1

if init **then** { initialize internal register }
begin $r := - a_{in} / b_{in}$; init := false ; $b_{out} := b_{in}$; **end**
else { update a_{in} }
begin $a_{out} := a_{in} + r * b_{in}$; $b_{out} := b_{in}$; **end**

Figure 4 : Program of the cells (ADM array)

In this way, we generate a combination factor whose module is always smaller than 1. Sorensen [1984] has shown that this method is numerically stable for a wide class of matrices. It is easy to modify the program of the cells of the Ahmed-Delosme-Morf array in order to include pairwise pivoting techniques. The first time they operate, rectangular cells determine which combination is to be generated according to the values of their inputs, that is, whether the corresponding two rows should be interchanged or not. After initialization, they apply the transformation (whose encoding is) stored in their internal register. The additional information costs only one bit. Another solution is to use Givens rotations, see Quinton and Robert [1989].

3. Spacetime-minimal systolic arrays

In this section, we derive complexity results for the mapping of Gaussian elimination onto 2D systolic arrays. We design several systolic solutions whose space-time complexity are better than those of the ADM array described in section 2. We use the method of Cappello [1989], Bermond et al [1988], Louka and Tchuente [1989] to derive spacetime-minimal systolic arrays for Gaussian elimination. The systolic computation is modeled by a directed acyclic graph (DAG), with nodes corresponding to computed values, and arcs denoting dependencies.

The computation DAG is taken to be fixed and given. Such DAGs can be extracted automatically from uniform recurrent equations or regular iterative arrays (see Kung [1988], Quinton [1987]). The time to compute a DAG is determined when we assign a timing function, or schedule, to the nodes, subject to the constraints that
• a node can be computed only when its predecessors (the nodes which it depends upon) have been computed at previous steps
• no processor can compute two different nodes at the same time-step.
Following Cappello [1989], we have the following two definitions:

Definition 1: A schedule is time-minimal when the number of time-steps is equal to the length of a longest path in the DAG (which is clearly the minimal time needed to achieve the computation).

Definition 2 : A systolic array is spacetime-minimal when it is scheduled in minimal time and when it uses the minimal number of processors among all possible minimal-time solutions.

3.1. Scheduling the DAG for Gaussian elimination

Let Ax= b be a linear system of size n. We transform it into the triangular system Ux = c using Gaussian elimination without pivoting. A is overwritten by its LU decomposition A = LU, and b is overwritten by c. The triangularization scheme can be formally expressed as the following system of equations:

Input equations (1)

$1 \le i \le n,\ 1 \le j \le n,\ k=0 \quad \rightarrow \quad A(i,j,k) = a_{ij}$

$1 \le i \le n,\ j=n+1,\ k=0 \quad \rightarrow \quad A(i,j,k) = b_i$

Computation equations (2)

$1 \le k \le n,\ j=k,\ k+1 \le i \le n \quad \rightarrow \quad A(i,j,k) = A(i,j,k-1)/A(k,k,k)$

$1 \le k \le n,\ i=k,\ k \le j \le n+1 \quad \rightarrow \quad A(i,j,k) = A(i,j,k-1)$

$1 \le k \le n,\ k+1 \le i \le n,\ k+1 \le j \le n+1 \quad \rightarrow \quad A(i,j,k) = A(i,j,k-1) - A(i,k,k)*A(k,j,k)$

Output equations (3)

$1 \le k \le n-1,\ k+1 \le i \le n$	$\rightarrow$	$l_{ik} = A(i,k,k)$
$1 \le k \le n,\ k \le j \le n$	$\rightarrow$	$u_{kj} = A(k,j,k)$
$1 \le k \le n,\ j=n+1$	$\rightarrow$	$c_k = A(k,n+1,k)$

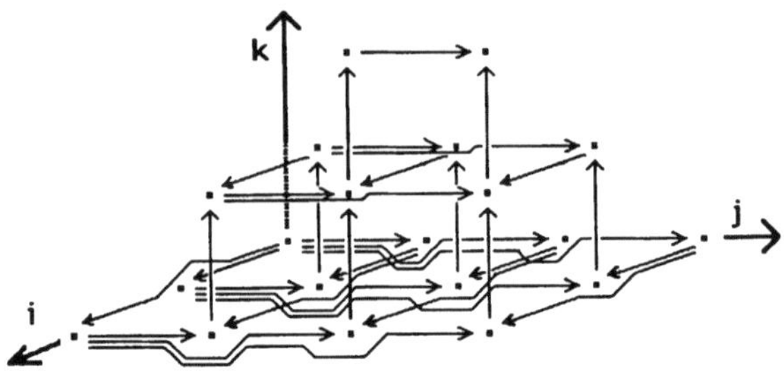

Figure 5 : Non-uniform dependence graph for Gaussian elimination

The computation domain is $D^n = \{(i,j,k),\ 1 \le k \le n,\ k \le i \le n,\ k \le j \le n+1\}$. The dependence graph G is represented in figure 5 for n=4. The dependences in the recurrence equations are not uniform: for instance vertex (i,k,k) depends upon vertex (k,k,k) for all i>k. In other words $\theta = (i-k,0,0)$ is a dependence vector for all i>k. We would like to uniformize the equations to avoid non-local dependences. A natural idea is to pipeline the circulation of the variables A(i,k,k) along the direction (0,1,0). In this way A(i,k,k) will start at vertex (i,k,k) and successively visit vertices (i,k+1,k), (i,k+2,k), ..., (i,n,k). Similarly, we pipeline the circulation of the variables A(k,j,k) along direction (1,0,0). Note that there are systematic ways of uniformizing equations whose domains are convex polyhedra, see van Dongen and Quinton [1988]. This leads to the new system of recurrence equations (input and output equations are left unchanged):

Computation equations (4)

$1 \le k \le n,\ j=k,\ k+1 \le i \le n$	$\rightarrow$	$A(i,j,k) = A(i,j,k-1)/U(i,j,k)$
$1 \le k \le n,\ i=k,\ k \le j \le n+1$	$\rightarrow$	$A(i,j,k) = A(i,j,k-1)$
$1 \le k \le n,\ k+1 \le i \le n,\ k+1 \le j \le n+1$	$\rightarrow$	$L(i,j,k) = L(i,j-1,k)$
$1 \le k \le n,\ j=k,\ k+1 \le i \le n$	$\rightarrow$	$L(i,j,k) = A(i,k,k)$
$1 \le k \le n,\ k \le j \le n+1,\ k+1 \le i \le n$	$\rightarrow$	$U(i,j,k) = U(i-1,j,k)$
$1 \le k \le n,\ i=k,\ k \le j \le n+1$	$\rightarrow$	$U(i,j,k) = A(k,j,k)$
$1 \le k \le n,\ k+1 \le i \le n,\ k+1 \le j \le n+1$	$\rightarrow$	$A(i,j,k) = A(i,j,k-1) - L(i,j,k)*U(i,j,k)$

The new dependence graph (still named G) is shown in figure 6. Note that all dependences are now local. The set of dependence vectors is

$$\Theta = \{\theta_a, \theta_l, \theta_u\}, \text{ with } \theta_a = (0,0,1),\ \theta_l = (0,1,0), \text{ and } \theta_u = (1,0,0)$$

For the sake of simplicity, we assume that each vertex of D^n has the same computation time, although the operations are not the same in all the vertices. A schedule (or timing function) is a mapping $t: D^n \rightarrow \mathbb{N}$ such that if computation at vertex $x \in D^n$ depends upon those at vertex $y \in D^n$ then $t(x) > t(y)$. Given t, the total execution time is $T(n) = \text{Max}\ \{t(x);\ x \in D^n\}$. We want to determine a schedule t such that T(n) is minimal. Note that finding such a schedule will only give a partial result: the original system of equations that specifies Gaussian elimination has been transformed through the uniformization of the recurrences, and we only deal with the complexity of the resulting uniform system. However, the uniform DAG G has been obtained in the most natural way, and all systolic arrays derived for Gaussian elimination use this DAG.

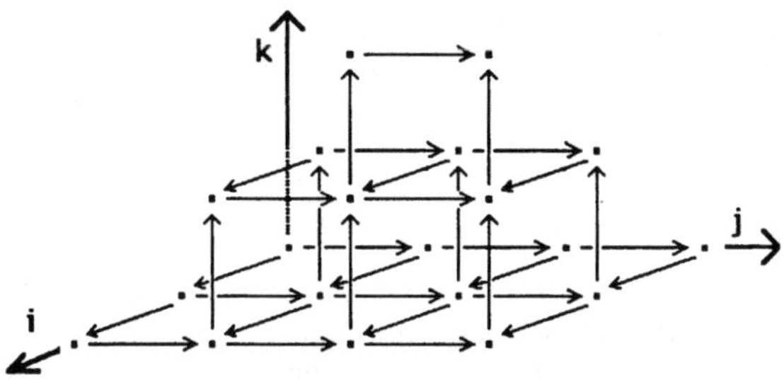

Figure 6 : Dependence graph after uniformization

If we search for an affine timing function $t(i,j,k) = \lambda_1 i + \lambda_2 j + \lambda_3 k + \alpha$, with λ_1, λ_2, λ_3 and α integers, we must have $t'(\theta) > 0$ for each dependence vector $\theta \in \Theta$, where t' is the linear part of t, which leads to $\lambda_1 \geq 1$, $\lambda_2 \geq 1$, $\lambda_3 \geq 1$. We choose the λ_i so that the execution time $T(n) = \max \{ t(x); x \in D^n \}$ is minimum, i.e. we take $\lambda_1 = \lambda_2 = \lambda_3 = 1$. Finally we choose α so that every vertex $x \in D^n$ is scheduled at a positive time: the minimum of t over D^n is for node (1,1,1), and we get $\alpha = -2$. We obtain $t(i,j,k) = i+j+k-2$ for $(i,j,k) \in D^n$.

The next step is to define an architecture that supports these calculations. This is obtained by defining an allocation function which maps D^n onto a *finite* set E: we want the number of processors |E| of the target architecture to be finite. Let alloc: $D^n \rightarrow E$ be the allocation function. We do not want to map on the same processor the computations of two distinct vertices that are scheduled at the same time; in other words:

$$\text{alloc}(x) = \text{alloc}(y) \Rightarrow t(x) \neq t(y) \text{ for all } x, y \in D^n, x \neq y \qquad \text{(A)}$$

In order to obtain designs with a good regularity and local interconnections, we take the allocation function to be linear: more precisely, we take alloc to be an orthogonal projection along a direction u, where u is a vector non-collinear to λ, so that condition (A) holds: if alloc(x) = alloc(y), then $x-y \in \text{Ker(alloc)}$, and x-y is collinear to u. Hence x-y is not collinear to the time vector $\lambda=(\lambda_1,\lambda_2,\lambda_3)$, therefore $x-y \notin \text{Ker } t'$ and $t(x) \neq t(y)$.

Once the timing function t and the allocation function a have been defined, the architecture is completely defined. Each processor π of alloc(D^n) executes the computation attached to the nodes of D^n that are mapped onto it. For the allocation function, we can choose u = (0,1,0), that is alloc(i,j,k) = (i,0,k), to obtain the Ahmed-Delosme-Morf array.

3.2. Time and space complexity

We already know that $t(i,j,k) = i+j+k-2$ is a timing function, since $t'(\theta)>0$ for each dependence vector $\theta \in \Theta$. We also know that this timing function is optimal among all linear timing functions. What we want to prove is that it is optimal among all possible timing functions:

Proposition 1 (Bermond et al. [1988]) : The schedule $t(i,j,k) = i+j+k-2$ is time-minimal. Moreover, it is the only time-minimal schedule. As a consequence, the optimal time for systolic Gaussian elimination is $T(n) = 3n-1$.

The Ahmed-Delosme-Morf is scheduled with the optimal timing function, hence their execution time 3n-1 is optimal. A natural question arises: what is the minimum number of processors that requires a systolic solution with the optimal scheduling ? When discussing allocation functions, we have insisted that the target design should be regular and locally connected, and therefore we have considered linear allocation functions. If we find a non-linear allocation function that requires fewer processors, then the resulting solution is not guaranteed to be as regular as the "pure" systolic schemes. However, finding the minimum number of processors for any allocation function will give us a lower bound of what can be achieved, and this will give us an indication of how efficient are our favorite solutions.

Proposition 2 (Bermond et al. [1988]) : The minimum number of processors M_n for the systolic triangularization of an n x (n+1) matrix [A,b] in optimal time T(n) = 3n-1 is $M_n = n^2/4 + O(n)$. More precisely $M_{2p-1} = p^2$ and $M_{2p} = p(p+1)$ for all $p \geq 1$.

In the following we present a spacetime-minimal systolic array for Gaussian elimination. Note that the Ahmed-Delosme-Morf array requires approximately twice this number of processors. To help the reader understand the derivation, we first present a solution which works in T(n) and uses (of the order of) $3n^2/8$ processors. Then we show how to modify this solution in order to derive a spacetime-minimal array.

3.3. Folding

We take a macroscopic point of view, as in figure 1, and we look carefully at table 1. Here a task represents a combination of columns and is assumed to last one time step. We see that processor P_1 executes its last task at time 6, that is one step before P_4 initiates its first computation T_{44}. So we can fold the array around its horizontal axis and assign to processor P_1 the set of tasks that were previously assigned to P_4 (in addition to its own tasks). Similarly, P_2 is given the set of tasks of P_5. In the general case, let n = 2p+1 be odd for the sake of simplicity. Then P_i, $1 \leq i \leq p$, is assigned the set of tasks $\{T_{ij};\ i \leq j \leq n+1\} \cup \{T_{i+p+1,j};\ i+p+1 \leq j \leq n+1\}$, and the last processor P_{p+1} is assigned the set of tasks $\{T_{p+1,j};\ p+1 \leq j \leq n+1\}$. This leads to the parallel algorithm illustrated in figure 7. Processor i, $1 \leq i \leq p$, executes task T_{ij} at time i+j-1, $i \leq j \leq n+1$, and task $T_{i+p+1,j}$ at time i+j+p, $i+p+1 \leq j \leq n+1$

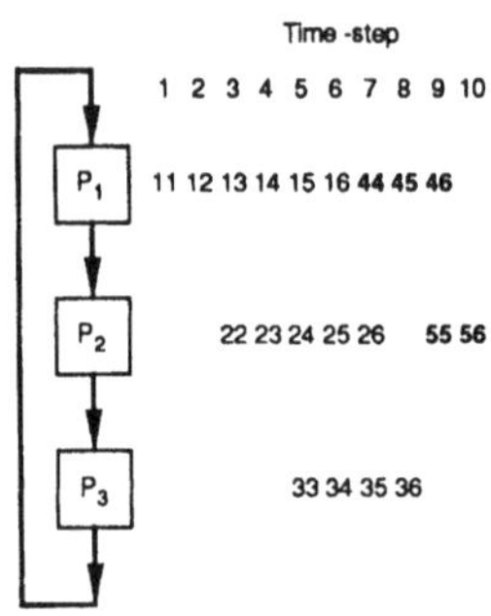

Figure 7 : Solving a problem of size n on a ring of $\lceil n/2 \rceil$ processors (n=5)

We move to a systolic design by the same space-time extension as in section 2.1. We obtain the array represented in figure 8 (Benaini , Robert and Tourancheau [1989]). Let n=2p+1: the new systolic array is composed of p+1 rows numbered 1 to p+1. The row i , $1 \leq i \leq p+1$, is composed of n-i+1 cells numbered P_{ik}, $1 \leq k \leq n-i+1$. The computation of node (i,j,k) is assigned to to processor $P_{alloc(i,j,k)}$, $1 \leq k \leq i,j \leq n$, where

$$alloc(i,j,k) = (i,k) \text{ if } 1 \leq k \leq p+1 \text{ and } alloc(i,j,k) = (i,k-p-1) \text{ if } p+1 \leq k \leq 2p+1.$$

The array has $\frac{(3p+2)(p+1)}{2} \approx \frac{3n^2}{8}$ cells as claimed. Note however that this saving of around 25% over the Ahmed-Delosme-Morf array does not come for free: the period of the array, i.e. the minimum time P between the solution of two consecutive instances of the problem, is now P=n+p+1, while P=n+1 for the Ahmed-Delosme-Morf array. There are now cycles in the dependence graph, and partitioning issues become more involved (Huang and Abraham [1984], Hwang and Cheng [1982], Kung and Lam [1984]).

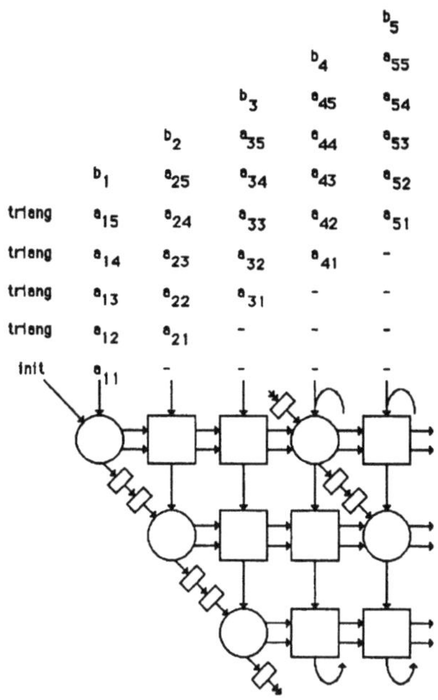

Figure 8 : A time-minimal solution with $3n^2/8 + O(n)$ processors (n=5)

3.4. A spacetime-minimal array

We are now ready for the design of the spacetime-minimal solution (Benaini and Robert [1990]. As the design is rather technical, we only outline the derivation process. In table 2, we illustrate the optimal allocation for a larger example: we let n=9. For each task T_{ik}, $1 \le k \le i \le n$, defined as

if i=k: T_{kk} : < **for** j := k **to** n+1 **do** $u_{kj} := a_{kj}$ >

if i>k: T_{ik} : < $l_{ik} := - a_{ik} / u_{kk}$; **for** j := k+1 **to** n+1 **do** $a_{ij} := a_{ij} + l_{ik} * u_{kj}$ >

we indicate as $t=t_1,t_2$ the time-step t_1 at which execution begins, and the time-step t_2 at which the last computation is performed. Two tasks T and T' can be assigned to the same processor if the first one completes execution before the second one is initiated, i.e. if $t_2(T) < t_1(T')$. We check in table 2 that our mapping is safe: this condition always holds when two tasks are assigned to the same cell.

We represent the resulting array in figure 9, without detailing further its operation. Some hints for the reader:

• rows 1 to 4 enter the array as indicated: they are not skewed as in the previous design. They progress leftwards until they are reflected by the anti-diagonal AD2. Then they move rightwards up to the right border, then they go down along the anti-diagonal direction. Such a scheme can be easily implemented with a few boolean control signals.

• row 5 to 9 move exactly as in the previous solution. They are input with a skewed scheme. There is a toroidal connection between the first and last rows of the array.

• the only problem is for the connections marked /1/ to /4/ in figure 9. Rows 1 to 4 use these connections to meet rows 5 to 9 as they did in the design of figure 8. These connections are not local. A possible solution to make them local is to fold the array two times, one along the main anti-diagonal AD1 and the second around the vertical axis (third column in the example).

Cell 5,1 task 5,1 t=5,14	Cell 6,1 task 6,1 t=6,15 task 6,6 t=16,20	Cell 7,1 task 7,1 t=7,16 task 7,6 t=17,21	Cell 8,1 task 8,1 t=8,17 task 8,6 t=18,22	Cell 9,1 task 9,1 t=9,18 task 9,6 t=19,23
Cell 5,2 task 5,2 t=7,15	Cell 6,2 task 6,2 t=8,16	Cell 7,2 task 7,2 t=9,17 task 7,7 t=19,22	Cell 8,2 task 8,2 t=10,18 task 8,7 t=20,23	Cell 9,2 task 1,1 t=1,10 task 9,2 t=11,19 task 9,7 t=21,24
Cell 5,3 task 5,3 t=9,16	Cell 6,3 task 6,3 t=10,17	Cell 7,3 task 7,3 t=11,18	Cell 8,3 task 2,1 t=2,11 task 8,3 t=12,19 task 8,8 t=22,24	Cell 9,3 task 2,2 t=4,12 task 9,3 t=13,20 task 9,8 t=23,25
Cell 5,4 task 5,4 t=11,17	Cell 6,4 task 6,4 t=12,18	Cell 7,4 task 3,1 t=3,12 task 7,4 t=13,19	Cell 8,4 task 3,2 t=5,13 task 8,4 t=14,20	Cell 9,4 task 3,3 t=7,14 task 9,4 t=15,21 task 9,9 t=25,26
Cell 5,5 task 5,5 t=13,18	Cell 6,5 task 4,1 t=4,13 task 6,5 t=14,19	Cell 7,5 task 4,2 t=5,14 task 7,5 t=15,20	Cell 8,5 task 4,3 t=7,15 task 8,5 t=16,21	Cell 9,5 task 4,4 t=10,16 task 9,5 t=17,22

Table 2 : Optimal allocation of tasks to processors

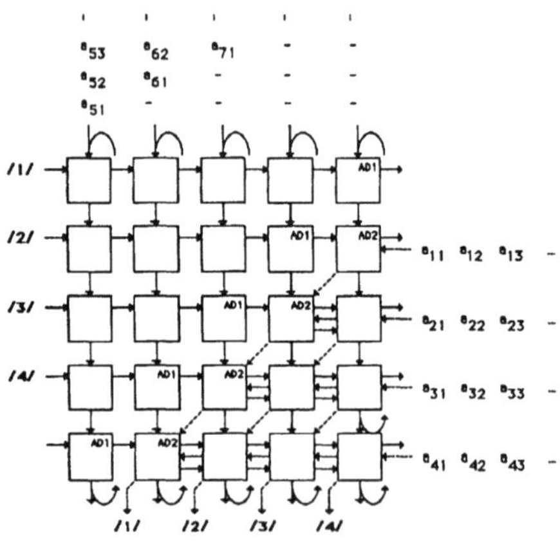

Figure 9 : The spacetime-minimal systolic array

4. Hypercube computing

In this section we deal with the implementation of Gaussian elimination on a hypercube. We first review some topological properties of hypercubes (section 4.1). We discuss broadcasting kernels (the movement of the same data from a given node to all other nodes) in section 4.2. Then we present centralized implementations of the Gaussian elimination algorithm, both pointwise and block variants (section 4.3). In section 4.4 we move to local algorithms, using only neighbor-to-neighbor communications. In section 4.5 we deal with complexity results.

4.1. Topological properties of hypercubes

Hypercubes grow by replication. Given two identical m-cubes (that is, with 2^m nodes), we obtain an (m+1)-cube by linking their corresponding nodes in a one-to-one fashion. Adding that a 0-cube consists of one node, we have a recursive definition of hypercubes.

Another definition is the following: an m-cube consists of 2^m nodes numbered by m-bit binary numbers, from 0 to 2^m-1, and interconnected so that there is a link between two processors if and only if their binary labels differ by exactly one bit. With the recursive construction it is easy to label the processors likewise. Indeed, given two m-cubes, it suffices to renumber the nodes of the first one as $0 \wedge a_i$ and those of the second one by $1 \wedge a_i$, where a_i is a binary number representing two similar nodes of the n-cubes and where $\wedge$ denotes the concatenation of binary numbers.

In a distributed memory system, we seek for a compromise between the number of neighbors of each processor and the largest distance between two processors. The degree of the connection graph is defined as the maximum number of neighbors of a given processor. The smaller the degree of the connection graph, the easier the realization of the architecture under given technological constraints. The other factor to be minimized is the diameter of the connection graph, i.e. the largest distance between two processors. It takes a time proportional to the diameter of the connection graph to exchange information between two remote processors. An architecture where $\delta = \Delta$ will achieve a good balance between the communication speed and the complexity of the topology network. Hypercubes achieve this equality, which explains why they are one of today's most popular design (e.g. iPSC of Intel Corp., T Series of FPS, n-CUBE, Connection Machine of Thinking Machines Corp).

Perhaps the most important advantage of the hypercubes is that many of the classical topologies such as rings, 2D-torus and 3D-torus can be embedded preserving proximity in them. By preserving proximity, we mean that any two adjacent vertices in the original graph are mapped to neighboring nodes in the hypercube.

Gray codes are the tool we need to embed a ring in an m-cube. A Gray code of dimension m is a finite sequence $G_m = \{g_0^{(m)}, g_1^{(m)}, \ldots, g_{2^m-1}^{(m)}\}$ of 2^m binary numbers which represent all m-bit binary numbers and so that $H(g_i^{(m)}, g_{i+1}^{(m)}) = 1$ for all i. Indices should be considered modulo 2^m, so that the Hamming distance between the last and first numbers is also 1. Thus, a Gray code defines a sequence of all the nodes of a hypercube so that any two successive nodes in the sequence are neighbors. In other words, a Gray code of dimension m enables us to embed a ring of 2^m nodes into an m-cube: we map the node i in the ring, $0 \le i \le 2^m$-1, into the node of the m-cube whose binary label is $g_i^{(m)}$, the i-th element of G_m.

There are many different ways in which Gray codes can be generated. The best known method is the construction of the binary reflected Gray code. We start with the sequence of two 1-bit numbers $G_1 = \{0, 1\}$. This is a Gray code of dimension 1. To build a Gray code of dimension 2, we take the same sequence and insert a 0 in front of each number. Then we take the same sequence in reverse order, and insert a 1 in front of each number. We obtain G_2 = {00, 01, 11, 10} which is a Gray code of dimension 2. Repeating the process, we take the sequence G_2, insert a 0 in front, then take the reverse of G_2 and insert a 1 in front. We obtain G_3 = {000, 001, 011, 010, 110, 111, 101, 100}, a Gray code of dimension 3. Repeating the process again, we obtain G_4 = {0000, 0001, 0011, 0010, 0110, 0111, 0101, 0100, 1100, 1101, 1111, 1110, 1010, 1011, 1001, 1000}. G_4 is the Gray code of dimension 4 which we have used to label the nodes of the 4-cube reproduced in figure 10, where we outline in bold the ring embedded in the cube. Formally, given G_i, denote by $G_i^{(R)}$ the sequence obtained by reversing its order, and by $0G_i$ (resp., $1G_i$) the sequence obtained from G_i by prefixing a 0 (resp., a 1) to each element of the sequence: G_{i+1} is generated as $G_{i+1} = \{0G_i, 1G_i^{(R)}\}$.

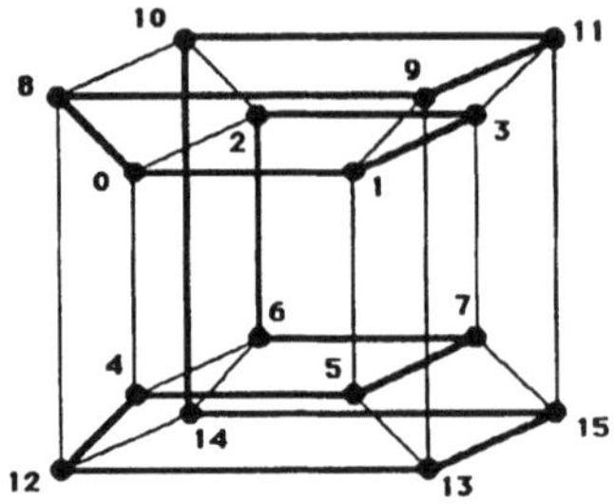

Figure 10 : Gray code numbering of a 4-cube

To embed a 2D-torus of size $2^u \times 2^v$ in an m-cube, where u+v=m, we simply map the node (i,j) of the grid, $0 \le i \le 2^u-1$, $0 \le j \le 2^v-1$, into the node of the hypercube whose binary label is $g_i^{(u)} \wedge g_j^{(v)}$, where $\wedge$ denotes concatenation. Observe that the set of nodes obtained by fixing one coordinate and letting the other vary, forms a subcube. In other words, the mapping of every row or column of processors in the grid forms a subcube of the cube.

4.2. Broadcasting

In a distributed memory system, processors communicate by message passing. Two neighboring processors can exchange data directly, but two remote processors can only exchange data by routing a message through intermediate processors. A very frequently encountered communication kernel is the broadcasting of a message, i.e. the distribution of the same data from one node to all the other nodes. For instance, when implementing Gaussian elimination, the processor holding the pivot column at a given step will have to distribute the information to all the other processors. We discuss in this section several implementations of the broadcast procedure on a hypercube.

Let processor 0 originate a broadcast in the m-cube. We cannot simply say that it sends the message to all its neighbors, which in turn will forward it to their own neighbors, and so on; such a process would be very difficult to control, and a lot of redundant information would be propagated. A simple way to perform a broadcast in the m-cube is to generate a spanning tree of height m for the routing: see figure 11.

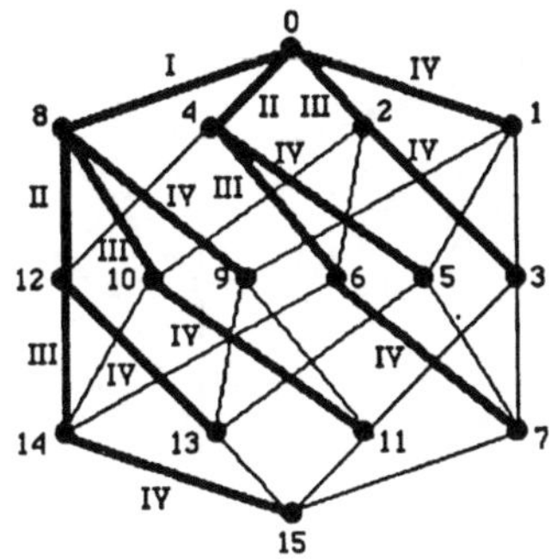

Figure 11 : A spanning tree embedded in a 4-cube
(Roman numbers represent broadcast steps)

Processors are numbered according to the standard binary reflected Gray code (denoted *abs*). The critical path in the spanning tree is of length m. Suppose that the syntax for sending and receiving a message between two neighbors is the following:

{ receiving } **receive**(hypercube_link,receive_buffer,length) ;
{ sending } **send**(hypercube_link,send_buffer,length) ;

where *hypercube_link* is the dimension along which the message is passed, i.e. the index of the bit which differs in the Gray code numbering of the two neighbors. The idea is that the processors receive on the link corresponding to their first bit equal to one, and they send on the links corresponding to the preceding zeros. There are m phases in the algorithm, numbered from m-1 to 0 in decreasing order. Time steps are numbered from 1 to m (step = m - phase). At phase i, those processors whose first bit equal to one is in a position greater than i send on link dimension i. See figure 11 where m=4 and the broadcast is initiated from processor 0, which is supposed to have a fictitious 1 in position m=4:

- at step 1, P_0 sends on link dimension m-1 =3, that is to P_8
- at step 2, P_0 and P_8 send on link dimension m-2 =2, that is to P_4 and P_{12} respectively
- and so on until step m where each even-numbered processor P_{2q} sends to P_{2q+1}.

The time to transfer L words between two adjacent processors is modelled by $\beta + L\,\tau_c$, where β is the communication start-up and τ_c the elemental transfer time. The time to broadcast L words using the above algorithm is then equal to $m\,(\beta + L\,\tau_c)$.

The standard broadcast does not make full use of the hypercube connectivity: we are using only 2^m-1 links while there are $m.2^{m-1}$ (bidirectional) links available in the m-cube. In fact each node uses only one of its links for sending and receiving. For most commercial hypercubes, however, all the communication links of a node can be activated in parallel. We can improve our first broadcast algorithm by simultaneously generating m different spanning trees in the m-cube, so as to make full use of the hypercube connectivity. As depicted in figure 12 for m = 4, we rotate the Gray code numbering of the nodes to generate m spanning trees which can be used to schedule m distinct broadcasts in parallel. Note that at a given step, we always communicate on different links for different broadcasts. Note also that at step m all the links of the m-cube are used. We split the data into m pieces and transmit them in parallel using the m broadcasts.

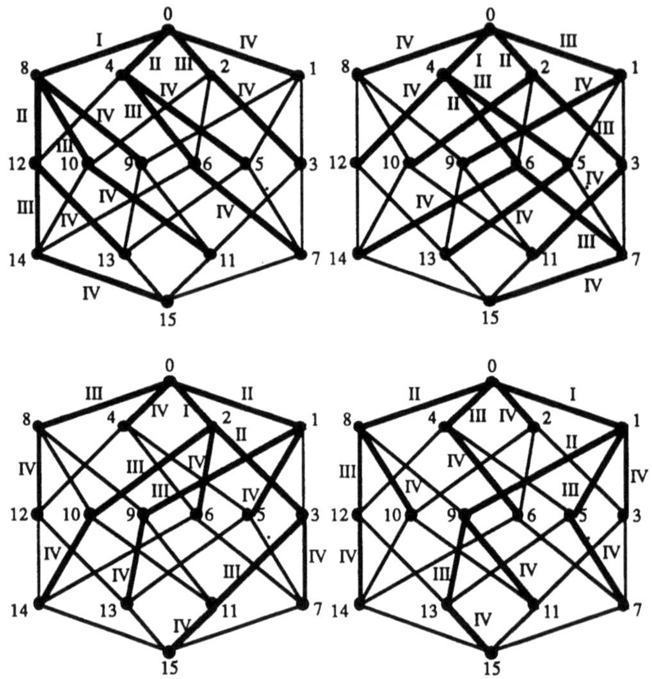

Figure 12 : 4 spanning trees embedded in a 4-cube (Roman numbers represent broadcast steps)

The time to broadcast L words using the new procedure is simply the time to broadcast L/m words using the first procedure, because the initial message has been split into m packets of the same length. We obtain the value $m\,(\beta + \frac{L}{m}\tau_c) = m\beta + L\tau_c$.

Another improvement to the standard procedure is to pipeline the data in the communication processes (Johnsson and Ho [1989], Saad [1986b]). The data set is partitioned into several packets and these packets are sent along the links in succession. At step j, the first packet reaches the nodes which are at distance j from the root, the second packet reaches the nodes at distance j-1 and the j-th packet reaches the neighbors of the root. If there are ν packets of length L/ν, the total time will be $T_m(\nu) = (m+\nu-1)\,(\beta + \frac{L}{\nu}\tau_c)$: it takes $m\,(\beta + \frac{L}{\nu}\tau_c)$ for the first packet to reach the last node, plus $(\nu-1)\,(\beta + \frac{L}{\nu}\tau_c)$ for the remaining packets. We have

$$T_m(\nu) = ((m-1)\beta + L\tau_c) + (\nu\beta + \frac{(m-1)L\tau_c}{\nu}),$$

so that the optimal number of packets ν_{opt} satisfies $\nu\beta = \frac{(m-1)L\tau_c}{\nu}$,

hence $\nu_{opt} = \sqrt{\frac{(m-1)L\tau_c}{\beta}}$, leading to a time $T_m(\nu_{opt}) = \left(\sqrt{L\tau_c} + \sqrt{(m-1)\beta}\right)^2$.

Note that we assume here that each processor can activate all its communication links in parallel. However, communication is unidirectional, from low-numbered nodes to high-numbered nodes, so that half of the full connectivity is used if we assume that links are bidirectional (simultaneous sending and receiving on the same link).

It is tempting to include pipelining in the rotating broadcast procedure. However, this is not possible, because the m generated spanning trees are not edge-independent. We have to resort to a more complex construction termed the m-Edges disjoint Spanning Binomial Tree (m-ESBT) in the literature, Johnsson and Ho [1989]. The construction of the m-ESBT is illustrated in figure 13:

- first take a standard Spanning Binomial Tree rooted at node 0, SBT_0 in figure 13 (a)
- rotate it just as before to build m-1 other spanning trees using the left rotation operator: formally, if $s=(s_{m-1},\ldots,s_0)$ is any node of the m-cube, the left rotation (of the bits) of s is $R(s)=(s_{m-2},\ldots,s_0,s_{m-1})$. The d-th left rotation of s, $2\le d\le m-1$, is defined by $R^d(s)=R(R^{d-1}(s))$. Let us call the SBT obtained with the d-th left rotation the d-th spanning tree.
- as we said above, these m spanning trees are not edge-disjoint. Thus translate the d-th spanning tree by negating the (d-1)-th bit of each node, $1\le d\le m-1$, and translate the original spanning tree SBT_0 by negating the most significant bit: see figure 13 (a).
- now node 0 is a leaf in each of the m resulting spanning trees. Merge the m trees into a single one, of height m+1, rooted in 0: figure 13 (b). This last tree is the m-ESBT.

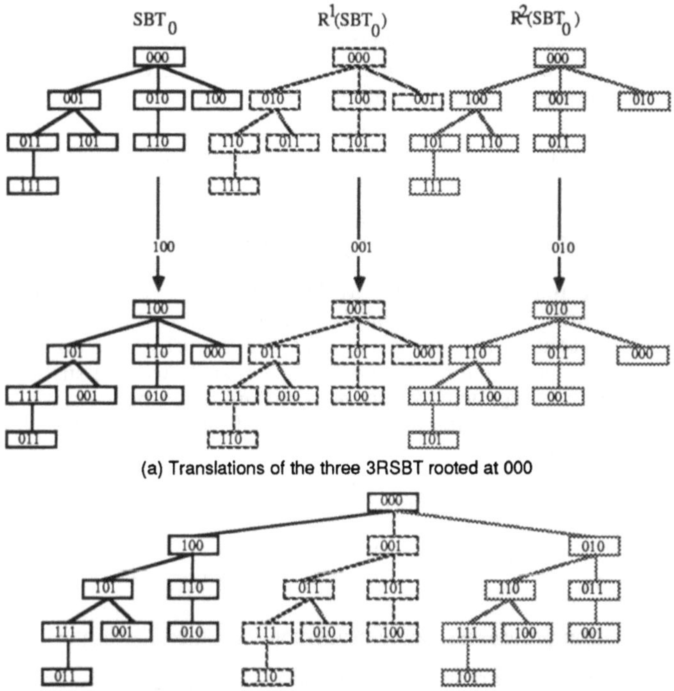

(a) Translations of the three 3RSBT rooted at 000

(b) Merge at 000 of the three translated 3RSBT

Figure 13 : Construction of the m-EBST (m=3)

Note that we use here bidirectional links: for instance 100 sends to 110 in the leftmost subtree of figure 13, while 110 sends to 100 in the rightmost subtree. To compute the communication time, we come back to the expression

$$T_m(\nu_{opt}) = \left(\sqrt{L\tau_c} + \sqrt{(m-1)\beta}\right)^2$$

We simply replace the length L of the message by L/m, and the longest path m by m+1 in this expression; we obtain the value

$$T_m(\nu_{opt}) = \left(\sqrt{\frac{L\tau_c}{m}} + \sqrt{m\beta}\right)^2$$

$T_m(\nu_{opt})$ is within a factor 2 of the universal lower bound $m\beta + (\frac{L}{m}+m-1)\tau_c$ for broadcasting L words in an m-cube: it necessarily takes m $(\beta+\tau_c)$ for the first word to reach the last node. At that time the last node can have received at most m words, one per link. We need to add at least (L-m)/m transmission units for receiving the L-m remaining words on the m channels.

4.3. Centralized Gaussian elimination

The first solution for implementing Gaussian elimination on a distributed memory machine is to emulate a shared-memory machine. In this section we deal with a centralized implementationwhere data needed by all processors will be broadcast through the hypercube.

First of all we have to allocate the coefficients of the matrix to the processors. Two possible strategies come to mind: we can distribute either the rows or the columns of the matrix to the processors. But recall the generic elimination algorithm with partial pivoting, where A is an n x n matrix (we do not consider the right-hand side):

```
for k := 1 to n-1 do
    execute Tkk { preparation of k-th transformation - routine prep(k) }
        < find p such that |apk| = max { akk, ..., ank } ; pivot(k) := p ;
        interchange apivot(k),k and akk ;  c := 1 / akk ;
        for i := k+1 to n do aik := aik * c >
    for j := k+1 to n do { update trailing columns - routine apply(k,j) }
        execute Tkj
            < interchange apivot(k),j and akj ;
            for i  := k+1 to n do aij = aij - aik * akj >
```

Following Geist and Heath [1986], Moler [1986] and Saad [1986b], we consider column-oriented algorithms. If a processor holds a whole column of the matrix, it can generate transformation T_{kk} (search for the pivot and scaling of the column) without any communication. On the other hand, if rows of the matrix were distributed among the processors, the execution of T_{kk} would require communication between all the processors. There are several strategies for distributing the columns of the matrix A among the processors. Assume that the size n of the problem is divisible by the number of processors p, and number the processors from 0 to p-1. The first strategy is to assign consecutive columns to a given processor (full block allocation). For instance, with n=16 and p=4:

	P_0	P_1	P_2	P_3
columns:	1,2,3,4	5,6,7,8	9,10,11,12	13,14,15,16

As the elimination progresses, some processors become idle: in the example, at step 5 processor 0 has finished its work. In the general case, at step $\frac{qn}{p}$ + 1, q processors will be definitely idle, resulting in a decrease of efficiency. The second strategy is to assign the columns with the same number modulo p to a given processor (wrap allocation). With the same example:

	P_0	P_1	P_2	P_3
columns:	1,5,9,13	2,6,10,14	3,7,11,15	4,8,12,16

This strategy is more attractive, since all processors are kept active until the very end of the computation. Let *alloc* denote the allocation function: alloc(j) = k if column j is assigned to processor k; for instance we obtain

- $\text{alloc}(j) = \lceil \frac{j\,p}{n} - 1 \rceil \bmod p$ $\quad 1 \leq j \leq n$, for the full block strategy
- $\text{alloc}(j) = (j - 1) \bmod p$ $\quad 1 \leq j \leq n$, for the wrap strategy.

Once the allocation strategy is chosen, we can design a very simple centralized implementation. At each step, the processor which holds the pivot column in its local memory

broadcasts it to all the other processors, so that they can use it to update their own internal columns. Hence each step k will consist of two phases:
• a communication phase: a broadcast originated by the processor which holds column k in its local memory, that is the processor i such that alloc(k) = i
• a computation phase: each processor independently updates its internal columns.
We have the following algorithm:

Broadcast Gaussian Elimination Algorithm :
{ program of processor P_i }
for k=1 **to** n-1 **do**
    if alloc(k) = i { P_i holds the pivot column } **then**
        execute T_{kk} ;
        broadcast column k and pivot(k) to the other processors
else
        receive column k and pivot(k)
    endif
    { perform the eliminations }
    execute T_{kj} for all columns j≥k+1 such that alloc(j) = i
endfor

We plot in figure 14 the performance (expressed in Mflops) of the centralized algorithm, obtained on a FPS T Series hypercube (Gustafson, Hawkinson, and Scott [1986]), using each of the three available levels for programming the VPU. This is mainly to show that good performance can be obtained with a very simple code: hypercube programming can be rewarding ! Note that it is possible to derive a block-version of the centralized algorithm, see Robert and Tourancheau [1989a].

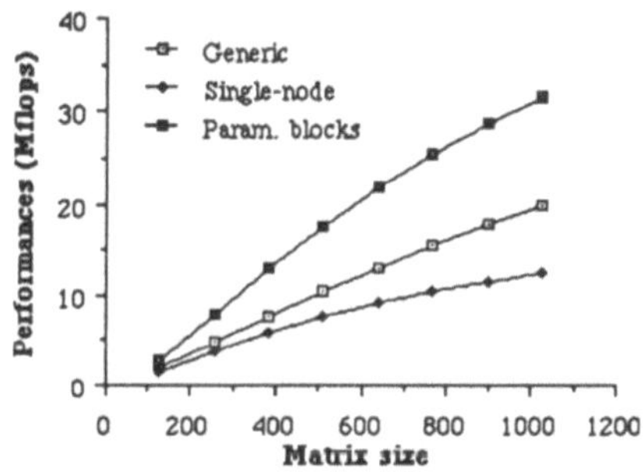

Figure 14 : Gaussian elimination (broadcast algorithm, 16 processors)

4.4. Local pipelined algorithms

In this section, we use the simplest subtopology of the cube, namely the ring, to derive pipeline algorithms that use only neighbor-to-neighbor communications. Local algorithms where the hypercube is configured as a 2D-grid can be found in Robert and Tourancheau [1989b].

We configure the hypercube as an oriented ring of p processors numbered from 0 to p-1. Processor P_i receives messages from its predecessor P_{i-1} and sends messages to its successor P_{i+1} (the subscripts are taken modulo p). At step k, the processor holding the pivot column, say P_i, sends it to its right neighbor P_{i+1}, which in turn sends it to P_{i+2}, and so on until the pivot column reaches the left neighbor of P_{i-1}. The main advantage of this asynchronous strategy is that communication and computation can be overlapped. A processor can start updating its internal columns as soon as it has received the pivot column and transmitted it to its neighbor. Their is no need to wait for all the processors to receive the information before starting the computation. The pipeline ring algorithm is as follows (processor indices are taken modulo p) :

Pipelined Ring (PR) Algorithm :

```
{ program of processor Pi }
for k=1 to n-1 do
    if alloc(k) = i then
        execute Tkk
        send column k to Pi+1
    else
        receive column k from Pi-1
        if (i+1 mod p) ≠alloc(k) then send column k to Pi+1
    endif
    { perform the eliminations }
    execute Tkj for all columns j≥k+1 such that alloc(j) = i
endfor
```

It might seem more natural at first to perform the eliminations as soon as the pivot column is available, and propagate the pivot column to the next processor after the arithmetic is completed, but this is not as efficient since the next processor will be idle waiting for the column to arrive while the processor P_i is computing.

We use a block-r allocation strategy, i.e. we distribute blocks of r consecutive columns to the processors in a wraparound fashion. For instance with n = 24, p =4 and r = 3:

	P_0	P_1	P_2	P_3
columns:	1,2,3	4,5,6	7,8 9	10,11,12
	13,14,15	16,17,18	19,20,21	22,23,24

The allocation function by blocks of size r, $1 \le r \le \frac{n}{p}$, is defined by

$$\text{alloc}(j) = \lceil \frac{j}{r} - 1 \rceil \bmod p \qquad 1 \le j \le n$$

Of course if r = 1 we retrieve the wrap allocation and if $r = \frac{n}{p}$ the full block one.

First consider the PR algorithm with the full block allocation, that is $r = \frac{n}{p}$. We depict its execution in figure 15. We see in the figure that communication and arithmetic do overlap. Now we use the same representation in figure 16 for the wrap allocation (r =1). We see that the communication time T_c is less for the full block allocation than for the wrap one, because of the idle time introduced every time that the pivot columns at two consecutive steps are not held by the same processor. This is a general result: the larger r, the smaller T_c. On the other hand, the arithmetic obeys the same rules as in the centralized case: the larger r, the greater the arithmetic time T_a. In other words, T_a is minimum for the wrap allocation (r=1), while T_c is minimum for the block allocation (r=n/p). The smaller r, the more balanced the computations, but the more expensive the communications. As the total time takes into account these two opposite laws, we must be prepared to find a compromise for the choice of the best block size r. As evidenced by figure 17, the best value is found to be r=4 for a matrix of order 1024 on the FPS T Series. We return in next section to the influence of data allocation strategies on algorithm performance, and we analytically determine the best allocation as a function of p, n and some machine-dependent parameters.

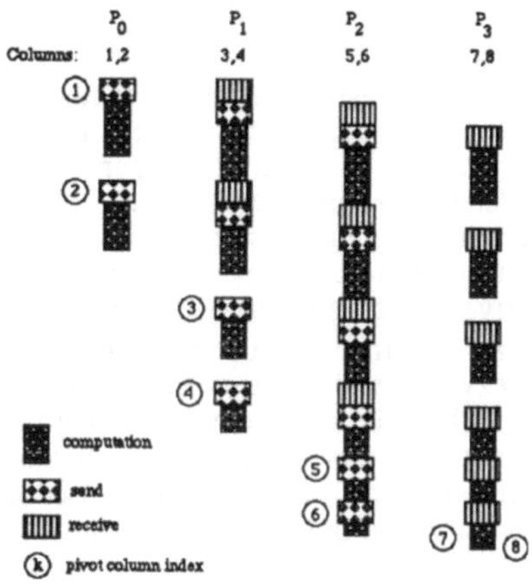

Figure 15 : Pipelined ring algorithm, full block allocation

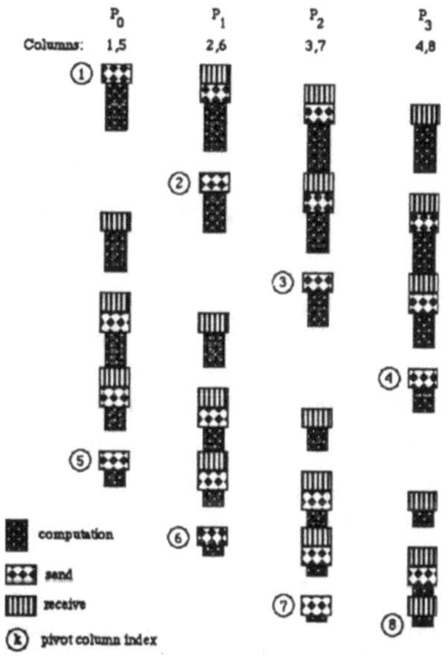

Figure 16 : Pipelined ring algorithm, wrap allocation

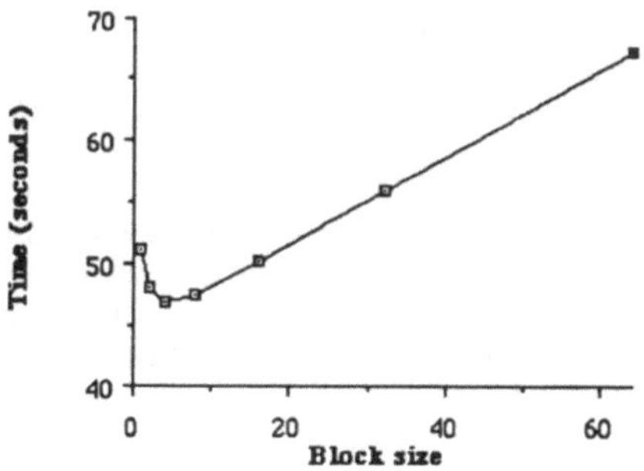

Figure 17 : Execution times for the PR algorithm, various block sizes (n=1024, p=16)

4.5. Complexity results

In this section we deal with complexity results for the implementation of Gaussian elimination on distributed memory machines. We focus on the oriented ring as a target machine, and we want to derive an analytical model for evaluating the performances of the data allocation strategies which we have considered in the previous section.

We consider an oriented ring of p processors numbered from 0 to p-1. Each processor has a private memory and can communicate by a message passing protocol with its two neighbors. Processor P_i receives messages from P_{i-1} and sends messages to P_{i+1} (the subscripts are taken modulo p). The machine operates in an asynchronous MIMD mode. Communications are by rendezvous, and there is no overlap between computations and communications. Moreover, for the sake of convenience, we now assume that a processor cannot simultaneously receive data from its left neighbor and send (different) data to its right neighbor. This assumption, although not mandatory, will facilitate our derivation of lower bounds. In our analytical analysis, we do not take pivoting into consideration. We consider a slightly modified version of the pipelined ring algorithm:

```
{ program of processor Pi }
for k=1 to n-1 do
    if alloc(k) = i then
        send column k to Pi+1
    else
        receive column k from Pi-1
        if (i+1 mod p) ≠ alloc(k) then send column k to Pi+1
    endif
    perform the eliminations: execute Tkj for all columns j≥k+1 s.t. alloc(j) = i
endfor
```

where T_{kj} is the following task: $< a_{kj} := a_{kj} / a_{kk}$
for $i = k+1$ **to** n **do** $a_{ij} := a_{ij} - a_{ik} * a_{kj} >$

Let τ_a be the time for an arithmetic operation such as a multiply-add. The time to transfer L words between two neighbor processors is modelled as $\beta + L\,\tau_c$, where β is the communication start-up and τ_c the elemental transfer time. We consider block-r allocation functions, i.e. we distribute blocks of r consecutive columns to the processors in a wraparound fashion. Column j, $1 \le j \le n$, is allocated to processor q=alloc(j), $0 \le q < p$, such that $\mathrm{alloc}(j) = \lceil \frac{j}{r} - 1 \rceil \bmod p$. The problem is to determine the allocation function which minimizes the total execution time, defined as the sum of the arithmetic time T_a, and of the communication time T_c (which includes idle time).

Clearly, we have restricted the problem to a special case. We have imposed constraints both on the algorithm and on the allocation function. This is only a first step towards full complexity results for the distributed implementation of Gaussian elimination. Our hypotheses are consistent with the following assumptions specified by Gerasoulis and Nelken [1988] and Saad [1986a]:

- equidistribution of the data: the number of processors p is a divisor of the problem size n, and each of the p processors holds exactly n/p columns of the system matrix in its local memory
- locality assumption: a processor may modify only the data which are stored in that processor
- completion assumption: a processor must complete all computations involving the most recently received message before beginning computations involving the next received message.

We assume that the number of processors is proportional to the size of the problem: we let $p = \alpha\, n$, where $\alpha \in\,]0,1]$ is a fixed number, and we let n go to infinity. The reason for choosing p proportional to n is as follows: we know from Saad [1986a] that communications predominate if $p = O(n^2)$, so that we should not use a number of processors greater than the problem size. On the other hand, if p is negligible compared to n, communications have no influence on the asymptotic execution time. The situation under consideration is when p and n are proportional, since the costs of the computations and of the communications are then of the same order $O(n^2)$: we have $T_a = O(n^3/p)$ and $T_c = O(n^2)$. Finally, for our asymptotic analysis, we assume $\beta = 0$, as in Ipsen, Saad and Schultz [1986] and Saad [1986a]. Even though β can be in practice much higher than τ_c, we still have messages of length O(n), so that the contribution from the start-ups is negligible.

Given the ratio $\alpha = p/n$ and the machine-dependent parameters τ_a and τ_c, we want to derive the value of the block size r which maximizes the efficiency. We summarize the results obtained by Robert, Tourancheau and Villard [1989]:

• the parallel arithmetic time is $T_a = \frac{n}{6p}\left(4n^2 + 3rpn - r^2p^2\right)\tau_a + o(pn)$

Note that T_a is a non-decreasing function of r (differentiate T_a with respect to r). The larger the blocks, the less balanced the workload distribution.

• the parallel communication time is $T_c = \left(pn + n^2 + \frac{n^2}{2r} - \frac{p^2r}{2}\right)\tau_c + o(np)$

We check that T_c is a non-increasing function of r (differentiate T_c with respect to r). The larger the blocks, the more reduced the volume of communications; this corroborates figures 15 and 16.

We know from the previous results that, given $\alpha = n/p$, T_a is an increasing function of r whereas T_c is a decreasing one. In other words, T_a is minimum for the wrap allocation (r=1), while T_c is minimum for the full block allocation (r=n/p). The smaller r, the more balanced the computations, but the more expensive the communications. We have:

Theorem (Robert, Tourancheau and Villard [1989]) : Let $\alpha = p/n$ be the ratio of the processor number to the problem size, and let $\rho = \tau_c / \tau_a$ be the ratio of the elemental costs of communication to arithmetic. The efficiency for a block-r allocation is

$$e_{\alpha,r} = \frac{1}{1 + \lambda_1 \alpha + \lambda_2 \alpha^2 + \lambda_3 \alpha^3} + o(1)$$

where $\lambda_1 = \frac{3r}{4} + \frac{3\rho}{2} + \frac{3\rho}{4r}$, $\lambda_2 = \frac{3\rho}{2} - \frac{r^2}{4}$, and $\lambda_3 = -\frac{3\rho r}{4}$

Given α and ρ, it is easy to compute the value of r which maximizes the efficiency. Note that for small α, the efficiency can be approximated as

$$e_{\alpha,r} \approx \frac{1}{1 + \left(\frac{3r}{4} + \frac{3\rho}{2} + \frac{3\rho}{4r}\right)\alpha}$$

(we drop the terms in α^2 and α^3). The efficiency is then maximum for $r_{opt} \approx \sqrt{\rho}$. This is a key result for all situations where the number of processors is significantly less than the size of the problem that we want to solve, a situation very likely to happen in practice.

We point out the good adequacy of these theoretical results with numerical experiments run on a hypercube. Let n=480 and use various numbers of processors, from 4 to 16, so that α varies from 1/120 to 1/30. The best experimental value of r is always r=4. We report in table 3 the execution times for r=3, 4 and 5. On the other hand, we estimate $\rho = 14.13$ for the FPS hypercube and messages of average length 240. We report this value in the expression of the efficiency e_α, and we compute the value of r which maximizes it. This is the "optimal value" reported in table 3. Note that for small α, the theory predicts that $r_{opt} = \sqrt{\rho} \approx 3.76$.

Block size r	p = 4	p = 8	p = 16
3	35.58	19.97	12.37
4	35.53	19.96	12.26
5	35.71	20.14	12.56
optimal value of r	3.80	3.85	3.97

Table 3 : Time (seconds) for n = 480

5. Conclusion

In this paper we have reviewed the implementation of the Gaussian elimination algorithm on distributed memory architectures such as systolic arrays and general-purpose hypercubes.

First we have introduced the well-known 2D systolic array of Ahmed-Delosme-Morf. Next we have presented space-time complexity results: we have derived time-minimal systolic arrays for Gaussian elimination that use a minimal number of processors, i.e. $n^2/4 + O(n)$ for a problem of size n. The minimal processor requirement is therefore half of that of the ADM array. The design of spacetime-minimal arrays represents an interesting and important contribution to the knowledge of the systolic model

• from the theoretic point of view, we can prove optimality results which will contribute to our basic understanding of the limits and potential of systolic computation

• from the practical point of view, we can compare any existing solution against the one which minimizes both time and space. Therefore, if other design criteria (including technical implementation constraints) are to be taken into account, we have a good basis for performance evaluation.

Then we have moved to the hypercube as a target machine. We have discussed the performances of various allocation functions for different subtopologies of the hypercube and different communication protocols (centralized versus local). We have reported both experimental and theoretical results. It is important to point out that many factorization algorithms share the same generic structure as Gaussian elimination:

> **for** k = 1 **to** n-1
> 1. prepare a transformation using column k of A:
>   task T_{kk} : PREP(...,k,...)
> 2. apply this transformation to columns k+1 to n
>   **for** j = k+1 **to** n
>   task T_{kj} : APPLY(...,k,j,...)
>
> **endfor**

Hence most of the results presented in this paper can be straightforwardly extended to a wide class of algorithms.

Acknowledgments

This work has been supported by the Project C3 of the French Council for Research CNRS, and by the ESPRIT Basic Research Action 3280 "NANA" of the European Economic Community. Part of the material presented in this chapter is a shortened version of chapters 4, 5, 7 and 8 of the book *The impact of vector and parallel architectures on the Gaussian elimination algorithm*, by Yves Robert, Manchester University Press (1990).

The author would like to thank the co-authors of the published material that has been used in this paper, namely Abdelhamid Benaini, Bernard Tourancheau and Gilles Villard.

References

1. Ahmed, H.M., Delosme, J.M. & Morf, M., "Highly concurrent computing structures for matrix arithmetic and signal processing", *Computer* **15**, 1 (1982), 65-82
2. Benaini, A. & Robert, Y., "Spacetime-minimal systolic arrays for Gaussian elimination and the algebraic path problem", to appear in *Parallel Computing*. Also available as technical Report 90-09, LIP-IMAG, Ecole Normale Supérieure de Lyon (1990)
3. Benaini A., Robert Y. & Tourancheau B., "A new systolic architecture for the algebraic path problem", in *Systolic array processors*, J. McCanny et al. eds, Prentice Hall (1989), 73-82
4. Bermond, J.C., Peyrat, C., Sakho, I. & Tchuente, M., "Parallelization of Gauss elimination algorithm on systolic arrays", Technical Report 430, Laboratoire de Recherche en Informatique, Université de Paris-Sud (1988)
5. Bromley, K., Kung, S.Y. & Swartzlander, E. (eds), *International conference on systolic arrays*, IEEE Computer Society Press (1988)
6. Cappello P.R., "A spacetime-minimal systolic array for matrix product", in *Systolic array processors*, J. McCanny et al. eds, Prentice Hall (1989), 347-356
7. Chuang, H.Y.H. & He, H., "Design of problem-size independent systolic arrays systems", in *Int. Conf. Computer Design ICCD 84*, IEEE Press (1984), 152-156
8. van Dongen, V. & Quinton, P., "Uniformization of linear recurrence equations: a step towards the automatic synthesis of systolic arrays", in *International Conference on Systolic Arrays*, K. Bromley et al. eds., IEEE Computer Society Press (1988), 473-482
9. Geist, G.A. & Heath, M.T., "Matrix factorization on a hypercube multiprocessor", in *Hypercube Multiprocessors 1986*, M.T. Heath ed., SIAM Press (1986), 161-180
10. Gerasoulis, A. & Nelken, I., "Gaussian elimination and Gauss-Jordan on MIMD architectures", Report LCSR-TR-105, Department of Computer Science, Rutgers University (1988)
11. Gustafson, J.L., Hawkinson, S. & Scott, K., "The architecture of a homogeneous vector supercomputer", in *Int. Conf. Parallel Processing ICPP 86*, The Pennsylvania State University Press (1986), 649-652
12. Huang, K.H. & Abraham, J.A., "Algorithm based fault-tolerance for matrix operations", *IEEE Trans. Computers* **33**, 6 (1984), 518-528
13. Hwang, K. & Cheng, Y.H., "Partitioned matrix algorithm for VLSI arithmetic systems", *IEEE Trans. Computers* **31**, 12 (1982), 1215-1224
14. Ipsen, I.C.F., Saad, Y. & Schultz, M.H., "Complexity of dense linear system solution on a multiprocessor ring", *Lin. Alg. Appl.* **77** (1986), 205-239
15. Johnsson, S.L. & Ho, C.T., "Optimum broadcasting and personalized communication in hypercubes", *IEEE Trans. Computers* **38**, 9 (1989), 1249-1268
16. Kung, H.T., "Why systolic architectures?", *Computer* **15**, 1 (1982), 37-46
17. Kung, H.T. & Lam, M.S., "Fault-tolerance and two-level pipelining in VLSI systolic arrays", *J. Parallel and Distributed Computing* **1**, 1 (1984), 32-63
18. Kung, H.T. & Leiserson, C.E., "Systolic arrays for (VLSI)", in *Introduction to VLSI systems*, C.A. Mead and L.A. Conway, Addison-Wesley (1980), chapter 8.3

19. Kung, S.Y., *VLSI array processors*, Prentice-Hall (1988)
20. Kung, S.Y., Whitehouse, H.J. & Kailath T. (eds), *VLSI and modern signal processing*, Prentice-Hall (1985)
21. Louka B., Tchuente M., "An optimal solution for Gauss-Jordan elimination on systolic arrays", in *Systolic array processors*, J. McCanny et al. eds, Prentice Hall (1989), 264-274
22. McCanny, J., McWhirter, J. & Swartzlander, E.E. Jr. (eds), *Systolic array processors*, Prentice-Hall (1989)
23. Moler, C., "Matrix computation on distributed memory multiprocessors", in *Hypercube Multiprocessors 1986*, M.T. Heath ed., SIAM Press (1986), 181-195
24. Moore, W., McCabe, A. & Urquhart, R. (eds), *Systolic arrays*, Adam Hilger (1987)
25. Ortega, J.M. & Romine, C.H., "The ijk forms of factorization methods II. Parallel systems", *Parallel Computing* **7** (1988), 149-162
26. Quinton, P., "The systematic design of systolic arrays", in *Automata Networks in Computer Science*, F. Fogelman-Soulie et al. eds., Manchester University Press (1987), 229-260
27. Quinton, P., "Mapping recurrences on parallel architectures", in *Int. Conf on Supercomputing ICS 88*, L.P. Kartashev and S.I. Kartashev eds., International Supercomputing Institute (1988), III 1-8
28. Quinton, P. & Robert, Y., *Algorithmes et Architectures Systoliques*, Masson (1989)
29. Robert, Y. & Tourancheau, B., "Block Gaussian elimination on a hypercube vector multiprocessor", *Revista de Matematicas Aplicadas* **10** (1989a), 49-69
30. Robert, Y. & Tourancheau, B., "Impact of the architecture topology on data allocation strategies for Gaussian elimination on the hypercube", in *4-th Conf. on Hypercube Concurrent Computers and Applications HCCA4* (1989b)
31. Robert Y., Tourancheau B. & Villard, G., "Data allocation strategies for the Gauss and Jordan algorithms on a ring of processors", *Information Processing Letters* **31**, 1 (1989), 21-29
32. Saad, Y., "Communication complexity of the Gaussian elimination algorithm on multiprocessors", *Lin. Alg. Appl.* **77** (1986a), 315-340
33. Saad, Y., "Gaussian elimination on hypercubes", in *Parallel Algorithms and Architectures*, M. Cosnard et al. eds., North-Holland (1986b), 5-18
34. Saad, Y., & Schultz, M.H., "Topological properties of hypercubes", *IEEE Trans. Computers* **37**, 7 (1988), 867-872
35. Sorensen, D.C., "Analysis of pairwise pivoting in Gaussian elimination", Technical Report MCS-TM-26, Argonne National Laboratory (1984)

Parallel Algorithms for General Sparse Systems

Iain S Duff

Rutherford Appleton Laboratory
Chilton, Oxfordshire OX11 0QX, England.
and
CERFACS
42 Ave G. Coriolis, 31057 Toulouse, France.

Abstract

We discuss several algorithms for the solution of sparse linear systems on parallel architectures. We will be concerned primarily with shared memory machines and will consider both direct and iterative methods of solution. The direct methods are based on Gaussian elimination and exploit parallelism coming from both the sparsity and from a full matrix kernel. The iterative method is a block projection method using a direct solver within the blocks. We then discuss the development of two codes, one for the solution of general sparse equations, the other for sparse symmetric indefinite systems. Some results on vector multiprocessor supercomputers are presented to illustrate the behaviour of our approaches.

Keywords: sparse matrix, sparse equations, vectorization, parallelism, multifrontal, block iterative, Cimmino, augmented system, MA28, symmetric indefinite, constrained optimization, block pivots.

1 Introduction

For most large-scale scientific computations there is a great deal of potential parallelism which, we contend, can be thought of as occurring at three levels. The first is at the system level where parallelism could even be exploited on a continuous problem prior to discretization. It is usually a form of domain parallelism, the most common example being domain decomposition in the solution of partial differential equations. We are not here concerned with this level of parallelism. The second level comes from the parallelism inherent in a sparse matrix, where the connectivity of the matrix gives the possibility of independent or semi-independent subproblems. We discuss this parallelism in Section 2.

One of the most common calculations in scientific kernels is the multiplication of two dense matrices. This is fortunately one of the easiest operations that permits the efficient exploitation of vectorization and parallelism because of regular data access and natural vector operations. It is this form of parallelism that we consider at the third level of our hierarchy. Matrix-matrix kernels of this type are standardized and defined within the Level 3 BLAS (Basic Linear Algebra Subprograms), Dongarra, Du Croz, Duff, and Hammarling (1990), and most manufacturers provide highly efficient machine dependent versions of these kernels. The important feature of the BLAS is that the call to the routines from Fortran has been standardized so that code utilizing them is entirely portable since Fortran definitions exist for use on machines without their own versions. A large range of linear algebra codes, for example for the solution of dense systems of linear equations, have been designed to utilize

NATO ASI Series, Vol. F 77
Computer Algorithms for Solving Linear Algebraic Equations
Edited by E. Spedicato

these kernels through the use of block algorithms. The most prominent example of these can be found in the LAPACK project (Anderson, Bai, Bischof, Demmel, Dongarra, DuCroz, Greenbaum, Hammarling, McKenney, and Sorensen, 1990) that uses both the Level 3 BLAS and the matrix-vector kernels of the Level 2 BLAS (Dongarra, Du Croz, Hammarling, and Hanson 1988).

We study some approaches to the solution of sparse linear equations that either exploit parallelism or at least provide the potential of doing so. Although many of our remarks apply to distributed-memory architectures, we are primarily concerned with shared-memory multiprocessors and it is on these machines that we have done our experimental work. Our main aim is to develop codes that are efficient on vector multiprocessor supercomputers.

We first discuss, in Section 2, some projects at CERFACS aimed at developing parallel sparse direct codes for general systems. In Section 3, we examine the use of a block iterative method based on partitioning the system and using a direct solver within the blocks. In Section 4, we introduce two new sparse codes being developed at the Rutherford Appleton Laboratory. We are primarily concerned with the efficient vectorization of these codes but our design should make them much more amenable to parallelism than the codes they are replacing. The codes are discussed in more detail in Sections 5 and 6. We make some concluding remarks in Section 7.

2 Parallel direct solvers

The main technique that we have been using at CERFACS to take advantage of parallelism in the solution of general sparse systems using direct methods has been based on the multifrontal approach of Duff and Reid (1983, 1984). The most important feature of this approach is that it takes advantage of two levels of parallelism, both our second level due to sparsity and the third level using level 3 BLAS. The natural parallelism of a sparse system is exposed through viewing the factorization of the matrix using an elimination tree where elimination operations associated with the nodes of the tree depend only on operations at descendant nodes in the tree. The other, and often more important, level is that all elimination operations are performed on small dense submatrices so that we can take advantage of parallelism by utilizing similar techniques to those employed in the solution of full systems of equations, namely block algorithms employing the level 3 BLAS (Dongarra, DuCroz, Duff, and Hammarling 1990).

The multifrontal method for solving

$$\mathbf{Ax} = \mathbf{b}, \tag{2.1}$$

where $\mathbf{A}$ is large and sparse is a direct method of solution based on Gaussian elimination. $\mathbf{A}$ is decomposed into triangular factors according to

$$\mathbf{PAQ} = \mathbf{LU}, \tag{2.2}$$

where $\mathbf{P}$ and $\mathbf{Q}$ are permutation matrices, $\mathbf{U}$ is an upper triangular matrix, and $\mathbf{L}$ is unit lower triangular. The decomposition (2.2) is then used to solve the system (2.1) through the forward elimination

$$\mathbf{Ly} = \mathbf{Pb}, \tag{2.3}$$

followed by the back substitution

$$U(Q^T x) = y. \tag{2.4}$$

We are here concerned with the most time-consuming part, the factorization (2.2). The kernel operation in forming the **LU** factors is just

$$a_{ij} = a_{ij} - a_{ik}[a_{kk}]^{-1}a_{kj}, \tag{2.5}$$

where $k < i, j \leq n$, n is the order of the system being solved, and a_{kk} is the pivot at stage k, $(k = 1,...n-1)$. Note that, in the sparse case, most of the a_{ik} and a_{kj} will be zero and so these operations need not be included in the calculations (2.5). The traditional way of doing this is to store the a_{ij} as sparse vectors so that the nonzero entries are stored contiguously (in array `A`, say) and their column indices in an integer array (`IND`, say). Then we need access only the nonzero entries using indirect addressing. For example, if we wish to update a row of the matrix held in positions `I1` to `I2` of `A` and `IND` by adding a multiple of another sparse vector which may have already been expanded into array `W`, then we may use a loop of the following kind

```
      DO 10 I = I1,I2
         A(I) = A(I) - AMULT*W(IND(I))
   10 CONTINUE
```

The most important feature of multifrontal codes is that the nonzeros in the pivot row and column have already (in an assembly step) been gathered into a small dense submatrix so that the operations (2.5) can be effected using regular direct addressing. Even though vector computers such as CRAYs and Alliants provide gather/scatter vector instructions, these operations are irregular and require additional memory traffic, degrading the performance.

A second important feature of the approach is the organization of the elimination steps in an assembly tree providing a partial ordering for the elimination sequence. For example, if we are factorizing a tridiagonal system, the diagonal entries 1, 3, 5, 7,... may be chosen as pivots simultaneously and independently since none directly affects the value of another. After this step (or rather steps), the diagonal entries 2, 6, 10, 14, ... can be used as pivots (again independently and simultaneously). The next set of pivots are from diagonals 4, 12, 20, ... Continuing in this way, if n is equal to a power of 2 less one, the elimination can be represented as a binary tree with $\log_2(n+1)$ levels and $(n+1)/2$ leaf nodes. In the general case, the tree is not a balanced binary tree but each node of the tree still corresponds to some eliminations on a full submatrix, and the edges of the tree still represent dependencies inasmuch as the elimination operations at any node can proceed as soon as the operations on the sons of the node are complete. The operations at each node of the tree correspond to one or more steps of Gaussian elimination on full matrices and so can be implemented using Level 2 and Level 3 BLAS.

At CERFACS, we have used these two features of the multifrontal approach to obtain good performance on a range of parallel computers. Much of this work is described in detail by Amestoy and Duff (1989, 1991) and in the thesis of Amestoy (1990).

We exploit the natural parallelism of the assembly tree by assigning separate processes to each node, keeping a list of executable processes (initialized to the leaf nodes), and allowing processors to execute processes on the list as soon as they are idle. This certainly provides plenty of parallelism in the early stages of the elimination, but as we progress to the root the number of parallel processes declines and only one process is present at the root itself. We see the overall effect of this in Table 2.1, where the code of Amestoy has been used. In each case the same Fortran code has been used although parallelism has been controlled through user-controllable parameters. The matrix used in these experiments was the matrix BCSSTK15 from the Harwell-Boeing Collection (Duff, Grimes, and Lewis 1989). This matrix is from a structural analysis application and has order 3948 with 117816 nonzero entries. These results indicate that, for typical sparse problems, the level of parallelism that can be obtained from the tree alone is very disappointing. Indeed, if we use the work of Duff and Johnsson (1989) to obtain theoretical estimates then we find that the results in Table 2.1 are due to the inherent lack of parallelism rather than any deficiency in the code.

Machine	Mflops	Speedups due to parallelism			
	1 proc	3 procs	4 procs	6 procs	8 procs
IBM 3090E	44	1.9			
IBM 3090J	60	1.9	2.0	2.1	
CRAY-2	176	1.7	1.8		
CRAY Y-MP	216	1.9	2.1	2.3	
Alliant FX/80	8	1.7	1.8	1.9	1.9

Table 2.1. Speedups due to tree only on matrix BCSSTK15 from Harwell-Boeing collection.

We resolve this problem by exploiting parallelism within the node, making use of the earlier observation that each nodal process corresponds to several steps of Gaussian elimination on a full submatrix. To obtain parallelism at this level we use a block algorithm for the decomposition and employ Level 2 and Level 3 Basic Linear Algebra Subprograms as their kernel (Dongarra, Du Croz, Hammarling, and Hanson, 1988, and Dongarra, Du Croz, Duff, and Hammarling, 1990, respectively). The efficiency and parallelism comes both from within efficient (machine dependent) implementations of the BLAS and also from being able to execute different calls to the BLAS in parallel. We use implementations similar to those described by Daydé and Duff (1989, 1990) who employ similar algorithms to those in the LAPACK project (Anderson, Bai, Bischof, Demmel, Dongarra, DuCroz, Greenbaum, Hammarling, McKenney, and Sorensen, 1990). We show, in Table 2.2, speedup figures on the same problem as earlier.

Machine	Mflops	Speedups due to parallelism			
	1 proc	3 procs	4 procs	6 procs	8 procs
IBM 3090E	44	2.4			
IBM 3090J	60	2.5	3.0	3.8	
CRAY-2	176	2.0	2.3		
CRAY Y-MP	216	2.7	3.3	4.1	
Alliant FX/80	8	2.4	3.0	3.9	4.3

Table 2.2. Speedups due to tree and nodes on matrix BCSSTK15 from Harwell-Boeing collection.

The main reason why this dual parallelism is effective is that the frontal matrices are invariably larger near the root and so the main benefit from nodal parallelism (through Level 3 BLAS) occurs just when the parallelism from the tree becomes less. It is this nice balance that produces the good speedups indicated in Table 2.2, where it should be noted that all speedups are relative to a high uniprocessor performance.

The decoupling of the indirect addressing operations (in the assembly) and the direct addressing operations (in the eliminations) enables us to measure the relative performance of each for a specified problem on a particular architecture and we show, in Table 2.3, such figures for a matrix arising from a nine-point discretization of the Laplacian operator on a 70×70 grid on a range of machines. Obviously it makes sense to move work from indirect to direct addressing.

Machine	Assembly	Elimination	Ratio
Alliant FX/80	0.3	3	10
IBM 3090E	1.5	23	15
ETA 10P	0.7	28	43
CRAY-2	2.1	79	37
CRAY Y-MP	2.2	139	63

Table 2.3. Megaflop rates for assembly (indirect addressing) and elimination (direct addressing).

One way of doing this and controlling the ratio of the work in assembly (corresponding to tree edges) and the nodal work (full eliminations) is to collapse the tree by combining adjacent nodes. Although this can lead to more entries in the factors and more operations of the kind (2.5), a judicious use of this tree amalgamation can lead to greater overall efficiency on vector and parallel architectures by trading off a reduction in costly indirect addressing assembly operations for the faster directly addressed elimination operations. This was used by Duff and Reid (1983) to improve vectorization on the CRAY-1 and has been explored in depth by Amestoy (1990). A crucial observation is that, at low levels of node amalgamation, the reduction in indirect operations significantly exceeds the increase in arithmetic. The best point for the trade off depends on the target machine but is easily controlled in our portable code by a single parameter governing the amount of node amalgamation.

Node amalgamation coupled with the use of Level 3 BLAS can lead to highly efficient code on vector machines. Amestoy (1990) has illustrated, by extensive experiments and analysis, that similar techniques can be used to obtain good parallelism without sacrificing the vector performance. For a wide range of matrices, uniprocessor performance of sometimes over half the peak performance of the machine is obtained. A typical performance rate is shown in the second column of Tables 2.1 and 2.2 with speedups over these figures, due to parallelism, of the order shown in Table 2.2. It should be stressed that the number of actual floating-point operations required to factorize the matrix is near minimum since a minimum degree ordering has been used to determine the pivotal sequence used in constructing the elimination tree. Thus these Megaflops are all doing useful and necessary work in solving the problem. A higher rate could obviously be obtained by solving the system as a full matrix but the execution time would be much longer. Furthermore, we re-emphasise that the same code is run on each target machine, the only differences being in the values of user-controllable parameters and the use of tuned BLAS, transparent at the level of our Fortran source code.

Present work in this area includes better data management algorithms and the porting and tuning of the multifrontal code to the BBN TC 2000. Further work is also under way to improve the software design and interface so that the codes used by Amestoy (1990) can be made publicly available.

Other approaches being developed at CERFACS include the work on the D2 algorithm by Davis (Davis and Yew 1990). The D2 algorithm exploits parallelism by obtaining maximal independent pivot sets but does not use any full matrix algebra. The algorithm combines the symbolic and numerical phases into a single phase. Its key feature is a non-deterministic parallel pivot search that builds a set of, say m, *independent pivots* which when permuted to the diagonal form a diagonal $m \times m$ submatrix. The pivot search is followed by a parallel rank-m sparse update that takes advantage of the independence between the m pivots but uses sparse algebra and does not use the Level 2 and 3 BLAS. These two steps alternate until a switch is made to dense matrix code or until the matrix is factorized.

To a varying degree, both the multifrontal algorithm and D2 are designed to take advantage of the vector and parallel characteristics of typical high-performance computers. Their different design goals, however, can lead to a wide range in relative performance for different classes of matrices.

D2 uses a parallel unsymmetric ordering that is merged with the numerical phase, so it does not need to make any a priori assumptions about numerical accuracy. An algorithm such as the multifrontal method that has separate symbolic and numerical phases must make this assumption, and its performance can be degraded if numerical pivoting causes too much fill-in and extra work. The multifrontal method uses the pattern of $A+A^T$ throughout the factorization, while D2 uses the unsymmetric pattern of the original matrix A. Therefore D2 should have less fill-in for matrices with a highly asymmetric nonzero pattern. Of course, although the fill-in and number of arithmetic operations may be less, the use of sparse algebra in the inner loops will offset this gain so that we might expect the performance to be somewhat similar for all except very unsymmetric patterns. The balance between these two opposing factors is illustrated by the results in Table 2.4.

More recent work involves the extension of multifrontal techniques to unsymmetric structures. Although this work is still at a fairly early stage, some encouraging results are already being obtained (Davis and Duff 1991). In this work, the idea is to combine the efficiency of full matrix working within a frontal scheme with an efficient ordering for unsymmetric systems thus obtaining both benefits that the two previous algorithms were trying to achieve. Many of the ideas carry over from the symmetrically structured case although the underlying graph becomes a directed acyclic graph rather than a tree. Because it is really not sensible in the general unsymmetric case to perform a symbolic factorization and then perturb our elimination graphs according to subsequent numerical pivoting, we perform a numerical factorization at the same time as the original analysis when generating the elimination graph. We show some very preliminary results from this work in Table 2.4.

Matrix			
Name	LNS3937	SHERMAN5	GRE1107
Order	3937	3312	1107
Entries	25407	20793	5644
Asymmetry	0.15	0.26	1.00
Entries in LU			
Symmetric multifrontal	292231	177818	181143
D2	705240	308405	68305
Unsymmetric multifrontal	533829	152011	53324
MA28	417603	163468	44980
Millions of flops			
Symmetric multifrontal	21.6	16.0	24.0
D2	174.7	39.7	3.6
Unsymmetric multifrontal	98.3	13.9	3.2
Time (secs on Alliant FX/80)			
Symmetric multifrontal	12.0	8.9	7.2
D2	101.7	40.9	4.5
Unsymmetric multifrontal	129.9	20.3	10.2

Table 2.4. Performance of codes on some Harwell-Boeing matrices.

In Table 2.4, we have compared the three direct codes we have just discussed on three matrices from the Harwell-Boeing test set. The three matrices have been chosen to illustrate the effect of different levels of asymmetry, where we define the asymmetry by

$$\frac{\text{Number of pairs such that } a_{ij}=0, a_{ji}\neq 0}{\text{Total number of off-diagonal entries}}$$

so that symmetrically structured matrices have an asymmetry value of 0.0. We have called the code of Amestoy and Duff "Symmetric multifrontal" and that of Davis and Duff "Unsymmetric multifrontal". For an added comparison, we have included the number of entries in **LU** produced by a general unsymmetric code MA28 (Duff 1977, and Section 5). We note that, on the most symmetric example (LNS3937), the symmetric multifrontal code performs best in all measures in the table. It is significantly better than the other methods in the number of entries in **LU** because it is usually better to restrict pivoting to the diagonal for such systems and the other codes do not do this. Because of the relaxation of sparsity pivoting controls to enable D2 to obtain reasonably sized independent sets of pivots, this method can do quite badly although when the matrix is very unsymmetric (GRE1107) it becomes much better than the symmetric multifrontal method and competitive with the other unsymmetric methods. D2 could then be an attractive method because of its greater degree of parallelism. The relatively poor times of the unsymmetric multifrontal method reflect its early stage of development. Thus it is perhaps best not to dwell on the precise figures in Table 2.4 but rather to use them to gain an appreciation of the relative merits of our approaches over a range of systems.

3 Block iterative methods

Recently, there has been much activity in preprocessing the coefficient matrix with another matrix or matrices so that iterative methods will have good convergence. The preprocessing or preconditioning matrix is chosen so that the solution to the original problem can be easily recovered from the solution to the preconditioned one. Another related approach that we discuss here is to subdivide the problem, solve the subproblems using direct methods, and recover the solution to the overall system by iterating over the subproblems. We describe such a block iterative technique in this section.

Our method is based on a block version of Cimmino's method (Sloboda 1988), where the blocks are obtained by partitioning **A** into block rows. We choose to solve the underdetermined systems corresponding to the blocks by a direct method which uses a sparse indefinite solver on a symmetric augmented system obtained from the underdetermined system. We accelerate our basic scheme using conjugate gradients and find that this is strongly influenced by the partitioning of the original system. We further improve the convergence by preconditioning the conjugate gradient acceleration. The work is described in detail by Arioli, Duff, Noailles, and Ruiz (1990). We present here a brief outline.

We obtain the blocks by partitioning the system $\mathbf{Ax}=\mathbf{b}$ as

$$\begin{pmatrix} \mathbf{A}^1 \\ \mathbf{A}^2 \\ \cdot \\ \cdot \\ \mathbf{A}^p \end{pmatrix} \mathbf{x} = \begin{pmatrix} \mathbf{b}^1 \\ \mathbf{b}^2 \\ \cdot \\ \cdot \\ \mathbf{b}^p \end{pmatrix} \tag{3.1}$$

where $1 \leq p \leq n$.

If we define by $\mathbf{P}_{\mathcal{R}(\mathbf{A}^{i^T})}$ the projector onto the range of $\mathbf{A}^{i^T}$, and the pseudo-inverse of $\mathbf{A}^i$ as $\mathbf{A}^{i^+}$, then we can describe the block Cimmino algorithm in the following way :

Algorithm 3.1 (Block Cimmino Method)

Choose $\mathbf{x}^{(0)}$

repeat until convergence

begin

compute in parallel

$$i=1,\ldots,p \quad \delta^{i\,(k)} = \mathbf{A}^{i+}\mathbf{b}^{i} - \mathbf{P}_{R(\mathbf{A}^{iT})}\mathbf{x}^{(k)} = \mathbf{A}^{i+}(\mathbf{b}^{i} - \mathbf{A}^{i}\mathbf{x}^{(k)}) \tag{3.2}$$

end parallel

$$\mathbf{x}^{(k+1)} = \mathbf{x}^{(k)} + \omega\sum_{i=1}^{p}\delta^{i\,(k)}$$

end

When solving the subproblems (3.2), we use the augmented system

$$\begin{bmatrix} \mathbf{G} & \mathbf{A}^{iT} \\ \mathbf{A}^{i} & \mathbf{0} \end{bmatrix} \begin{bmatrix} \mathbf{u}^{i} \\ \mathbf{v}^{i} \end{bmatrix} = \begin{bmatrix} \mathbf{0} \\ \mathbf{b}^{i} - \mathbf{A}^{i}\mathbf{x} \end{bmatrix} = \begin{bmatrix} \mathbf{0} \\ \mathbf{r}^{i} \end{bmatrix} \tag{3.3}$$

with solution

$$\mathbf{v}^{i} = -(\mathbf{A}^{i}\mathbf{G}^{-1}\mathbf{A}^{iT})^{-1}\mathbf{r}^{i}$$

$$\mathbf{u}^{i} = \mathbf{G}^{-1}\mathbf{A}^{iT}(\mathbf{A}^{i}\mathbf{G}^{-1}\mathbf{A}^{iT})^{-1}\mathbf{r}^{i} \ .$$

We use the augmented system approach because it is more stable than the normal equations (Arioli, Duff, and de Rijk 1989). It is also trivial to include a scaling matrix in the augmented systems, as indicated by the matrix $\mathbf{G}$ in (3.3). The solution to (3.2) would have $\mathbf{G}=\mathbf{I}$, but by including other matrices for $\mathbf{G}$ the iterative scheme can be accelerated. We solve the augmented systems with the sparse symmetric linear solver MA27 from the Harwell Subroutine Library (Duff and Reid 1982). The solver MA27 is a multifrontal method that computes the $\mathbf{LDL}^T$ decomposition of a permutation of the augmented matrix, using a mixture of 1×1 or 2×2 pivots chosen during the numerical factorization. The resulting $\mathbf{L}$ and $\mathbf{D}$ factors are then used to solve the augmented systems by forward and backward substitution in the usual way.

For the block Cimmino algorithm, with $\omega=1$, the iteration matrix is of the form

$$\mathbf{I}-\sum_{i=1}^{p}\mathbf{A}^{iT}(\mathbf{A}^{i}\mathbf{A}^{iT})^{-1}\mathbf{A}^{i}=\mathbf{I}-\mathbf{H}, \tag{3.4}$$

where $\mathbf{H}$ is symmetric and positive definite if the $\mathbf{A}^{i}$ all have full row rank. Thus it is appropriate to use the method of conjugate gradients to solve the iteration equations (Hageman and Young, 1981). An important aspect of using conjugate gradient acceleration is that we can show that the method is independent of the ω in algorithm 3.1 (Arioli, Duff, Noailles, and Ruiz 1990), which is why we set it equal to 1 above. Arioli *et al.* (1990) also show that the matrix $\mathbf{G}$ in (3.3) can be viewed as a preconditioner for the conjugate gradient iteration.

From a point of view of theoretical analysis and to assist in understanding the influence of different partitionings, it is advantageous to examine the case of a two-block partitioning. There is little loss of generality in doing this since, as we shall illustrate later, any partition can be viewed as a two-block one. In a two-block partitioning, we partition the matrix $\mathbf{A}$ in two blocks

$$\mathbf{A} = \begin{bmatrix} \mathbf{B}^1 \\ \mathbf{B}^2 \end{bmatrix} ,$$

where $\mathbf{B}^1$ and $\mathbf{B}^2$ have n_1 and n_2 rows respectively. We assume without loss of generality that $n_1 \geq n_2$. With such a partitioning, the iteration matrix can be written in the form

$$\begin{pmatrix} \mathbf{0}_{n_1 \times n_1} & \mathbf{S} \\ \mathbf{S}^T & \mathbf{0}_{n_2 \times n_2} \end{pmatrix}, \tag{3.5}$$

where the off-diagonal matrix $\mathbf{S}$ is related to the principal angles between the subspaces $R(\mathbf{B}^{1^T})$ and $R(\mathbf{B}^{2^T})$, and there are at most $2n_2$ nonzero eigenvalues. The finite termination property of the conjugate gradient algorithm suggests that our method in general works better for small n_2, and in exact arithmetic takes not more than $2n_2$ steps for convergence.

Although our algorithm is quite general, the efficiency is helped by matrix structure and in particular the method can benefit from the structure of block tridiagonal matrices. Matrices of this structure are very common when discretizing partial differential equations. Consider, for instance, a block tridiagonal matrix partitioned in block rows $\mathbf{A}^i$ such that each $\mathbf{A}^i$ has $k \times l$ rows, $k \geq 2$, l being the size of the blocks. It can be reordered as

$$\begin{bmatrix} \mathbf{B}^1 \\ \\ \\ \mathbf{B}^2 \end{bmatrix} = \begin{bmatrix} \mathbf{A}^1 \\ \mathbf{A}^3 \\ \mathbf{A}^5 \\ \mathbf{A}^2 \\ \mathbf{A}^4 \end{bmatrix} = \begin{bmatrix}
* & * & & & & & & & & & & & \\
* & * & * & & & & & & & & & & \\
 & * & * & * & & & & & & & & & \\
 & & & & * & * & * & & & & & & \\
 & & & & & * & * & * & & & & & \\
 & & & & & & * & * & * & & & & \\
 & & & & & & & & & * & * & * & \\
 & & & & & & & & & & * & * & * \\
 & & & & & & & & & & & * & * \\
 & & * & * & * & & & & & & & & \\
 & & & * & * & * & & & & & & & \\
 & & & & & & & * & * & * & & & \\
 & & & & & & & & * & * & * & &
\end{bmatrix}, \tag{3.6}$$

so that we have effectively the two-block partitioning shown earlier but with a good degree of independence evident within $\mathbf{B}^1$ and $\mathbf{B}^2$.

We can vary the number of zero eigenvalues in the iteration matrix; setting it close to zero by taking the same size for the $\mathbf{A}^i$ in $\mathbf{B}^1$ as for the $\mathbf{A}^i$ in $\mathbf{B}^2$ ($n_1 \sim n_2$), or setting it as big as possible ($n_1 \gg n_2$) by taking large blocks $\mathbf{A}^i$ in $\mathbf{B}^1$, and defining small interface blocks in $\mathbf{B}^2$ of size $2 \times l$ (the minimum required for making the $\mathbf{A}^i$ in $\mathbf{B}^1$ structurally orthogonal). A good partitioning strategy is to compromise between reducing the size of the interface block $\mathbf{B}^2$, with the aim of reducing the number of CG iterations, and maintaining the degree of parallelism of the method.

We will use three different test problems. The first two problems come from simulations of real flows developed at CERFACS by the fluid dynamics team and the remaining one is from the literature.

Problem 1, from a model developed by Françoise Perrel, concerns the study of a body entering the atmosphere at a high mach number. It is based on a finite-volume discretisation of the Navier Stokes equations coupled with chemistry. This leads to block tridiagonal matrices of order 20700 where the blocks are of order 345. The matrix is unsymmetric and non-diagonally dominant and, in this case, we do not require the solution to be very accurate.

Problem 2, from a model developed by Per Weinerfelt, is from the study of a two-dimensional wing profile at transonic flow (with no chemistry effects). It is still based on finite-volume discretisation, but this time of the Euler equations. This leads to unsymmetric but diagonally dominant block tridiagonal matrices of order 7680 with each block of order 240. This time the test problem we are solving is close to the steady-state solution and a very accurate solution is required.

Problem 3 comes from the paper by Kamath and Sameh (1989) and is used also in similar work by Bramley and Sameh (1990). It is obtained from the finite-difference solution on a 64 by 64 grid of the equation

$$-u_{xx} - u_{yy} + 1000\, e^{xy}\, u_x - 1000\, e^{xy}\, u_y = g,$$

where g is calculated assuming the solution $u = x+y$.

Since we are using an 8 processor Alliant, we consider the three following partitionings, which are a subset of those considered by Arioli, Duff, Noailles, and Ruiz (1990).

Partitioning 1 has eight equal sized blocks and so $n_1 = n_2$ in (3.5) when the partitioning is viewed as a two-block one.

Partitioning 2 has the smallest possible size for the blocks in the interface. This should decrease the number of iterations at the possible cost of a reduction in parallelism.

Partitioning 3 is the one used by Kamath and Sameh (1989) and is defined by taking equal-sized blocks of the smallest possible size (two times the number of rows in one block of the block tridiagonal structure). This partitioning gives good load balancing, but it has a large interface block. It has the added advantage, however, that it minimises the overall fill-in in the factorization of the different augmented systems because they are all of the smallest size.

All the tests are performed on an Alliant FX/80 at CERFACS in double precision (64 bit words) with machine precision 2.2 10^{-16}. For the convergence criterion, we use the scaled residual defined by

$$\frac{\| \mathbf{A}\, \mathbf{x}^{(k)} - \mathbf{b} \|_\infty}{\| \mathbf{A} \|_\infty \| \mathbf{x}^{(k)} \|_1 + \| \mathbf{b} \|_\infty}$$

at step k, and we require this value to be less than or equal to 10^{-7} for problem 1 and less than 10^{-14} for problems 2 and 3. For a full discussion of this stopping criterion see Arioli, Duff, and Ruiz (1991).

	Iterations	Storage Kwords	Time secs (FX/80)	Speedup
Order: 20700				
Entries: 511050				
Part. 1	24	5249	163	4.0
Part. 2	18	6728	299	4.5
Part. 3	33	1743	47	6.1
Order: 7680				
Entries: 113760				
Part. 1	16	782	16	4.9
Part. 2	16	1049	23	4.5
Part. 3	19	547	13	6.1
One part	1	5249	799	–

Table 3.1. Effect of partitioning on the Block Cimmino method on an Alliant FX/80 (8 processors).

The results in Table 3.1 illustrate the power of our method and the effect of the partitioning. In most cases, we converge very quickly to the desired solution. The results support the comments made when we introduced the partitionings. The number of iterations is least for partitioning 2 where n_2 of (3.5) is smallest. However, in this partitioning the blocks in $\mathbf{B}^1$ are larger than in partitionings 1 and 3 so that the storage and time requirement of the direct method part of the computation are high. Sometimes the reduction in storage for the factors when we work with smaller blocks means that the iterations take less time although there are more of them, particularly in problem 1. The overall effect is that partitioning 3 is the fastest although a better direct solver could well alter the balance. For the direct solution of the subproblems, a new version of MA27 (see Section 6) that recognizes and exploits the zeros on the diagonal and behaves much better in terms of fill-in is being developed (Duff, Gould, Reid, Scott, and Turner 1991) and should improve all the results. Also, as we will see later, partitioning 3 cannot be used with the preconditioner that we will examine later in this section. Good speedups are obtained, particularly with partitioning 3. The entry "One part" for the second problem gives the results from running a direct code on an augmented system corresponding to the whole problem. Naturally, if the original system was square and nonsingular it would normally be much better to employ the direct method on the original coefficient matrix. There was insufficient memory on our Alliant to run "One part" on problem 1!

We now introduce a preconditioning associated with a two-block partitioning. The aim is to make the two subspaces $R(\mathbf{B}^{1\,T})$ and $R(\mathbf{B}^{2\,T})$ orthogonal, so that the eigenvalues of the iteration matrix (3.5) will all be equal to zero. The solution would then be obtained in one iteration only.

The simplest way to make two sets of rows in a matrix orthogonal is to multiply their overlapping part by zero. But obviously, if we do this, the resulting system can be rank deficient. We instead divide these overlapping columns by α, leading to a diagonal matrix $\mathbf{G}$ with α^2 in positions corresponding to the overlapping columns and 1 elsewhere. For example block columns 3 and 4 of the matrix in (3.6) would be so scaled to partially decouple the blocks $\mathbf{A}^1$ and $\mathbf{A}^2$.

Of course, with such a preconditioner, we expect some numerical instability in the factorization of the augmented systems because of bad scaling. Also, the overlapping part must not involve all the columns of the matrix, because in that case the preconditioning would be equivalent to a scaling of the complete matrix, which would not modify at all the principal angles between the two spaces. This means that we cannot use this preconditioning with partitioning 3.

For problem 3, we examine a preconditioning using the scaling matrix $\mathbf{G}$ of this type. A value of α equal to 10 was found best so we show the results from using this value compared with the unscaled problem (α equal to 1) in Table 3.2.

	Number iterations	Time (secs) Alliant FX/80
PART 1		
$\alpha = 1$	193	26
$\alpha = 10$	95	19
PART 2		
$\alpha = 1$	182	35
$\alpha = 10$	105	28

Table 3.2. Effect of scaling (preconditioning) on Problem 3 (Order 4096, Number of entries 20224).

From the results in Table 3.2 we notice that the scaling substantially reduces the number of iterations required showing that the subspaces are, as intended, more independent. The reason that the time for solution is not reduced by the same proportion is that the subproblems become more ill-conditioned because of the scaling and thus are more difficult to solve. The increase in time in such instances is caused partly by the fact that ill-conditioning causes the actual pivotal order to be further from that predicted by the analysis of the multifrontal MA27. If partitioning 3 is run on the same problem then 379 iterations are required and the execution time is 55 seconds on the Alliant FX/80. Additionally, we cannot use the preconditioner with this partition.

4 Two sparse codes

At the Rutherford Appleton Laboratory, we are presently concerned with substantial modifications of two of our most popular sparse codes from the Harwell Subroutine Library: MA28 (Duff 1977) for the solution of general sparse unsymmetric equations, and MA27 (Duff and Reid 1982) for the solution of sparse symmetric indefinite problems.

The project to develop the MA28 suite will produce a new package, MA48, that will contain all the functionality of the old but will be much more efficient on vector machines and will provide some additional features to the earlier code. These include switching to dense code, a more reliable iterative refinement option, a much faster routine for factorizing matrices with the same structure as matrices already factorized, and faster solve and pattern analysis algorithms. The previous code was written in Fortran 66 and, by using Fortran 77, the new code is cleaner and easier to maintain. We examine this project more closely in Section 5.

The second project is a substantial revision of the multifrontal code MA27 for the solution of symmetric indefinite problems. The revisions in this project, MA47, include better vectorization using the Level 2 and Level 3 Basic Linear Algebra Subprograms and new algorithms so that matrices of the form

$$\begin{pmatrix} \mathbf{H} & \mathbf{A}^T \\ \mathbf{A} & \mathbf{0} \end{pmatrix} \tag{4.1}$$

can be efficiently factorized. The MA27 code did not fully exploit the zero (2,2) block in (4.1). Systems with coefficient matrices of the form (4.1) occur very commonly in constrained optimization and least squares, an example being given by equation (3.3) in the previous section. We examine this project more closely in Section 6.

In both cases, care has been taken to reduce the overhead of indirect addressing and perform operations in full arithmetic as much as possible. We use the Level 3 BLAS routines or indeed full matrix code for most of the numerical computations. This gives the prospect of enhanced performance on many machines since optimized versions are widely available. The new designs also allow scope for a greater degree of parallelism than with the earlier codes.

5 THE MA48 PACKAGE

The MA28 package for solving sparse sets of linear equations, using an **LU** factorization of the sparse coefficient matrix **A**, was written in 1976 and was available in the Harwell Subroutine Library from 1977 (Duff 1977). Although there have been some significant improvements to the package since then, its basic design has remained unchanged and thus reflects the state-of-the-art of over a decade ago. We are currently designing a new package, called MA48, altering the user interface, the facilities offered and algorithms used, and coding an implementation that better exploits the capabilities of architectures of recent machines.

Before we discuss the new package, let us first briefly review the design of MA28. The main entry implements an ANALYSE-FACTORIZE phase so that the pattern of the matrix and its numerical values are analysed and the **LU** factors are generated and output. It uses a pivoting strategy based on the algorithm of Markowitz with threshold pivoting. An option exists for first permuting the matrix to block triangular form. A separate FACTORIZE entry performs the factorization of a new matrix with the same sparsity structure as the first, using information from the earlier ANALYSE-FACTORIZE. It is assumed that the pivot sequence from the first factorization is still appropriate (if not, a fresh ANALYSE-FACTORIZE is needed). Finally, the factors can be used to solve sets of equations in the SOLVE entry. Modifications in a revision in 1983 included the use of drop-tolerances in ANALYSE-FACTORIZE and an iterative refinement option in SOLVE.

An unsatisfactory feature of the MA28 ANALYSE-FACTORIZE is that the user has to guess how many fill-ins will be generated and provide the corresponding amount of storage. For some problems this can be very much greater than that needed to store the original matrix. Since ANALYSE-FACTORIZE is much slower than FACTORIZE (typically 4 to 7 times slower), we have decided that it is reasonable to change to an ANALYSE code that needs a subsequent FACTORIZE call, particularly if we can reduce the heavy storage requirements of ANALYSE-FACTORIZE. We keep only the Markowitz ordering, and so only need hold the active matrix until a switch to full code occurs. We throw away the pivotal rows and columns once they have been used and do not even generate them for the full part. We also have options for restricting pivot choices to certain rows and columns — an important tool for handling some kinds of structure. As with the earlier code, the input matrix can be singular or rectangular. We see the substantial gains in storage requirements over a range of Harwell-Boeing matrices in Table 5.1, with the ANALYSE phase often requiring little additional storage to that needed for the input matrix.

Order	Entries	MA28	MA48
185	1005	4880	2114
216	876	3505	1054
343	1435	7258	2661
512	2192	13803	5354
1107	5664	49234	19133
147	2449	5777	2836
1176	18552	22615	18552

Table 5.1. Number of entries stored by ANALYSE.

Largely because we can terminate the ANALYSE phase as soon as we switch to full code, the execution times are much reduced. As we see in Table 5.2 (and the relative performance is typical of a much wider range of matrices), the new ANALYSE code is up to nearly ten times faster than the old, thus giving the unusual benefit of significant savings in both storage and time. In Table 5.2, "Threshold" is the density of the reduced matrix when the code switches to full code. A value of 1.0 means that the reduced matrix is full, while we use 2.0 to indicate that no switch occurs. In the ANALYSE phase nothing is of course done when we switch to full code except a completion of the permutation arrays. Of course we then need to do a subsequent FACTORIZE, the phase which we now examine.

Order	1107	1176
Entries	5664	18552
MA28	6.00	1.7
MA48		
Threshold		
2.0	5.82	1.50
1.0	4.98	0.80
0.5	2.22	0.53
0.2	1.29	0.28
0.1	1.04	0.19

Table 5.2. Variation of ANALYSE times with the density at which the switch to full code is made (CRAY-2 uniprocessor secs).

We have decided to include a FACTORIZE option that uses the input row order but allows partial pivoting by columns because this means that a fresh ANALYSE is not needed if the entries change too much for the original pivot order to be satisfactory. We construct and use a mapping vector for the permutation from the user's ordering to the ordering of the factors with all their fill-in, thus avoiding a potentially expensive sort. A fast sparse triangular solve is used (for example, Gilbert and Peierls, 1988) giving a significant performance improvement over our earlier code. As a further aid to vectorization, we also include in FACTORIZE the option of switching to full code when the reduced matrix reaches a specified density. Even with pivoting included and only the mapping information in addition to the input matrix, this entry is usually faster than the MA28 FACTORIZE, which does not have pivoting and requires full information on the factors. We also provide a FACTORIZE entry in the new code that uses the same pivotal sequence as an earlier FACTORIZE.

In FACTORIZE when the switch is made to full code, the routine SGEFA from the Cray scientific library SCILIB is used. This subroutine is coded in assembler for high performance on the CRAY-2 but has exactly the same interface as the similarly named routine from LINPACK so that portability is not compromised. In Table 5.3, the switch is made only when the reduced matrix is full but even then there are significant gains over the MA28 code. If we switch over at lower densities there are further gains and the storage penalty is not great until we switch at quite low densities. The "Slow" entry shows greater improvement because the switching cuts the overhead of the integer processing used in generating the factors. This integer overhead does not vectorize as well as the sparse factorization itself. At the optimal switch over density, the new "Fast" entry can be over ten times as fast as the MA28 code.

Order	Entries	MA28	MA48	
			Slow	Fast
185	1005	0.041	0.038	0.007
216	876	0.021	0.025	0.006
343	1435	0.061	0.061	0.011
512	2192	0.158	0.163	0.024
1107	5664	0.855	0.788	0.105
147	2449	0.052	0.077	0.011
1176	18552	0.258	0.202	0.031

Table 5.3. FACTORIZE times on uniprocessor CRAY-2. Here the new code switches to dense code when the reduced matrix is full.

Order Entries	1107 5664			1176 18552		
	Time			Time		
	Slow	Fast	Storage	Slow	Fast	Storage
Threshold						
2.0	0.978	0.129	84.8	0.359	0.045	39.9
1.0	0.788	0.105	78.0	0.202	0.031	33.5
0.5	0.290	0.064	74.8	0.144	0.031	39.8
0.2	0.177	0.076	101.7	0.114	0.059	81.4
0.1	0.189	0.121	154.7	0.180	0.140	162.8

Table 5.4. Variation of FACTORIZE with the density at which the switch to full code is made (CRAY-2 uniprocessor times).

The SOLVE entry again employs iterative refinement but uses the theory of Arioli, Demmel, and Duff (1989) to obtain a good stopping criterion and an accurate estimate of the error in the solution. We show the effectiveness of this over a wide range of test cases. We have also streamlined the SOLVE algorithm for better vectorization and have more than doubled the speed on the CRAY-2, for example. Note also that, because we now access the matrix by columns the direct, as opposed to the transpose, solution is more efficient on the CRAY-2 because of better vectorization of the inner loop. The full code used by SOLVE after the switch over is the optimized assembler code SGESL from the CRAY Library SCILIB. There is a smaller advantage in switching at lower densities in SOLVE than in the other entries.

Order	Entries	MA28	MA48 $Ax = b$	MA48 $A^T x = b$
185	1005	0.0020	0.0010	0.0016
216	876	0.0020	0.0011	0.0016
343	1435	0.0034	0.0017	0.0027
512	2192	0.0054	0.0027	0.0043
1107	5664	0.0126	0.0068	0.0097
147	2449	0.0017	0.0009	0.0015
1176	18552	0.0126	0.0057	0.0094

Table 5.5. SOLVE times on uniprocessor CRAY-2. Here the new code switches to full code when the reduced matrix is full.

6 The MA47 package

The Harwell code MA27 performs the direct solution of a set of linear equations (3.1) whose matrix $\mathbf{A}$ is sparse, symmetric and possibly indefinite. An initial ANALYSE chooses the pivot sequence but does no numerical factorization. FACTORIZE and SOLVE are similar to those in MA48 with the FACTORIZE entry checking the numerical suitability of pivots and altering the proposed sequence if necessary. We will summarize the improvements considered by Duff, Gould, Reid, Scott, and Turner (1991) on the basis of an experimental code and describe some of the further improvements planned to be incorporated in MA47, the replacement for MA27.

MA27 is a multifrontal code and owes its efficiency to performing the elimination operations within a full symmetric matrix (called the frontal matrix) with a row for each nonzero of the pivot column (or columns for 2×2 pivots). It chooses its pivotal sequence by assuming that the matrix is definite (so that any diagonal entry is suitable as pivot) and represents each frontal matrix by the list of row indices involved. For nondefinite matrices, the resulting sequence is used only as a guide. During actual factorization, a pivot step is not performed if the pivot fails to satisfy a numerical threshold criterion. As a consequence, later frontal matrices may be larger than was anticipated when the pattern was analysed; any of the extra rows are potentially pivotal, as well as the row(s) anticipated to be pivotal at this stage. Progress may be made by using a 2×2 pivot with a large enough off-diagonal entry for an appropriate threshold criterion to hold.

In some nondefinite cases, the pivot sequence chosen in this way is far from optimal. Suppose the frontal matrix has the form

$$\begin{pmatrix} \mathbf{A}_1 & \mathbf{A}_2 \\ \mathbf{A}_2^T & \mathbf{0} \end{pmatrix} \tag{6.1}$$

and a 2×2 pivot with one entry in each block is available. Since using a 2×2 pivot is mathematically equivalent to using each of its off-diagonal entries in turn as 1×1 pivots, it is easily seen that such a

2×2 pivot preserves the zero block. Now the zero block can be large, so this property can lead to significant computational savings. Indeed, $\mathbf{A}_1$ may have order 1, in which case the generated matrix consists only of the zero submatrix and need not be kept at all. There is no possibility of the ANALYSE phase of MA27 choosing such pivots since it assumes frontal matrices are full. In MA47 we therefore track which diagonal entries are certainly zero and hold generated matrices in the form

$$\begin{pmatrix} \mathbf{0} & \mathbf{A}_3 & \mathbf{A}_4 \\ \mathbf{A}_3^T & \mathbf{A}_5 & \mathbf{A}_6 \\ \mathbf{A}_4^T & \mathbf{A}_6^T & \mathbf{0} \end{pmatrix} \tag{6.2}$$

of which (6.1) is a special case.

The success of this strategy is illustrated by the results from Duff *et al.* (1991) shown in Table 6.1. The problems have matrices of the form

$$\begin{pmatrix} \mathbf{D} & \mathbf{N} \\ \mathbf{N}^T & \mathbf{0} \end{pmatrix}, \tag{6.3}$$

where $\mathbf{N}$ is a Netlib matrix (Gay, 1985) and $\mathbf{D}$ is a diagonal matrix with diagonal entries that are either 1 or 0. The problems represent constrained least-squares problems. It can be seen, from the results in Table 6.1, that the improvements can be dramatic.

Problem		Capri	Share1b	FFFFF800	E226	Beaconfd
Number of rows in **N**		466	253	1028	472	295
Number of cols in **N**		271	117	524	223	173
Number of entries		2190	1387	7247	3050	3633
Predicted flops	Original	136381	34658	1433091	196208	158850
	Revised	35355	9632	139376	11557	31990
Actual flops	Original	246713	48862	2849355	440361	387451
	Revised	37844	10556	164579	11582	43077

Table 6.1. A comparison of the proposed strategies for the ANALYSE phase of MA27.

7 Conclusions

We have given a brief review of several techniques for the solution of sparse systems of equations that exploit parallelism or at least provide the potential of so doing. Space does not permit us to go into great depth on any of these methods so we strongly encourage the interested reader to explore the various approaches further by consulting the references that follow.

Our main emphasis in all methods is to use the natural parallelism from sparsity while also benefiting from sub-calculations using full matrix algebra and the Level 3 BLAS. To a greater or lesser extent all the methods that we have discussed do this and their relative performance will depend both on the target machine and the structure of the problem under consideration. As always there is no definite "best buy" although we hope that readers will obtain a deeper understanding of the issues involved. Through this, it is possible to get a feeling for the best approach in any particular case in advance of or without the need for any extensive and possibly expensive experimentation.

References

Amestoy, P.R. (1990). Factorization of large unsymmetric sparse matrices based on a multifrontal approach in a multiprocessor environment. PhD Thesis. CERFACS, Toulouse.

Amestoy, P.R. and Duff, I.S. (1989). Vectorization of a multiprocessor multifrontal code. *Int. J. Supercomputer Applics.* **3** (3), 41-59.

Amestoy, P.R. and Duff, I.S. (1991). Efficient and portable implementation of a multifrontal method on a range of MIMD computers. *SIAM Symposium on sparse matrices,* May 1989, Salishan, Oregon. To appear.

Anderson, E., Bai, Z., Bischof, C., Demmel, J., Dongarra, J., DuCroz, J., Greenbaum, A., Hammarling, S., McKenney, A., and Sorensen, D. (1990). LAPACK: A portable linear algebra library for high-performance computers. LAPACK Working Note 20. Report CS-90-105, Computer Science Department, The University of Tennessee, Knoxville, Tennessee. To appear in Proceedings of Supercomputer '90, ACM Press, New York.

Arioli, M., Demmel, J.W., and Duff, I.S. (1989). Solving sparse linear systems with sparse backward error. *SIAM J. Matrix Anal. and Applics.* **10**, 165-190.

Arioli M., Duff I.S., Noailles, J., and Ruiz, D. (1990). A block projection method for sparse matrices. Report RAL-90-093, Rutherford Appleton Laboratory, Chilton, Didcot, Oxfordshire OX11 0QX. Report TR/PA/90/31, CERFACS, Toulouse. To appear in *SIAM J. Sci. Stat. Comput.*

Arioli M., Duff I.S., and de Rijk P.P.M. (1989). On the augmented system approach to sparse least-squares problems. *Numerische Math.* **55**, 667-684.

Arioli, M., Duff, I.S., and Ruiz, D. (1991). Stopping criteria for iterative solvers. Technical Report, CERFACS, Toulouse. To appear.

Bramley, R. and Sameh, A. (1990). Row projection methods for large nonsymmetric linear systems. Report, 957. Center for Supercomputing Research and Development, University of Illinois at Urbana-Champaign. To appear in *SIAM J. Sci. Stat. Comput.*

Davis, T.A. and Yew, P-C. (1990). A stable nondeterministic parallel algorithm for general unsymmetric sparse LU factorization. *SIAM J. Matrix Anal. and Applics.* **11**, 383-402.

Davis, T.A. and Duff, I.S. (1991). Design and implementation of an unsymmetric multifrontal algorithm. Technical Report, CERFACS, Toulouse. To appear.

Daydé, M. and Duff, I.S. (1989). Use of Level 3 BLAS in LU factorization on the CRAY-2, the ETA 10-P, and the IBM 3090/VF. *Int. J. Supercomputer Applics.* **3** (2), 40-70.

Daydé, M. and Duff, I.S. (1990). Use of Level 3 BLAS in LU factorization in a multitasking environment on three vector multiprocessors, the CRAY-2, the IBM 3090 VF, and the Alliant FX/80. Technical Report RAL-90-083, Rutherford Appleton Laboratory, Chilton, Didcot, Oxfordshire OX11 0QX. Technical Report TR/PA/90/30, CERFACS, Toulouse.

Dongarra, J.J., Du Croz, J., Duff, I.S., and Hammarling, S. (1990). A set of Level 3 Basic Linear Algebra Subprograms. *ACM Trans. Math. Softw.* **16**, 1-17.

Dongarra, J.J., Du Croz, J., Hammarling, S., and Hanson, R.J. (1988). An extended set of Fortran Basic Linear Algebra Subprograms. *ACM Trans. Math. Softw.* **14**, 1-17.

Duff, I. S. (1977). MA28 – a set of Fortran subroutines for sparse unsymmetric linear equations. AERE R8730, HMSO, London.

Duff, I. S., Gould, N. I. M., Reid, J. K., Scott, J. A., and Turner, K. (1991). Factorization of sparse symmetric indefinite matrices. Report RAL-90-066, Rutherford Appleton Laboratory, Chilton, Didcot, Oxfordshire OX11 0QX. To appear in *IMA J. Numer. Anal.* **11** (2)

Duff, I. S., Grimes, R. G., and Lewis, J. G. (1989). Sparse matrix test problems. *ACM Trans. Math. Softw.* **15**, 1-14.

Duff, I. S. and Johnsson, S. L. (1989). Node orderings and concurrency in structurally-symmetric sparse problems. In *Parallel Supercomputing: Methods, Algorithms and Applications*. G F Carey (ed.). John Wiley & Sons, Ltd., Chichester, New York, Brisbane, and Toronto, 177-189.

Duff, I. S. and Reid, J. K. (1982). MA27 – A set of Fortran subroutines for solving sparse symmetric sets of linear equations. AERE R10533, HMSO, London.

Duff, I. S. and Reid, J. K. (1983). The multifrontal solution of indefinite sparse symmetric linear systems. *ACM Trans. Math. Softw.* **9**, 302-325.

Duff, I. S. and Reid, J. K. (1984). The multifrontal solution of unsymmetric sets of linear systems. *SIAM J. Sci. Stat. Comput.* **5**, 633-641.

Gay, D. M. (1985). Electronic mail distribution of linear programming test problems. *Mathematical Programming Society COAL Newsletter*.

Gilbert, J. R. and Peierls, T. (1988). Sparse partial pivoting in time proportional to arithmetic operations. *SIAM J. Sci. Stat. Comput.* **9**, 862-874.

Hageman, L. A. and Young, D. M. (1981). *Applied iterative methods*. Academic Press, New York and London.

Kamath, C. and Sameh, A. (1989). A projection method for solving nonsymmetric linear systems on multiprocessors. *Parallel Computing* **9**, 291-312.

Sloboda F. (1988). A projection method of the Cimmino type for linear algebraic systems, Technical Report n. 16, Dipartimento di Matematica, University of Bergamo, Italy.

THE SOLUTION OF PARTIALLY SEPARABLE LINEAR EQUATIONS ON PARALLEL PROCESSING SYSTEMS

L. C. W. Dixon
Numerical Optimisation Centre
Hatfield Polytechnic
Hatfield, Herts. AL10 9AB
England

Abstract

In this paper we consider the task of forming and solving the sets of linear equations that arise in the context of nonlinear problems. In particular we consider the fields of optimisation, ordinary differential equations and partial differential equations. We show that in each the system formation can be handled efficiently by automatic differentiation and quote results for the sparse doublet and sparse triplet implementations in Ada. Results are given that show that this can be very effectively performed on a parallel processing computer. The more recent reverse automatic differentiation approach is also discussed.

On the question of the solution of the set of equations, it is shown that in general the truncated Newton method outperforms more traditional optimisation codes and that when the optimisation problem arises in a two-dimensional finite element context the use of the SIMD/DAP processor is very effective. In contrast, it will be shown that obtaining effective speed up of the solution of sparse sets of equations on small MIMD machines of the Sequent or Transputer network variety is much more difficult.

In the context of large sparse optimisation it is demonstrated that the use of accurate dot products can greatly reduce the number of iterations when using a conjugate gradient based algorithm.

Keywords

Automatic differentiation, truncated Newton, conjugate gradients, parallel processing, linear equations, optimisation, ordinary differential equations.

NATO ASI Series, Vol. F 77
Computer Algorithms for Solving Linear Algebraic Equations
Edited by E. Spedicato

1. Introduction

In this paper we will be mainly concerned with the formation and solution of sets of linear equations that arise when solving fundamentally nonlinear problems. Given that we are considering the solution of large sets of equations this discussion will be carried out in the context of parallel computation.

Viewed in this way we will see that similar problems arise in a number of different applications and that similar approaches can be applied in each. As illustrations the formation and solution of sets of linear equations that arise in nonlinear problems in

(1) unconstrained optimisation
(2) constrained optimisation
(3) systems of ordinary differential equations
(4) systems of partial differential equations

will all be given.

If we consider as our first example the solution of the unconstrained optimisation problem,

$$\text{Min } f(\underline{x}) \qquad \underline{x} \in R^n \tag{1.1}$$

then most classical algorithms consist of a two stage iteration process in which a sequence of points $x^{(k)}$ is calculated, by first selecting a search direction $\underline{\Delta}x$ and then a step length α.

$$\underline{x}^{(k+1)} = \underline{x}^{(k)} + \alpha\, \Delta x \tag{1.2}$$

Using the result that the solution of (1.1) is characterised by

$$\nabla f(x^*) = 0 \tag{1.3}$$

a Taylor expansion about the point $x^{(k)}$ is used to derive the Newton prediction

$$\nabla f(\underline{x}^{(k)} + \underline{\Delta}x) = \nabla f(x^{(k)}) + \nabla^2 f(x^{(k)})\underline{\Delta}x + O(\Delta x)^2 \tag{1.4}$$

so setting the new gradient to zero gives

$$\nabla^2 f . \Delta x = -\nabla f \tag{1.5}$$

a set of n linear equations in n unknowns, where ∇f and $\nabla^2 f$ are either calculated or approximated at $x^{(k)}$.

Similarly if we take the equality constrained optimisation problem

$$\begin{aligned} &\text{Min } f(x) \\ &\text{s.t. } e_i(x) = 0 \qquad i = 1\ldots E < n \end{aligned} \tag{1.6}$$

then we can form the Lagrangian function

$$L = f - \sum_i \lambda_i e_i \tag{1.7}$$

and use the result that the solution of (1.6) is characterised by n+E equations in the n+E unknowns (x,λ), namely

$$\begin{aligned} \nabla L &= 0 \\ e &= 0. \end{aligned} \tag{1.8}$$

A similar linear expansion now yields the equations

$$\begin{aligned} \nabla^2 L\Delta x - J\Delta\lambda &= -\nabla L \\ J^T \Delta x &= -e \end{aligned} \tag{1.9}$$

as the constrained equivalent of (1.5), where J is the Jacobian of the constraint vector e.

Alternatively we could recall the penalty function

$$P(x,r) = f(x) + r\sum e_i^{\,2}(x) \tag{1.10}$$

and use the S.U.M.T. approach to minimise P(x,r) for a sequence of decreasing values of r obtaining a set of n equations in n unknowns

$$\nabla^2 P\ \Delta x = -\nabla P. \tag{1.11}$$

Similar situations arise in many other application areas. In the solution of systems of first order ordinary differential equations

$$\dot{y} = \phi(y,t) \tag{1.12}$$

numerical stability for large step sizes $h = \Delta t$ is usually obtained by the use of an implicit formula where the y on the right hand size is evaluated at the new time step. For instance the crudest implicit method, the implicit Euler method is given by

$$y(t+h) = y(t) + h\phi(y(t+h),\ t+h).$$

as

$$\phi(y(t+h),t+h) = \phi(y(t),t) + \frac{\partial\phi}{\partial y}(y(t+h)-y(t)) + O(h)$$

$$(I-h\frac{\partial\phi}{\partial y})y(t+h) = y(t) + h\phi(y(t),t) - h\frac{\partial\phi}{\partial y}y(t) + O(h^2)$$

we obtain

$$(I-h\frac{\partial\phi}{\partial y})\Delta y = h\phi + O(h^2) \tag{1.13}$$

This is another set of linear equations which plays a role in ODEs which is equivalent to (1.5) in optimisation.

In each case the set of linear equations (1.5), (1.9), (1.11), (1.13) arises inside a more major calculation. Although we may represent each set of equations by the standard form

$$Ax = b \tag{1.14}$$

the values of the matrix A and the vector b are not known but need to be calculated, and the calculation of these values followed by the solution of the equations dominate the rest of the calculation.

The paper will therefore be concerned mainly with these two dominant tasks. The calculation of the derivatives involved in (1.14) will be considered in Section 2. The implications of parallel computation in their calculation and the calculation of appropriate search directions in a parallel computing context will be discussed in Section 3. In that section it will be argued that the accurate solution of (1.14) is rarely justified when solving nonlinear problems.

2. The calculation of derivatives

2.1 Gradient and Hessian of a function f(x)

Let us consider the unconstrained optimisation problem (1.1); we will assume that the objective function f(x) is computable and that it can be expanded as a series of elementary operations of one or two variables

$F_i(x_j,x_k)$.

So the objective function is given by

For $i = n+1, \ldots, n+M$

$$x_i = F_i(x_j,x_k) \qquad j,k < i \qquad (2.1)$$

next i.

$$f = x_{M+n}$$

The elementary operations consist of addition, subtraction and multiplication of two variables supplemented by functions of one variable i.e. sine, cosine, log, exponential, power, reciprocal etc.

Any computable objective function can be expanded into such a form if Boolean switches are ignored. For the purpose of this paper such switches will be ignored.

If we consider the well known Rosenbrock function

$$f(x) = 100(x_1^2 - x_2)^2 + (1-x_1)^2 \qquad (2.2)$$

This could be expanded as

$$\begin{aligned} x_3 &= x_1^2 \\ x_4 &= x_3 - x_2 \\ x_5 &= x_4^2 \\ x_6 &= 100x_5 \\ x_7 &= 1-x_1 \\ x_8 &= x_7^2 \\ f = x_9 &= x_6 + x_8 \end{aligned} \qquad (2.3)$$

So we see that this simple two dimensional problem contains seven operations. It is not unusual that for many practical problems $M>n^2$ or even n^3. This example also indicates that the order of the operations is often flexible and that the calculation may be indicated by a graph, which connects each operation with its data and emphasises any parallel computation possibilities in its calculation.

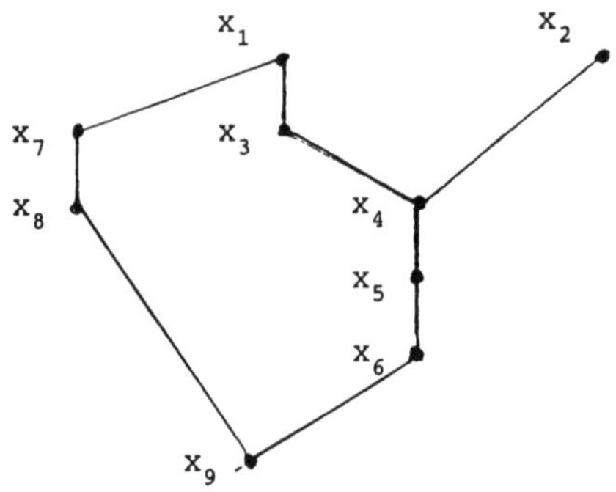

Figure 1. Graph of Rosenbrock's function

This graph emphasizes that the operations labelled 7 and 8 could have been undertaken at any stage in the calculation without altering the result. It also indicates that at most two processors could be used and that at least five sequential steps are required.

If we now consider any least squares problem

$$f(x) = \sum_{k=1}^{K} s_k{}^2(x) \tag{2.4}$$

then each subfunction $s_k(x)$ could have a graph like Rosenbrock's function and the subfunctions could be performed as independent calculations and the graph only linked at the end.

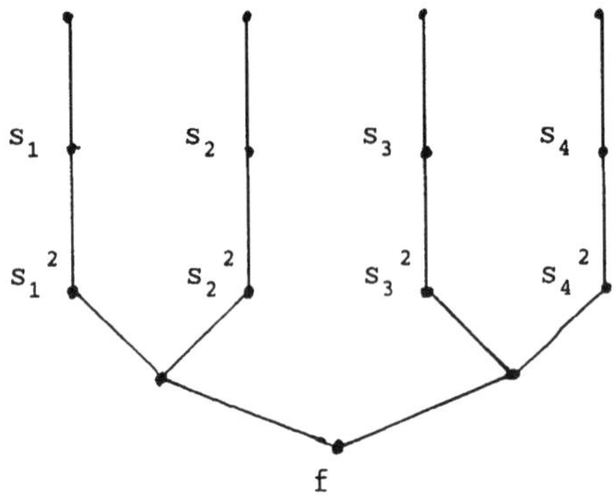

Figure 2

Any function with a number of parallel subtrees will be termed partially separable following Griewank and Toint (1981). Frequently but not necessarily each of the subfunctions s_k will depend upon a relatively small number of the independent variables x_i, i=1, ..., n.

Now suppose we wish to find the gradient vector ∇f. How should we proceed? There are at least **four** alternative approaches.

M1 Analytic Differentiation: the formula for the gradient vector are derived analytically and entered into the computer. This was the method usually adopted by practitioners until quite recently. For Rosenbrock's function this is fairly simple as n=2 and the gradient vector is given by

$$\nabla f = \begin{pmatrix} 400\, x_1(x_1^2 - x_2) - 2(1-x_1) \\ \\ -200(x_1^2 - x_2) \end{pmatrix} \tag{2.5}$$

but as n and M increase this becomes increasingly tedious and mistakes are liable to occur.

M2 Numerical Approximation: one sided or central difference formula may be used to estimate the derivatives i.e. the approximation

$$\nabla f_i = (f(x+h\hat{a}_i) - f(x))/h \tag{2.6}$$

could be adopted.

This is of course simple to program and is often successfully used, it is however expensive as calculating ∇f now requires at least nM operations and is also dependent on choosing a suitable value of h.

M3 Symbolic Differentiation: many mathematical aids such as MACSYMA will accept a FORTRAN listing of an objective function and produce a FORTRAN listing of the gradient. The difficulty of this approach is that such symbolic codes often produce inefficient FORTRAN and to date are often very restrictive on the complexity (M) of the problem that can be tackled. Griewank (1988) gives the listing of the gradient of the Helmholtz energy function given by MACSYMA. Such a listing then has to be evaluated at each iterative point $x^{(k)}$, and when the formula is lengthy it is not clear that this has any advantages over automatic differentiation.

<u>**M4** Automatic Differentiation;</u> a full description of automatic differentiation is given in Rall (1981). Automatic differentiation is essentially the computer implementation of a new algebra, which is implemented by defining a new data type and overwriting the meaning of operators for that data type. The simplest such data type, the doublet, consists of the n+1 numbers $U=(u,\nabla u)$. Then given one or two doublets the meaning of the elementary operations is overwritten, so we obtain:

$$\left.\begin{aligned} W &= U+V \rightarrow (u+v,\ \nabla u+\nabla v) \\ W &= U-V \rightarrow (u-v,\ \nabla u-\nabla v) \\ W &= U*V \rightarrow (u\cdot v,\ u\nabla v+v\nabla u) \\ W &= U^2 \rightarrow (u^2,\ 2u\nabla u) \\ W &= U^{-1} \rightarrow (u^{-1},\ -u^{-2}\nabla u) \\ W &= \log U \rightarrow (\log u,\ u^{-1}\nabla u). \end{aligned}\right\} \qquad (2.7)$$

As each of the operations in (2.1) is of this form we can obtain the derivatives by forward automatic differentiation in terms of doublets:

For $i = n+1, \ldots, M+n$

$$(x_i, \nabla x_i) = F_i((x_j, \nabla x_j), (x_k, \nabla x_k)) \qquad j,k < i$$

next i

$$(f, \nabla f) = (x_{M+n}, \nabla x_{M+n})$$

This approach is accurate but is bounded above by (2n+1)M operations due to the fact that a full vector ∇w is created at each step even though ∇w is usually sparse.

<u>**M4.1** Sparse Doublet and Sparse Triplet Arithmetic:</u> an efficient implementation in ADA is described in Dixon, Maany, Mohseninia (1989) where the vector ∇w is stored in sparse form i.e. (Number of nonzero elements, position, value) so $\nabla x_1 = (1,1,1)$ even if n=1000.

This method can be readily extended to the calculation of the second derivatives by introducing the algebra of sparse triplets in which $T = (u,\ \nabla u,\ \nabla^2 u)$ and both ∇u and $\nabla^2 u$ are stored in sparse form. We will illustrate this by showing the calculation for Rosenbrock's function and then the Helmholtz energy function. As n=2 for Rosenbrock's function the illustration will ignore sparsity.

x_1	$(x_1;\ 1,0;\ 0,0,0)$
x_2	$(x_2;\ 0,1;\ 0,0,0)$
$x_3 = x_1^2$	$(x_1^2;\ 2x_1,0;\ 2,0,0)$
$x_4 = x_3 - x_2$	$(x_4;\ 2x_1,-1;\ 2,0,0)$
$x_5 = x_4^2$	$(x_5;\ 4x_1x_4,\ -2x_4;\ 4x_4 + 8x_1^2, 2, -4x_1)$
$x_6 = 100x_5$	(all multiplied by 100)
$x_7 = 1 - x_1$	$(x_7;\ -1,0;\ 0,0,0)$
$x_8 = x_7^2$	$(x_8;\ -2x_7,0;\ +2,0,0)$
$x_9 = x_6 + x_8$	$(x_9;\ 400x_1x_4 - 2x_7, -200x_4;$ $400x_4 + 800x_1^2, +2, 200, -400x_1)$

which is the correct solution. It will be noted that even on this very simple problem 19 of the 54 entries in the 9 triplets calculated are zero.

The Helmholtz energy function introduced by Griewank (1988) as a standard example for automatic differentiation is given by:

$$f = RT \sum_{i=1}^{n} x_i \log \frac{x_i}{1-b^T x} - \frac{x^T A x}{\sqrt{8}\, b^T x} \log \frac{1+(1+\sqrt{2})b^T x}{1+(1-\sqrt{2})b^T x} . \qquad (2.8)$$

This can be divided into certain subfunctions, and in Table 1 we list the number of operations involved in the calculation of f, ∇f and $\nabla^2 f$ in sparse doublet and sparse triplet arithmetic.

From the Table 1.1 we can see that very many terms contribute $O(n^2)$ operations to $\nabla^2 F$ but that the term $x^T Ax$ dominates F, whilst the term

$$\sum_{i-1}^{n} x_i \log x_i/(1-b^T x)$$

	Function	Operations in F	∇F	$\nabla^2 F$
	$b^T x$	2n	n	0
	$1-b^T x$	1	0	0
	$1+(1+\sqrt{2})b^T x$	4	n	0
	$1+(1-\sqrt{2})b^T x$	4	n	0
	$1/(1-b^T x)$	1	n+2	n^2+3
	$1/1+(1-\sqrt{2})b^T x$	1	n+2	n^2+3
n times	$x_i/(1-b^T x)$	1	3n	$6n^2$
n times	$\log x_i/(1-b^T x)$	log	2n	$3n^2$
n times	$x_i \log x_i/(1-b^T x)$	1	3n	$6n^2$
n times	So the sum becomes	(2+log)n	(8n)n	$(15n^2)n$
	$x^T Ax$	$3n^2$	$4n^2$	n^2
	$1/b^T x$	1	n+2	n^2+3
	$x^T Ax/b^T x$	1	3n	$6n^2$
	log( )	log	2n	$3n^2$
	$\frac{x^T Ax/b^T x}{\log(\ )}$	1	3n	$6n^2$
	other	3	3n	$3n^2$
Total (approx)		$(n+1)\log + 3n^2$	$12n^2 + 17n$	$15n^3 + 22n^2$

Table 1.1

dominates $\nabla^2 F$ and appears to make it an $O(n^3)$ operation. As Maany observed when performing our tests in ADA, this term can be rearranged as

$$\sum_{i=1}^{n} x_i \log x_i/(1-b^T x) = \sum_{i=1}^{n} x_i(\log x_i - \log(1-b^T x))$$

$$= \left(\sum_{i=1}^{n} x_i \log x_i\right) - \left(\sum_{i-1}^{n} x_i\right) \log (1-b^T x)$$

and when performing these operations in this order we have the following operations:

$\log x_i$	log	1	1
$x_i \log x_i$	1	1	1
So $\sum x_i \log x_i$	n	n	n
$\sum x_i$	n	n	0
$\log(1-b^T x)$	log	2n	$3n^2$
$\left(\sum x_i\right)\log(1-b^T x)$	1	3n	$6n^2$
So the sum becomes	(n+1)log + 3n	9n	$9n^2 + 3n$
& the overall total	(n+1)log + $3n^2$	$4n^2$+ 26n	$31n^2$

Table 1.2. Revised operations count

We note that the n^3 term in the calculation of $\nabla^2 F$ no longer occurs and all three calculations are now dominated by order n^2 terms. We may first conclude from this that the order in which a function calculation is performed can dramatically affect the cost of calculating the derivatives by sparse doublet and sparse triplet methods.

Dimension	Sparse Doublet	Sparse Triplet
5	1.68	33
10	3.20	49
20	5.27	46
30	5.55	41
40	5.69	46
50	5.95	48
60	6.23	48
200	7.23	
500	8.15	

Table 2. The relative cost of calculating ∇f by sparse doublet and $\nabla^2 f$ by sparse triplet to the cost of calculating f.

We may note too that as reported by Dixon, Maany and Mohseninia (1989) and shown in Table 2, the times taken by the ADA implementation confirmed the constant nature of the ratios of the number of operations required to evaluate f, ∇f and $\nabla^2 f$; the difference in the actual ratios obtained do however imply that the cost of the index operations in the link lists cannot be ignored.

M 4.2 Reverse Automatic Differentiation; In Griewank (1988) a proof is given that by using reverse automatic differentiation whenever f can be calculated in M elementary operations then ∇f can be calculated in less than 5M elementary operations whatever the dimension n.

Recalling the notation for calculating f given in (2.1), reverse automatic differentiation is simply:

For $i = 1, \ldots, n$

$$\bar{x}_i = 0$$

For $i = n+1, \ldots, n+M$

$$x_i = F_i(x_j, x_k) \qquad j,k < i$$

$$\bar{x}_i = 0$$

$$f = x_{n+M}$$

$$\bar{x}_{n+M} = 1.$$

For $i = n+M, \ldots, n+1$

$$\bar{x}_j = \bar{x}_j + \frac{\partial F_i}{\partial x_j} \cdot \bar{x}_i$$

$$\bar{x}_k = \bar{x}_k + \frac{\partial F_i}{\partial x_k} \cdot \bar{x}_i$$

$$\nabla f_i = \bar{x}_i. \qquad i = 1, \ldots, n.$$

This reduction in the number of operations required to calculate ∇f in reverse rather than forward automatic differentiation is balanced by the need to store the computational graph so that it is available for the reverse sweep. One of my colleagues, Bruce Christianson (1990), has written an implementation of this in ADA as a standard "subroutine" compatible with our ADA optimisation codes. His results for the Helmholtz energy function are given in the following table 3. The method can be readily extended to obtain the product $\nabla^2 f.u$ for any given direction and these results are also given

Helmholz Energy Function	f ordinary	f by graph	∇f by graph	$\nabla^2 fu$ by graph
n = 20	-	7	8	16
50	11	53	52	120
100	40	320	300	660

Times in Seconds

Table 3 Reverse Automatic Differentiation

2.2 Jacobian of a vector function s_k (x)

As well as needing to calculate the gradient and Hessian of a scalar function f(x) we often need to calculate the Jacobian of a vector function s_k. This occurs in constrained optimisation where s_k could be the vector of active constraints, in ordinary differential equations where it might be the vector ϕ (1.12) or in the least squares minimisation of

$$f(x) = \sum_{k=1}^{K} s_k^2(x).$$

When each s_k is a separate calculation then we can use either sparse doublet or reverse automatic differentiation to efficiently calculate the gradient of each s_k. The sparse doublet form is particularly effective when each s_k only depends on a few variables (i.e. the Jacobian is sparse). This is the case for the Lokta-Volterra predator prey function described in Byrne and Hindmarsh (1987).

Using the sparse doublet implementation in ADA, Parkhurst (1990) obtained the following results:

$$r_1 = \frac{\text{Evaluation of } (\underline{f},\ J) \text{ using Sparse Doublets}}{\text{Evaluation of } (\underline{f}) \text{ using ordinary arithmetic}}$$

$$r_2 = \frac{\text{Evaluation of } (\underline{f}) \text{ using Sparse Doublets}}{\text{Evaluation of } (\underline{f}) \text{ using ordinary arithmetic}}$$

Dimension	50	200	800	1800	3200	5000
r_1	52.39	54.10	51.14	50.42	49.67	50.19
r_2	5.05	4.74	4.73	4.75	4.72	4.77

Table 4. Performance of Sparse Doublet Automatic Differentiation

Preliminary results with Christianson's implementation of reverse automatic differentiation indicate that this implementation gives a value of r_1 of approximately 19. If we assume that the s_k are separate calculations and that the total number of operations in calculating s_k is M_k, then the total operations for calculating ∇s_k is qM_k (where q is theoretically less than 5 for reverse automatic differentiation).

Now as $M = \sum M_k$ then the total number of operations for calculating J is less than $\sum qM_k$ which equals qM.

However, assuming that the s_k are separate calculations is not always fair as frequently there will be common strands in the calculations. If we drew a computational tree for the calculation of s_k such common strands must commence at the start of the tree and will probably involve a subset NEL_c of the variables and lead into a subset K_c of the subfunctions and will involve M_c elementary operations. Let us assume the common strand ends with a value x_c. There can of course be more than one common strand. In sparse doublet arithmetic we would proceed forward through this common strand once and would use less than $q \times NEL_c \times M_c$ operations to obtain ∇x_c.

In reverse automatic differentiation we would reverse through the common strand K_c times and would thus appear to possibly need $q \times K_c \times M_c$ operations. This destroys the simple bound relating the operations involved in the calculation of $\underline{s}$ with those required by the calculation of J.

However suppose now we identified the common strand, and it would have to be identified if it is to be utilised in the calculation of $\underline{s}$, then we could start reverse differentiation at the end of the strand and calculate ∇x_c by one pass in $q\ M_c$ operations.

Now suppose we calculate back in reverse mode for the gradient of s_k, we will reach a value i=c and have $\bar{x}_c = \frac{\partial s_k}{\partial x_c}$ and can update those $\bar{x}_i$, i<n contained in NEL_c by $\bar{x}_i = \bar{x}_i + (\nabla x_c)_i \bar{x}_c$.

This linking operation involves $(NEL_c)(K_c)$ elementary operations, so we have replaced q $K_c M_c$ operations by $qM_c + (NEL_c)(K_c)$ operations. There is therefore a saving provided

$$q(K_c-1)M_c > (NEL_c)(K_c)$$

which will frequently be true. Similar considerations apply in the calculation of the Hessian matrix of a function. If we consider the calculation of the gradient of Rosenbrock's function by reverse differentiation it is simply

$$\bar{x}_9 = 1$$

$$\bar{x}_8 = \bar{x}_9 \qquad \bar{x}_6 = \bar{x}_9$$

$$\bar{x}_7 = 2x_7\bar{x}_8$$

$$\bar{x}_1 = -\bar{x}_7$$

$$\bar{x}_5 = 100\bar{x}_6$$

$$\bar{x}_4 = 2x_4\bar{x}_5$$

$$\bar{x}_3 = \bar{x}_4 \qquad \bar{x}_2 = -\bar{x}_4$$

$$\bar{x}_1 = \bar{x}_1 + 2x_1\bar{x}_3$$

The last step illustrates that not all the operations generated by reverse differentiation are elementary and we strictly need to divide this into two steps

$$\bar{x}_{1\frac{1}{2}} = 2x_1\bar{x}_3$$

$$\bar{x}_1 = \bar{x}_1 + \bar{x}_{1\frac{1}{2}}$$

The combined tree for the function and gradient calculation is therefore that given on the left

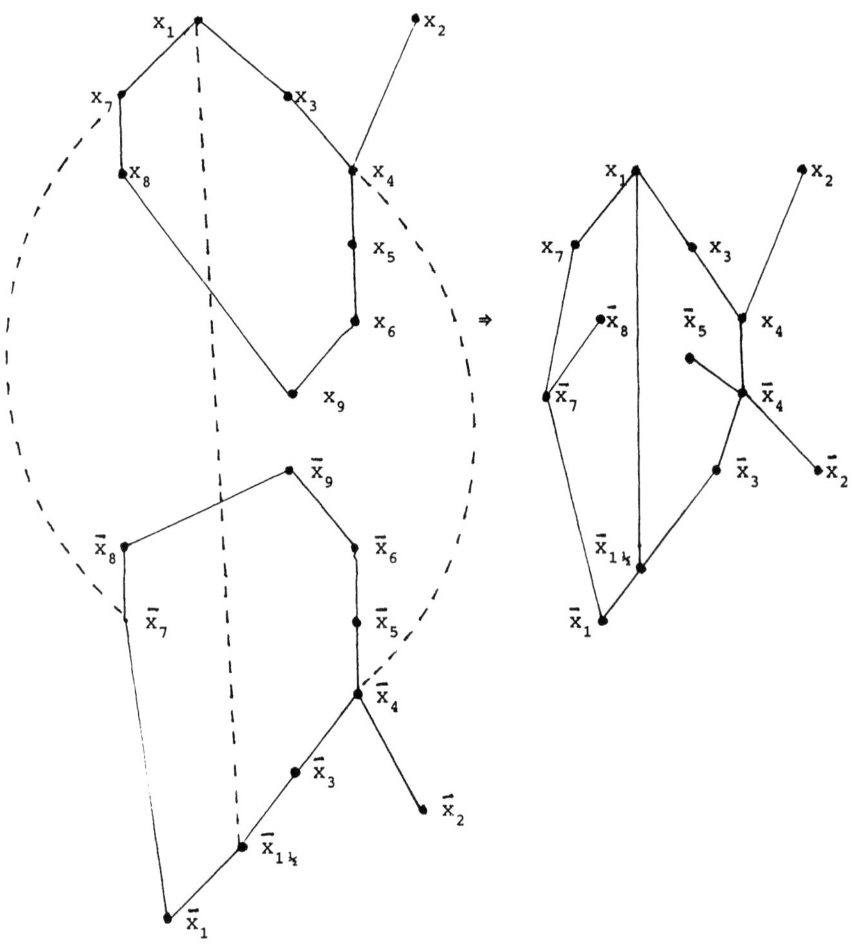

Figure 3

We may make the output a scalar, the directional derivative $\nabla f^T d = \bar{x}_1 d_1 + \bar{x}_2 d_2$ by adding an extra node at the bottom and then apply reverse differentiation again to obtain $\nabla^2 fd$. However we now need to reverse back from $\bar{x}_1$, $\bar{x}_2$ to x_1, x_2 and we note that this can only be done through the dotted lines linking the two graphs i.e. the nonlinear operations. This observation allows us to drastically reduce the tree as shown on the right and even now we have the common subtree in the calculation of $\bar{x}_2$, $\bar{x}_1$ namely that from x_1, $x_2 \to \bar{x}_4$. We may note too that the graph is symmetric about a horizontal line passing through the non-linearities. This is a general result for all such "Hessian" graphs.

Whether this approach to the Hessian calculation has any advantage over that described and implemented by Christianson still needs to be determined.

We conclude that two automatic methods now enable us to compute partial derivatives efficiently. In particular the sparse doublet/sparse triplet approach gives efficient methods for calculating the gradient and Hessian of partially separable scalar functions and of the Jacobian of vector functions.

The reverse differentiation approach requires the storage of the computational tree but then provides a very efficient way of calculating the gradient of any scalar function and the Jacobian of vector functions that do not contain significant subtrees. The method also provides a very efficient means of calculating $\nabla^2 f\, u$ for given u.

A method of handling common subtrees in reverse differentiation has been suggested above which should be used whenever $qM_c > NEL_c$.

Some interesting conjectures have been made on the reduction of the computational trees for Hessian calculations.

The contrasting values for sparse forward and backward calculation are

	Reverse	Sparse Forward
$\nabla f/f$	q	NEL
$\nabla^2 f/f$	q^2	½NEL(NEL+1)
J/f	qk*	NEL
$\nabla^2 fu/f$	2q (ref[13])	2.NEL.
	q < 5	

where NEL is the average number of nonzeros in the gradient vector ∇x_i i> n during the calculation, if the function is partially separable NEL is less than the number of variables in each subfunction.

* We have however shown that a better bound for J/f by reverse automatic differentiation exists.

3. The Effect of Parallel Computation

3.1 Experience using the ICL/DAP

At Hatfield Polytechnic our investigation into the benefits of using parallel computation commenced with the study of Kanu Patel who implemented unconstrained optimisation algorithms on two very different parallel processing machines - the Neptune machine at Loughborough University and the ICL-DAP at Queen Mary College. The ICL-DAP was an SIMD machine with 4096 processors, the Neptune an MIMD machine with four processors. His work on the Neptune was mainly concerned with the solution of small dimensional multi-extremal problems and is not particularly relevant to this paper.

As the DAP is an SIMD machine, each of the 4096 processors must either perform an identical arithmetic operation or be turned off. On this basis he chose to perform function evaluations in parallel and to then approximate the gradient and Hessian by difference formula, and thus implemented a parallel modified Newton algorithm, Patel (1982).

3.1.1 The Modified Newton Algorithm

Patel's Parallel Modified Newton Algorithm

Step 1 Initial guess $x^{(0)}$, step h, tolerance ε

Step 2 Calculate $f(x \pm ha_i \pm ha_j)$ all $j > i$ where a_i is a unit vector along i^{th} axis. This is an obvious parallel calculation.

Step 3 Calculate ∇f and $\nabla^2 f$ by central differences. He showed this could be performed in parallel on an SIMD machine.

Step 4 Solve the set of linear equations

$$(\nabla^2 f + \mu I)p = -\nabla f \tag{3.1}$$

He used the standard DAP equation solver.

Step 5 Perform a 4096 grid search in the two dimensional space spanned by p and $-\nabla f$ which sensibly replaces the line search given that 4096 evaluations take the same time as 1.

Step 6

$$x^{(k+1)} = \text{Arg min } f(x_i) \qquad i = 1, \ldots, 4096$$

using the very fast DAP "min" operation

Step 7 Return to 2 with $k = k+1$.

Essentially this algorithm uses $P = n^2$ processors and calculated f, ∇f, $\nabla^2 f$ in M steps and solved the Newton equations in $O(n^3/P)$ steps.

At the time when the relative speed of calculating 4096 parallel function evaluations on the DAP was 20 times faster than evaluating them sequentially on a DEC 1091, the comparison times to solve five 64 dimensional problems were obtained. Standard Modified Newton, Variable metric and Conjugate gradient codes were run on the DEC 1091 and the parallel Newton on the DAP.

Problem	Sequential Newton	Variable Metric	Conjugate Gradient	Parallel Newton
Quadratic	16.60	2.10	1.18	0.966
Extended Rosenbrock	140.98	80.78	11.36	36.626
Extended Powell	134.97	41.28	11.06	1.500
Trigonometric	2604	455	78.68	17.57
Extended Box	7196	1263	354.0	202.37

Table 5 Performance of the Parallel Newton Method

From these results it is easily seen that the parallel Newton algorithm greatly outperformed the sequential Newton code, often by far more than the natural value of 20. However, it was also obvious that on these functions the conjugate gradient code also outperformed the sequential Newton code and that a comparison of its times with those of the parallel code was not particularly encouraging.

Two decisions were made at that time which have greatly influenced our subsequent development. The first was that as the functions we were interested in at that time possessed considerable structure this should be used in the evaluation of f and in the solution of the set of equations. The second was that we needed to consider conjugate gradient type algorithms and problems with dimensions greater than 64.

3.1.2 The parallel conjugate gradient algorithm

The first problem studied was a quadratic function derived from Stones set of heat conduction problems namely:

$$f = \iint K_x \left(\frac{\partial T}{\partial x}\right)^2 + K_y \left(\frac{\partial T}{\partial y}\right)^2 - 2QT \, dA \tag{3.2}$$

To solve this problem Ducksbury (1984) split it into rectangular elements and approximated T in each element by the standard bi-linear mapping.

The function can now be expressed in the form

$$f = \sum_{\text{elements}}^{K} s_{el}$$

where

$$s_{el} = \iint_{\text{element}} K_x \left(\frac{\partial T}{\partial x}\right)^2 + K_y \left(\frac{\partial T}{\partial y}\right)^2 - 2QT \, dA \tag{3.3}$$

and s_{el} is a function of 4 variables, the values of T at the corners of the elements.

If M_{el} is the number of operations needed to evaluate s_{el}, then the total operation count for f is simply C M_{el} where C is the number of elements and is slightly less than n.

Again $\nabla f = \sum_{el} \nabla s_{el}$

$$\nabla^2 f = \sum_{el} \nabla^2 s_{el}$$

where ∇s_{el} only has four nonzeros and $\nabla^2 s_{el}$ 16 nonzeros.

The resulting set of linear equations was solved on the DAP using the conjugate gradient algorithm. The decision was taken to allocate each element to its natural processor on the DAP grid and to distribute the search direction p over the processors so each processor held the four components that influence its value of f. Given the conjugate gradient formula this only required data passing between neighbouring processors thus reducing the communication costs, the conjugate gradient code then only needs the calculation of

$$y = \nabla^2 f \; p = \sum_{el} \nabla^2 s_{el} \, p_{el} = \sum_{el} y_{el} \tag{3.4}$$

and for each element/processor this is simply 16 operations. In practice the scalar products $r^T r$, $y^T p$, $p^T p$ were also calculated using partial vectors distributed over the processors.

This problem is equivalent to the heat conduction problem tested by Stone who introduced a number of different cases based on the distribution of K_x and K_y.

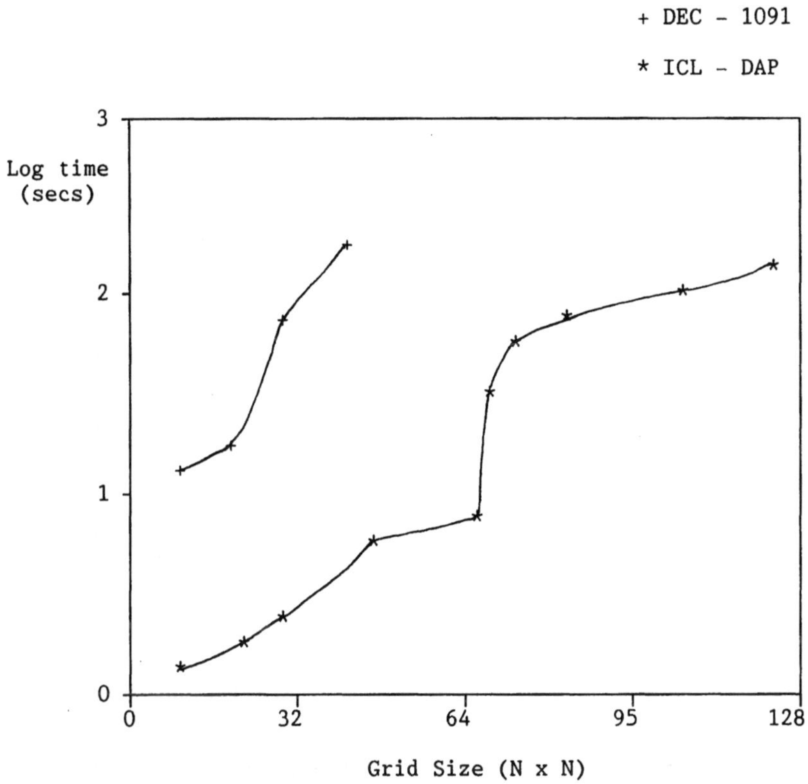

Figure 4.1 Performance of the Conjugate Gradient Codes

The simplest case with $K_x = K_y = 1$ leads to a well conditioned problem and the performance of the conjugate gradient codes on this problem are shown in Figure 4.1, where the log of the CPU time is plotted against the grid size.

Other distributions lead to less well-conditioned problems but the relative performance of the sequential and parallel codes is unaltered as shown in Figure 4.2.

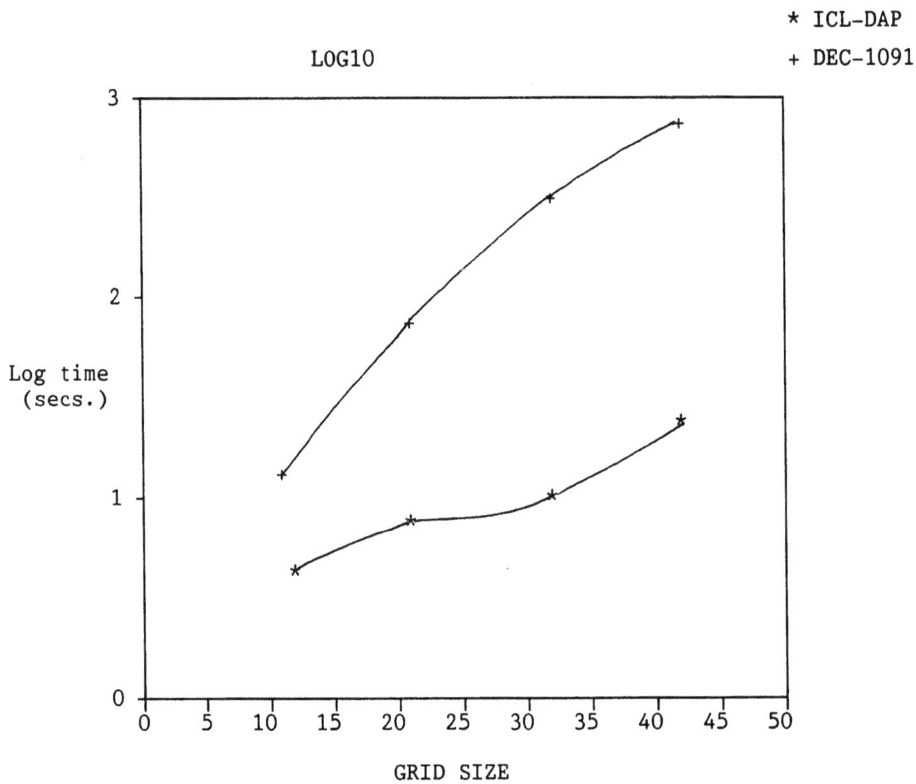

Figure 4.2 Performance of Sequential and Parallel Codes

The time required on the sequential machine precluded running the test for all the grid sizes, but there was a speed-up of over 100 even when only using a small part of the DAP.

This work was extended to the nonlinear domain by considering the differential equation

$$\frac{\partial^2 u}{\partial x^2} + \frac{\partial^2 u}{\partial y^2} - Ru \frac{\partial u}{\partial x} = 0 \qquad (3.5)$$

with boundary values such that the analytic solution was

$$u = \frac{6x_1}{Rx_2{}^2}$$

and converting it to a three dimensional map by introducing

$$a = \frac{\partial u}{\partial x}, \quad b = \frac{\partial u}{\partial y}$$

and minimising

$$f = \iint \left(a - \frac{\partial u}{\partial x} \right)^2 + \left(b - \frac{\partial u}{\partial y} \right)^2 + \left(\frac{\partial a}{\partial x} + \frac{\partial b}{\partial y} - Rua \right)^2 dA \quad (3.6)$$

Again the function and gradient evaluations were distributed over rectangular elements so that for a 64 x 64 grid there were 3 x 4096 unknowns before the specification of the boundary condition. This implies that the Hessian matrix is now a combination of 12 x 12 blocks. Again a parallel version of the Fletcher Reeves algorithm was used.

A typical result is shown in figure 5, where the number of processors used $P = (\text{grid size})^2$. It appears that the function/gradient cost is proportional to the $(\text{grid size})^2$ sequentially and is therefore constant for the parallel machine and that the number of iterations is approximately linear in the grid size, giving

sequential CPU α $(\text{grid size})^3$

parallel CPU α (grid size).

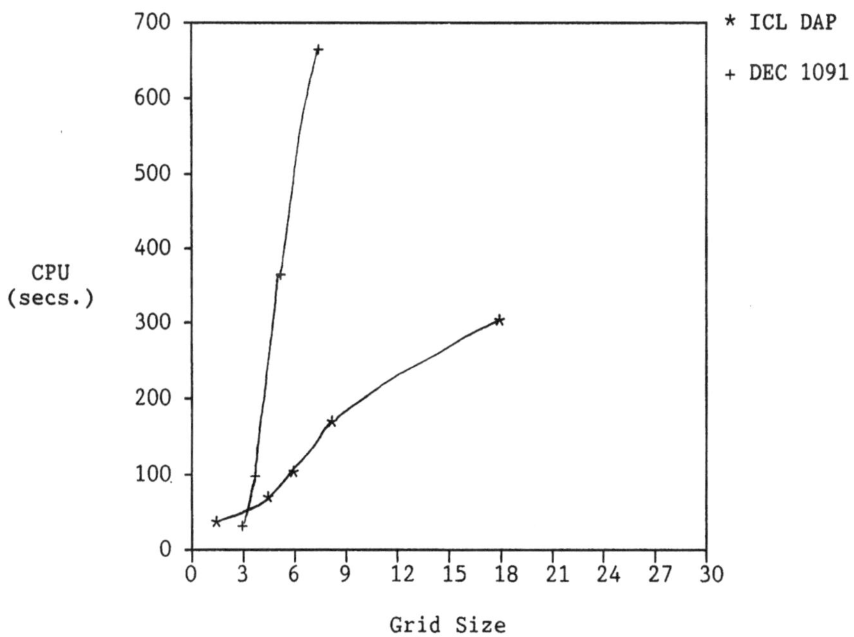

Figure 5 Performance of Parallel and Sequence Conjugate Gradient Codes

3.1.3 The Truncated Newton Algorithm

Before 1983 most existing optimisation routines that used the Newton's equation

$$r = \nabla^2 f(x^k)d + \nabla f(x^k) = 0$$

solved the equation as accurately as possible even though the current iterate was far from the optimal solution. For large problems a major part of the computing cost can be attributed to the solution of these equations.

The fact that it is not necessary to solve these equations accurately when far from the solution, was reported by Dembo, Eisenstat and Steihaug (1982, 1983). They introduced the idea of calculating an approximate solution when far from the solution and suggested solving the equations by the conjugate gradient method and also truncating the iteration before an accurate solution is obtained.

In our implementations at Hatfield this inner conjugate gradient iteration was terminated if

$$\text{(i)} \quad r_j^T r_j < \nabla f^T \nabla f * \min\left(\frac{.1}{k^2}, \nabla f^T \nabla f\right)$$

where ∇f is calculated at the kth iterate of the outer iteration and r_j at the jth iterate of the inner iteration or

$$\text{(ii)} \quad d_{j+1}^T d_{j+1} > D^2$$

a trust region imposed to limit the size of the step taken in the inner iteration.

The conjugate gradient method has the property that

$$d_{j+1}^T d_{j+1} > d_j^T d_j$$

so this limit will be reached and terminate the inner iteration unless the solution is within the trust region,

or

$$\text{(iii)} \quad d_j^T \nabla f > -.1||d_j|| \; ||\nabla f||$$

Code	Problem	N	No.F	No.G	EFE's	C.P.U.
TRUNEW	Wood	4	103	229	1019	0.48
OPCG	Wood	4	384	155	1004	0.80
OPVM	Wood	4	128	77	436	0.90
E04KDF	Wood	4	54	202	862	0.42
TRUNEW	Ext. Powell	4	39	82	367	0.28
OPCG	Ext. Powell	4	96	40	256	0.45
OPVM	Ext. Powell	4	9	31	173	0.46
E04KDF	Ext. Powell	4	18	90	378	0.31
TRUNEW	Ext. Powell	60	33	90	5433	1.27
OPCG	Ext. Powell	60	117	48	2997	1.87
OPVM	Ext. Powell	60	341	194	11981	59.85
E04KDF	Ext. Powell	60	42	1302	78162	15.31
TRUNEW	Ext. Powell	80	31	95	7631	1.81
OPCG	Ext. Powell	80	216	93	7656	3.79
OPVM	Ext. Powell	80	439	251	20519	134.71
E04KDF	Ext. Powell	80	43	1643	131483	32.84
TRUNEW	Ext. Rosenbrock	10	41	59	631	0.34
OPCG	Ext. Rosenbrock	10	99	33	429	0.54
OPVM	Ext. Rosenbrock	10	133	83	963	2.00
E04KDF	Ext. Rosenbrock	10	41	181	1851	0.53
TRUNEW	Ext. Rosenbrock	20	49	71	1469	0.56
OPCG	Ext. Rosenbrock	20	139	40	939	1.00
OPVM	Ext. Rosenbrock	20	198	113	2447	6.27
E04KDF	Ext. Rosenbrock	20	52	312	6292	1.21
TRUNEW	Ext. Dixon	80	21	148	11861	2.16
OPCG	Ext. Dixon	80	186	87	7146	3.04
OPVM	Ext. Dixon	80	201	102	8361	31.14
E04KDF	Ext. Dixon	80	28	929	66268	13.54
TRUNEW	Ext. Powell	2000	41	109	218041	24.98
OPCG	Ext. Powell	2000	125	49	98125	50.13
TRUNEW	Ext. Dixon	2000	29	851	1702029	206.42
OPCG	Ext. Dixon	2000	1064	520	1041064	393.47

Table 6 Comparison of the truncated Newton code (TRUNEW) with a conjugate gradient (OPCG), variable metric (OPVM) and modified Newton code (E04KDF).

which ensures that the resultant direction satisfies Wolfe's condition for finite termination.

These conditions when combined with an Armijo style line search that satisfies Wolfes Conditions II and III ensure that the truncated Newton method has finite termination to the neighbourhood of a stationary point.

Our first results comparing this algorithm with the standard optimisation codes were reported in Dixon & Price (1986, 1988) and confirmed the evidence reported by Dembo, Eisenstat and Steihaug that this method was much more efficient than Modified Newton, Variable Metric or Conjugate Gradient codes over a wide range of dimensions. For convenience these results are shown in Table 6.

In this code the matrix vector product

$$\nabla^2 fp$$

required in the truncated Newton method, was obtained by a difference approximation

$$\nabla^2 fp = (\nabla f(x+hp) - \nabla f(x))/h$$

as suggested by Dembo et al.

Ducksbury (1984) then implemented a parallel version of the Dembo-Steihaug Truncated Newton algorithm on the DAP and compared the relative performance of the parallel conjugate gradient algorithm with the parallel truncated Newton method on a large number of problems including ones based on the partial differential equation (3.5).

He noted that the Truncated Newton method consistently outperformed the conjugate gradient algorithm.

A typical result is one in which the sequential conjugate gradient code required in excess of one hour to solve a 39 x 39 grid (4111 unknowns) and the parallel conjugate gradient code on the DAP just 34 seconds this being a speed up of 104 over the DEC 1091 at a point where only 3/8 of the processors of the DAP were in use.

For a finer grid of 64 x 64 processors (12036 unknowns) the conjugate gradient code on the DAP required 50.76 seconds while the truncated Newton method only required 13.21 seconds.

Turning now to the Navier Stokes problem,

$$u\frac{\partial u}{\partial x_1} + v\frac{\partial u}{\partial x_2} = -\frac{\partial p}{\partial x_1} + \frac{1}{R}\nabla u^2$$

$$u\frac{\partial v}{\partial x_1} + v\frac{\partial v}{\partial x_2} = -\frac{\partial p}{\partial x_2} + \frac{1}{R}\nabla^2 v$$

$$\frac{\partial u}{\partial x_1} + \frac{\partial v}{\partial x_2} = 0$$

this was converted into a set of 6 first order equations by introducing the fields

$$a = u \; ; \quad b = \frac{\partial u}{\partial x_1} \; ; \; c = \frac{\partial u}{\partial x_2}$$

$$d = v \; ; \quad e = \frac{\partial v}{\partial x_1} \; ; \; f = \frac{\partial v}{\partial x_2} \text{ and } p$$

and noting that as the continuity equation is simply $b + f = 0$ the variable f can be eliminated.

The equations become

$$e_1 = ab + cd - \frac{1}{R}\left(\frac{\partial b}{\partial x_1} + \frac{\partial c}{\partial x_2}\right) + \frac{\partial p}{\partial x_1}$$

$$e_2 = ae + df - \frac{1}{R}\left(\frac{\partial b}{\partial x_1} + \frac{\partial c}{\partial x_2}\right) + \frac{\partial p}{\partial x_2}$$

$$e_3 = b + f$$

$$e_4 = \frac{\partial a}{\partial x_1} - b$$

$$e_5 = \frac{\partial a}{\partial x_2} - c$$

$$e_6 = \frac{\partial d}{\partial x_1} - e$$

$$e_7 = \frac{\partial d}{\partial x_2} - f$$

and the objective function is

$$I = \text{Min} \int \sum_1^7 e_k{}^2 dA$$

The minimization is carried out over the field variables and as the integrand only contains first derivatives the theory of the calculus of variables provides natural boundary conditions that will be satisfied on the boundary if any of the seven fields are not fully specified.

As we now have seven unknowns at each node point the storage requirement for $\nabla^2 I_{e1}$ for each finite element increases to 28 x 28 for the truncated Newton method and this was not available on the DAP when these tests were run, so the results quoted in Table 7 are for the conjugate gradient code.

	3 x 3	5 x 3	9 x 9	17 x 17
Value of I	0.2037	0.080917	0.03533	0.01730
CPU	2.56	5.39	14.11	13.51
Total CPU		36.37		
Iterations	82	164	417	389

Table 7 Performance on the Navier Stokes Equation

3.2 More Thoughts On Calculating The Search Direction

3.2.1 The Truncated Newton Method with Automatic Differentiation

Our first results obtained combining the concept of automatic differentiation with the truncated Newton algorithm used a crude Fortran implementation and were reported in Dixon & Price (1986, 1989).

In that paper it was noted that the vector matrix product $\nabla^2 f\ p$ could be obtained in two ways,

(1) Sparse triplets

First form $\nabla^2 f$ and then form the sparse matrix vector product $\nabla^2 fp$

(2) Structured triplets

Modify the definition of a triplet to be $(f, \nabla f, \nabla^2 fp)$ and alter the basic algebra appropriately.

It is important to notice that a structured triplet simply consists of two sparse vectors and a scalar and therefore requires far less store. Mohseninia implemented both versions in ADA and found that the structured version is much slower.

Problem	Dimension	Structured Sparse Triplets	Sparse Triplets
Extended Rosenbrock	2	.81	.62
	20	41.12	4.48
	40	98.61	11.66
	80	420.08	39.64
Extended Powell	4	.75	.57
	20	27.89	3.36
	40	112.3	9.02
	80	760.7	33.42
Extended Dixon	2	.19	.13
	20	26.06	2.96
	40	103.51	12.6
	80	546.13	42.31

Table 8 Comparison of structured and sparse triplet automatic differentiation.

We have therefore not pursued the concept of structured automatic differentiation further. It does however require far less store when this is important.

3.2.2 Accurate Arithmetic

While these results were fairly conclusive in implying that sparse triplets were preferable to structured triplets, they were worrying in that they implied that the codes were not behaving theoretically correctly on these extended functions.

These functions have the property that as their dimensions increase the number of distinct eigenvalues remains constant.

In 1974 Dixon had conjectured that for such extended problems the number of iterations of variable metric and conjugate gradient algorithms should be independent of n.

This was proved in Spedicato (1978). It was however clearly not occurring in the above results. The reason appears to be that while theoretically the value of the scalar product

$$\sum_{i=1}^{n} a_i b_i$$

is independent of the ordering of the elements in most computer languages this is not true and this breaks the symmetry of the calculations in the variable metric algorithm.

Dixon & Mills (1990) wrote a simple sort to make the scalar product independent of the order and probably more accurate and obtained results that reflected the theoretical result.

A typical result obtained using the variable metric algorithm on the extended Powell problem is given below

Dimension	Number of iterations using double precision arithmetic	Number of iterations using sorted dot products
4	39	39
8	64	39
16	25	39
32	67	39
64	52	39
128	165	39

Table 9. The effect of accurate arithmetic

Due to the result of this experience I would urge people to be very careful how they calculate scalar vector products, and matrix vector products to ensure that the arithmetic does not destroy the symmetric structure of the problem.

It is essential in the design of BLAS software for general purpose use that accuracy is not sacrificed for speed.

3.2.3 Maany's Test Set

To overcome the problems with the extended functions so often used to test algorithms in large dimensions Maany (1989) introduced a new family of test problems.

$$f(x) = 1.0 + \sum_{i=1}^{n} 0.5\left(\frac{i}{n}\right)^{k} x_i^{2}$$

$$+ \sum_{i=1}^{n-1} \beta(x_i x_{i+1} + x_i x_{i+1}^{2})^{2} + \sum_{i=1}^{2N} \delta\, x_i^{2} x_{i+N}^{4}$$

$$+ \sum_{i=1}^{N} \delta\left(\frac{i}{n}\right)^{K} x_i x_{i+2N}$$

Here the dimension is $n = 3N$, and the family contains three parameters β, δ, K.

With $K = 0$ the eigenvalue pattern is similar to that in an extended family but for other values of K this property is lost and the family gets more ill-conditioned as K increases.

If $\beta = \delta = 0$ the problem is quadratic and diagonal, when $\beta \neq 0$ the diagonals on either side of the main diagonal are introduced and $\delta \neq 0$ introduces four wide diagonals.

Maany tested the truncated Newton code including sparse triplet automatic differential on twelve cases drawn from this family for a range of dimensions up to $n = 3000$.

His results indicated the robustness of the truncated Newton code and the efficiency of the dynamic data structure used in ADA.

Three sets of his results on badly conditioned problems are detailed below:

No Preconditioning n = 3000						
	β	δ	K	No. of iter.	CPU time	% in AD
Case 10	1/16	1/16	2	8498	19001	12
Case 11	1/8	1/8	2	6482	15461	15
Case 12	0.26	0.26	2	17211	38072	11

Table 10.1. The truncated Newton Method and automatic differentiation on the Maany problem.

These results indicated that the percentage of time being spent in conjugate gradient code dominated that spent in automatic differentiation.

The results were therefore repeated using a diagonal pre-conditioner.

Diagonal Preconditioning n = 3000						
	β	δ	K	No. of iter.	CPU time	% in AD
Case 10	1/16	1/16	2	31	1220	84
Case 11	1/8	1/8	2	31	1226	84
Case 12	0.26	0.26	2	31	1318	84

Table 10.2. The truncated Newton Method and Automatic Differentiation on the Maany problem. Effect of preconditioning.

These results indicated that it was essential to use a preconditioner but that when doing so the calculation of the gradients even with a sparse automatic differentiation routine could dominate. We therefore wished to go on a parallel computing system to see how these results would improve.

Before doing so however it is appropriate to mention the results of Vespucci (1990) who has shown that if the conjugate gradient algorithm was replaced by that equivalent algorithm from the Abaffy-Broyden-Spedicato family which theoretically generates the same sequence of points x^k, then the number of outer and inner iterations is greatly reduced.

Two typical results one for a 40 dimensional modified extended Wood function with a complete set of distinct eigenvalues and one for a member of the Maany family with n = 150 and K = 3 indicate this effect.

		Maj. It.	Min. It.	Time
Modified Extended Wood n = 40	CG ABS/CG	412 137	14962 3218	16.1 7.5
Maany Function n = 150	CG ABS/CG	222 32	30082 1519	400 47

Table 11. The effect of using the conjugate gradient algorithm from the ABS family on iterations and computer time

These results emphasise the need for accurate arithmetic in CG algorithms. The results are not strictly comparable as the ABS/CG method stores a full set of n x n matrices H_i and it is indeed remarkable that the cost of updating such a matrix is dominated by the effect of rounding error in increasing the number of major and minor iterations.

In her paper she advocates replacing the conjugate gradient algorithm by a truncated LL^T iteration implemented within the ABS class but using the sparsity of matrix H_i.

A few of her results with this code are given in the following table:

Modified Extended Wood				
n = 40	CG	412	14962	16.1
	ABS/CG	137	3218	7.5
	ABS/LL^T	73	1967	1.8
n = 100	CG failed to converge in 1000 it.			
	ABS/CG	303	11704	161
	ABS/LL^T	130	7917	36
Maany Function				
n = 150 K = 3	CG	222	30082	300
	ABS/CG	32	1519	47
	ABS/LL^T	51	1651	15
n = 300 K = 3	CG	198	52601	115
	ABS/LL^T	75	6405	35
n = 900 K = 3	CG	138	116620	595
	ABS/LL^T	83	8149	41

Table 12: Use of the ABS/LL^T code

Because of the sparsity of H_i for the LL^T algorithm no difficulty was experienced in storing the matrix and the effectiveness of the new approach is obvious.

Analysis of the results does however indicate that if the use of accurate arithmetic reduced the number of iterations used by the CG algorithm to the number required by the ABS/CG algorithm then the time needed by the CG algorithm would be lowest.

3.3 Results using the Sequent Balance System

To obtain results using automatic differentiation on a parallel processor we needed to have access to a parallel machine that could run ADA and supported concurrent tasks. The Sequent Balance fulfilled this criteria and Professor Delves allowed us access to his machine at Liverpool University.

Dixon & Mohseninia (1987) implemented automatic differentiation (sparse triplets) and the truncated Newton method on this system. As the Sequent Balance contains far fewer processors than the DAP a number of elements needed to be allocated to each processor. Again as the processors are not allocated so that only nearest neighbour communication is possible, the effect of data communication is more obvious.

Typical times for the sparse triplet evaluation of f, ∇f, $\nabla^2 f$ on the Sequent, see Dixon & Mohseninia (1989), are shown below for the Olsen square cavity problem.

Elements	Number of Processors Used									
	1	2	3	4	5	6	7	8	9	10
8	37	20	12	10	11	12	13	8	-	-
32	156	85	62	44	38	35	26	22	24	26
128	670	352	227	175	146	124	107	92	87	81

Table 13 Parallel Sparse Automatic Differentiation

These results were really very satisfactory and indicated that dividing automatic sparse triplet differentiation into concurrent tasks is effective.

The effect of data communication becomes even more obvious however when considering the operations required within the truncated Newton code. The dominant operation of this part of the computation is the product of the sparse matrix $\nabla^2 f$ with the search direction p, and the cost of this multiplication for dimensions 512 and 1107 within the cavity driven flow problem on the Sequent Balance were

Time	Number of Processors									
	1	2	3	4	5	6	7	8	9	10
n=512	3.80	2.40	2.20	2.10	2.08	2.10	2.11	2.12	2.2	2.4
1107	9.50	6.00	5.00	4.5	4.5	4.3	4.1	4.4	4.4	5.2

Table 14 Parallel Matrix Vector Multiplication

These results are really quite disappointing and indicate that communication costs cannot be ignored when performing linear algebra on parallel systems.

However calculating $\nabla^2 fp$ is much less expensive than forming $\nabla^2 f$ and using the figures given in the previous section and assuming P = 8 we have as an overall effect

	P = 1	P = 8
Time in AD on sequential machine	.84	.12
Time in linear algebra	.14	.07
Other time	.02	.02
	1.00	.21

Giving an overall speed up of approximately 5.

3.4 Results using a Transputer Net

At about this time the Polytechnic took delivery of a small transputer network and we began tests to discover its capabilities and to construct a model of its behaviour.

Dim.	Number of Transputers							
	1		2		4		8	
	Act. Time	Exp. Time	Act. Time	Exp. Time	Act. Time	Exp. Time	Act. Time	Exp. Time
16	1	-	2	-	3	-	4	-
32	5	-	7	-	14	-	17	-
64	22	-	30	-	54	-	66	-
83	34	-	46	-	84	-	103	-
96	50	-	67	-	120	-	147	-
128	88	89.6	119	118.49	213	206.06	255	233.01

Table 15 REAL 64 Matrix-vector Multiplication

Calculation of expected timings in milliseconds

As the parallel linear algebra had proved disappointing on the Sequent, Jha (1990) began his investigation in this area and demonstrated that the operation Av could not be performed effectively in parallel if A had to be downloaded into the system (see table 15). In this table the measured times are given on 1, 2, 4 and 8 transputers and the expected time on 8 transputers using Jha's model of the communication time and computation time of the network.

Dim	Number of Transputers							
	1		2		4		8	
	No. of itr.	Time	Time	s.up	Time	s.up	Time	s.up
16	36	49	52	0.94	30	1.63	25	1.96
32	39	193	197	0.98	109	1.77	75	2.57
64	43	819	625	1.31	431	1.90	262	3.13
80	42	1236	832	1.46	610	2.03	384	3.22
96	42	1768	1097	1.61	760	2.33	549	3.23
128	46	3904	2039	1.92	1248	3.13	839	4.66

Table 16. Parallel Jacobi's Method on Network of Transputers.

However, when the matrix vector multiplication was embedded within Jacobi's iterative algorithm

$$x^{K+1} = (I - A)\, x^{K} + b \qquad\qquad A_{ii} = 1$$

an effective speed up was obtained (see table 16).

The difference is mainly due to the fact that A only had to be down loaded once. The results obtained were again in agreement with the constructed model, it should perhaps be stressed that the model predictions are dominated by the communication requirements and not by the computational time.

Dim	Number of transputers							
	1		2		4		8	
	Time	Model	Time	Model	Time	Model	Time	Model
128	3904	3894	2039	2096	1248	1220	839	834

Table 17.

Similar good speed ups were obtained when a parallel conjugate gradient code was applied to a full matrix with distinct eigenvalues (see table 18), but no speed up could be obtained by implementing a sparse matrix code in parallel as the much smaller amount of computation at each iteration was dominated by the communication costs (see Table 19).

Matrix Size	Number of Transputers										
	1		2			3			8		
	Time	Itr	Time	s.up	itr	Time	s.up	itr	Time	s.up	itr
16	45	26	27	1.67	26	19	2.37	26	19	2.37	26
32	358	63	198	1.81	63	116	3.09	62	91	3.93	62
64	3416	167	2013	1.70	167	976	3.50	166	621	5.50	163
80	8937	253	4669	1.91	256	2438	3.67	279	1269	7.042	234
96	17786	355	9128	1.95	355	4242	4.10	347	2829	6.29	388
128	39461	449	21483	1.84	479	10754	3.67	513	6188	6.377	513

Full code version

For matrix A^TA

* Times are in milliseconds

Table 18. Parallel Conjugate Gradient

Dim	No. of NNZ	No. of iter.	Number of transputers				
			1	2		4	
			Time	Time	s. up	Time	s. up
16	74	15	21	22	0.95	25	0.84
32	154	25	70	69	1.01	75	0.93
64	314	38	212	202	1.05	214	0.99
80	394	43	300	283	1.06	298	1.01
96	474	47	394	369	1.07	387	1.02
128	634	55	614	572	1.07	599	1.07
IN FULL MATRIX CODE							
128	634	55	5131	2745	1.87	1498	3.43

Table 19. Sparse Conjugate Gradient with Matrix of Distinct Eigenvalues

Eigenvalue to 1, ..., n, so CN = N.
Sparse code.
* Time in milliseconds

From our experience we would stress the difficulty of getting speed up for the solution of sparse linear equations on parallel systems, and the necessity to model the communication time carefully when predicting the performance of parallel algorithms.

4. Conclusions

In this paper we have demonstrated

(1) that sparse doublet and sparse triplet versions of forward automatic differentiation provide effective means for calculating gradient vectors, Jacobian and Hessian matrices that occur within nonlinear optimisation, ordinary differential equations and partial differential equation problems.

(2) that many partial differential equation problems can be converted into optimisation problems and then solved using the truncated Newton or conjugate gradient algorithms in a proven finite number of steps

(3) that when so converted these partially separable optimisation problems can be mapped and solved effectively on SIMD parallel processing machines of the DAP type

(4) that automatic differentiation of partially separable functions can be performed effectively on Sequent type parallel processors using Concurrent Tasking in ADA

More speculatively results have been presented that indicate that

(1) the use of accurate dot products may greatly reduce the number of iterations required in practice by conjugate gradient based solvers.

(2) the use of MIMD systems of the Sequent and Transputer network varieties do not easily give effective speed-up when solving sparse sets of linear equations.

(3) it is necessary to model the communication time implied by linear algebra algorithms if the performance of the parallel system is to reflect the model.

Acknowledgements

This paper was presented as the first two of three lectures at the NATO ASI held at Il Ciocco, Italy in September 1990. Thanks are expressed to NATO for inviting us to attend the meeting and to past and present members of the NOC whose results featured in the presentation, in particular K. Patel, P. Ducksbury, Z. Maany, M. Mohseninia, S. Parkhurst, M. Vespucci, D. Mills and M. Jha.

We would also wish to acknowledge the sources of funding that made these research projects possible, namely, The National Advisory Board, The Science and Engineering Research Council, CEC IT Task Force, U.S. Army, British Gas, British Aerospace, Rolls Royce and the Italian CNR.

References

1. Byrne, G. D. & Hindmarsh, A. C. (1987) Review Article: Stiff ODE Solvers: A Review of Current and Coming Attractions, Journal of Computational Physics 70, 1987, pp 1-62.
2. Christianson, B. (1990) Automatic Hessians by Reverse Accumulation. NOC, Hatfield Technical Report No.TR228, April 1990
3. Dembo, R., Eisenstat, S. C., and Steihaug, T. (1982) Inexact Newton Methods, SIAM Journal of Numerical Analysis, Vol 19, 1982, pp 400-408.
4. Dembo, R. and Steihaug, T. (1983) Truncated Newton Methods for Large Scale Optimisation, Mathematical Programming, Vol.26, 1983, pp.190-212.
5. Dixon, L. C. W. (1974) Nonlinear Optimisation: A Survey of the State of the Art, Hatfield Polytechnic, NOC TR42, 1973. Published in Software for Numerical Mathematics, Academic Press, pp 193-219, ed. D. J. Evans, 1974.

6. Dixon, L. C. W., Maany, Z. A and Mohseninia, M. (1989, 1990) Automatic Differentiation of Large Sparse Systems, presented at IFAC Symposium on Dynamic Modelling & Control of National Economies, Edinburgh, July 1989. Paper 1, Hatfield Polytechnic, NOC TR223. Published in Journal of Economic Dynamics & Control No.14(2), 1990.
7. Dixon, L. C. W. and Mills, D. (1990) The Effect of Rounding Error on the Variable Metric Method. Hatfield Polytechnic NOC TR229, April 1990.
8. Dixon, L. C. W. and Mohseninia, M. (1987) The use of the Extended Operations Set of Ada with Automatic Differentiation and the Truncated Newton Method. Hatfield Polytechnic NOC TR176, April 1987.
9. Dixon, L. C. W. and Mohseninia, M. (1989) Concurrent Optimisation on the Sequent Balance 8000. Hatfield Polytechnic NOC TR226, September 1989.
10. Dixon, L. C. W. and Price, R. C. (1986, 1988) Numerical Experience with the Truncated Newton Method. Hatfield Polytechnic, NOC TR169, May 1986. Published in JOTA, Vol.56, No.2, pp 245-255, February 1988.
11. Dixon, L. C. W. and Price, R. C. (1986, 1989) The Truncated Newton Method for Sparse Unconstrained Optimisation using Automatic Differentiation. Hatfield Polytechnic, NOC TR170, October 1986. Published in JOTA, Vol.60, No.2, pp 261-275, February 1989.
12. Ducksbury, P. G., (1984) An Investigation of the Relative Merits of Optimisation Algorithms on the ICL-DAP, Hatfield Polytechnic PhD Thesis, October 1984.
13. Griewank, A., (1988) On Automatic Differentiation, in Mathematical Programming 88, Kluwer Academic Publishers, Japan.
14. Griewank, A. and Toint, Ph.L. (1981) On the Unconstrained Optimization of Partially Separable Functions, in Nonlinear Optimization 1981, ed. M. J. D. Powell, Part 5, pp 301-312.
15. Jha, Manoranjan, (1990) Preliminary Results on some Parallel Linear Algebra Applications on Transputer Networks. Hatfield Polytechnic, NOC TR231, April 1990.
16. Maany, Z. A., (1989) The Performance of the Truncated Newton Conjugate Gradient Algorithm in Fortran and Ada. Hatfield Polytechnic NOC TR210, 1989.
17. Parkhurst, S. C. (1990) The Evaluation of Exact Numerical Jacobians using Automatic Differentiation. Hatfield Polytechnic NOC TR224, December 1990. Presented at The Eleventh Conference on Differential Equations, Dundee University, July 1990.
18. Patel, K. D. (1982). Implementation of a Parallel (SIMD) Modified Newton Algorithm on the ICL DAP, paper presented at the Progress in the use of Vector and Array Processors Workshop, University of Bristol, September 1982.
19. Rall, L. B., (1981) Automatic Differentiation: Techniques and Applications. Springer-Verlag, Berlin, Germany, 1981.
20. Spedicato (1978) On a Conjecture of Dixon and other Topics in Variable Metric Methods. Mathematical Programming 15, pp 123-9.
21. Vespucci, M. T. (1990) The Use of the ABS Algorithms in Truncated-Newton Methods for Nonlinear Optimisation. University of Bergamo. Presented at NATO Summer School, Il Ciocco, Italy, September 1990.

Author Index

Aasen, J.O. 9,13
Abaffy, J. 93,94,107,112, 113
Abraham, J.A. 260
Adachi, N. 93
Adams, L. 210,213
Ahmed, H.M. 251
Aho, A.V. 29,30,33,36
Amestoy, P.R. 279,281,282
Ammar, G.S. 38
Anderson, B.D.O. 38,43
Anderson, E. 278,280
Arioli, M. 64,66,73,74,76, 284,285,287,293
Ashkenazi, V. 80
Astrachancev, G.P. 133
Axelsson, O. 158,172,177, 180,182,189,201

Bachvalov, N.S. 133
Bai, Z. 278,280
Bank, R.E. 133,162,162
Barwell, V. 13
Bauer, F.L. 63,64,80
Beauwens, R. 172,177
Benaini, A. 260,261
Ben Bouzid, M. 172,177
Benson, M.W. 180
Berman, A. 81,170
Bermond, J.C. 257,259,260
Berry, M. 38
Bertocchi, M. 104
Bini, D. 27,31,34,38,39, 43,45,47,50
Bischof, C. 70,218,220,240
Bitmead, R.R. 38,43
Björck, Å. 17,71,72,73,74, 75,76,77,79,80,84,86,88
Bodon, E. 97,107,109
Borodin, A. 28,31,42
Braess, D. 133
Brakhage, H. 133
Bramble, J.H. 133,162,166, 196,198
Bramley, R. 287
Brandt, A. 133
Brent, R.P. 38,42,52
Briggs, F. 218,242
Brinkkemper, S. 172,180, 182
Bromley, K. 249
Broyden, C.G. 2,17,93,98, 99,102,106
Bunch, J.R. 13,37,74
Byrne, G.D. 311

Canny, J.F. 47
Cappello, P.R. 257
Carter, J.H. 51
Chan, T.F. 68,69,185,200
Chen, Y.T. 81
Chen, Z. 94
Cheng, Y.H. 260
Christianson, B. 311
Chun, J. 43,44
Clarke, M.R.B. 243,245
Coffmann, E.G. 220,222,246
Concus, P. 182,192,193,202
Coppersmith, D. 31
Cosnard, M. 226,227,229, 230,237,241,243,245,246, 248
Craig, E.J. 107
Csanky, L. 33,42
Cybenko, G. 15,38

Dahlquist, G. 146
Daniel, J. 99
Daoudi, M. 243,248
Davis, P. 38
Davis, T.A. 282
Daydé, M. 280
Delosme, J.M. 251
Dembo, R. 322
Demmel, J. 64,66,278,280, 293
Deng, N.Y. 94,112
Denning, P.J. 220,222,246
Dennis, J.E. 87
Deuflhard, P. 164
Di Benedetto, F. 38
Dixon, L.C.W. 2,112,306, 310,324,326,328,332
D'Jakonov, E.G. 133
Doi, S. 192
Dongarra, J. 217,218,220, 227
van Dongen, V. 258
Dryja, M. 196
Ducksbury, P.G. 318,324
DuCroz, J. 277,278,280
Duff, I.S. 64,66,190,192, 277,278,279,280,281,282, 283,284,285,287,288,290, 293,294,295
Dupont, T.F. 133,162

Eckhard, C. 61
Eijkhout, V. 192
Eisenstat, S.C. 211,322
Elfving, T. 17,80,82,84,87
Elman, H.C. 211
Engeli, M. 213
Eppstein, D. 32
Evans, D.J. 85,208

Fedorenko, R.P. 133
Feingold, D.G. 171
Fiedler, M. 170
Filippone, S. 192
Fletcher, R. 24,211
Foster, L.V. 68,69
Frederickson, P.O. 180
Freund, R. 85

Galantai, A. 94
Galil, Z. 32,33
Galligani, E. 209,211,213
Gao, Z. 104
de la Garza, A. 80
Gay, D.M. 295
Geist, G.A. 268
George, A. 13,27,35,36, 112,118
Gerasoulis, A. 219,225, 229,272
Gilbert, J.R. 36,37,292
Ginsburg, M. 213
Glowinski, R. 185,204
Gohberg, I.C. 39
Golub, G.H. 5,8,9,15,27, 28,34,35,47,81,82,87,88, 182,185,186,187,192,193, 202,233,234,235,236,240
Gould, N.I.M. 288,294,295
Gragg, W.B. 38,52
Greenbaum, A. 278,280
Griewank, A. 305,307,310
Grimes, R.G. 280
Gustafson, J.L. 269
Gustavson, F.G. 38,42,52, 227

Hackbusch, W. 35,50,133, 137,138,142,144,145,146, 147,150,151,152,154,155, 156,157,158,161,162
Hafsteinsson, H. 37
Hageman, L.A. 88,212,285
Hager, W.W. 64,171,178
Hammarling, S. 275,276,278
Hanson, R.J. 278,280
Hardy, G.H. 51
Hawkinson, S. 269
Heath, M.T. 79,268
Hegedus, C.J. 23
Hemker, P.W. 133
Hestenes, M.R. 23,82
Higham, N.J. 2,64,66
Hindmarsh, A.C. 311
Ho, C.T. 266,267
Hockney, R. 187
de Hoog, F.R. 38
Hopcroft, J.E. 29,30,33,36
Householder, A.S. 80
Huang, H.Y. 93,98
Huang, K.H. 260
Huang, Z. 94
Hwang, K. 218,242,260

Il'in, V.P. 172,180,182
Iohvidov, I.S. 39
Ipsen, I.C.F. 213,238

Jesshope, C. 187
Jha, M. 334
Johnsson, S.L. 266,267,280

Kahan, W. 82
Kailath, T. 38,43,44,45, 249
Kaltofen, E. 41,47
Kamath, C. 87,287
Karp, A. 227
Karp, R. 32,33
Kaufman, L. 13
Keller, J. 80
Keller-Gehrig, W. 42
Kettler, R. 158
Keyes, D. 205
Knuth, D.E. 32
Kolotilina, L.Y. 172,179, 180,181,183
Kowalik, J.S. 224,229,235
Kronsjö, L. 146
Kuck, D. 243,245
Kumar, S.P. 220,224,229, 235
Kung, H.T. 249,260
Kung, S.Y. 43,45,249,257
Kuznetsov, Y. 185,203

Lacroute, P.G. 240
Laderman, J. 31

Lakshman, Y. 47
Lam, M.S. 260
Laratta, A. 73,74,76
Läuchli, P. 84
Leinen, P. 164
Leiserson, C.E. 249
Lev-Ari, H. 43,44
Levinson, N. 14
Lewis, J.G. 280
Li, C. 85
Lions, P.L. 194
Lipton, R.J. 35,36,37
Liu, J. 27,35,36,112,118
Lord, R.E. 224,229
Lotti, G. 31
Louka, B. 257
Luk, F.T. 88

Maany, Z. 306,310,328
Maitre, J.F. 133
Mandel, J. 133
Manneback, P. 87,88
Manteuffel, T. 84
Marrakchi, M. 229,230,237
Martin, R.S. 9
McCabe, A. 249
McCanny, J. 249
McCormick, S. 35,50,133
McKenney, A. 278,280
McWhirter, J. 249
Mehrmann, V. 172
Meijerink, J.A. 84,172,190
Meurant, G. 182,185,186, 187,188,190,191,192,193, 201,202,204
Mills, D. 328
Mirsky, L. 61
Modi, J.J. 243
Moenck, R.T. 51
Mohseninia, M. 306,310,332
Moler, C. 268
Moore, W. 249
Morf, M. 43,45,251
Muller, J.M. 245,246
Munro, I. 28,31,42
Musicus, B.R. 38
Musy, F. 133

Nelken, I. 219,225,229,272
Nicolaides, R.A. 133
Nielsen, H.B. 84
Noailles, J. 284,285,287

Oettli, W. 65
O'Leary, D.P. 209
Ong, M.E.G. 166,210,213
Ostrowski, A.M. 170,171, 183

Paige, C.C. 72,74,75,77,82
Pan, V. 27,28,30,31,32,33, 34,36,37,39,40,41,42,43, 44,45,46,47,48,49,50
Parkhurst, S.C. 311
Parlett, B.N. 13
Pasciak, J.E. 133,162,166, 196,198
Patel, K.D. 316
Pease, M. 33
Peierls, T. 292
Pereyra, V. 17
Periaux, J. 185,204
Peters, G. 9
Peyrat, C. 257,259,260
Plemmons, R.J. 81,170
Polman, B. 158,171,177
Prager, W. 65
Preparata, F.P. 33
Price, R.C. 324,326
P̂tak, V. 170

Quinn, M.J. 32,33
Quinton, P. 257,258

Radicati, G. 194
Rall, L.B. 306
Ramachandran, V. 32,33
Reid, J.K. 13,187,278,281, 285,288,290,294,295
Reif, J. 36,37,49,50
Reusken, A. 133
de Rijk, P.P.M. 66,285
Robert, F. 172
Robert, Y. 217,226,227, 229,230,237,243,245,246, 248,257,260,261,269,273
Rodrigue, G. 195
Rose, D. 35,36,37
Ruggiero, V. 209,211,213
Ruiz, D. 284,285,287
Rutishauser, H. 213

Saad, Y. 186,193,218,238, 266,268,272,273
Saff, E.B. 172
Sakho, I. 257,259,260
Sameh, A. 87,227,243,245
Sarwate, D.V. 33
Saunders, M.A. 82,85

Schatz, A.H. 196
Schippers, H. 133
Schreiber, R. 34
Schultz, M.H. 186,211,238, 273
Scott, J.A. 288,294,295
Scott, K. 269
Semencul, A.A. 39
Sha, X.H. 31
Sherman, A.H. 163
Simon, J. 195
Skeel, R.D. 63,64,65
Sloboda, F. 284
Sonneveld, P. 133
Sorensen, D. 218,220,257
Spedicato, E. 93,94,97,99, 104,107,109,113,327
Steihaug, T. 87,322
Stewart, G.W. 8,62
Stiefel, E. 23,82,213
Strang, G. 38
Strassen, V. 28,31
Stüben, K. 158
Sun, J. 62
Swartzlander, E. 249
Swartzlander, E.E.Jr. 249

Tanabe, K. 81
Tarjan, R.E. 35,36,37
Tchuente, M. 257,259,260
Terrasson, G. 204
Toint, Ph.L. 87,88,305
Tourancheau, B. 260,269, 273
Trench, W.F. 39
Trottenberg, U. 35,50,133
Trystram, D. 229,230,237, 241
Turner, K. 288,294,295

Ullman, J.D. 29,30,33,36
Urquhart, R. 249

Valiant, L. 32
Van Loan, C.F. 5,8,9,15, 27,28,34,35,47,82,87,233, 234,235,236,240
Van der Sluis, A. 63
Van der Vorst, H.A. 84, 172,190,191
Varga, R.S. 79,81,171,172
Veldhorst, M. 229
Vespucci, M.T. 99,330
Viera, A. 43
Villard, G. 273

Wedin, P-Å. 62
Weiser, A. 163
Wesseling, P. 133,158
White, R.E. 208,209
Whitehouse, H.J. 249
Widlund, O. 185,196
Wiedemann, D.H. 41,42
Wilkinson, J.H. 2,3,5,7,8, 9,72,172
Winograd, S. 31
Wittum, G. 133,158
Wright, E.M. 51

Xia, Z. 94,104,109
Xu, J. 162,165,166,198

Yang, Z. 97
Yeremin, A.Y. 179,180,181, 183
Yew, P.C. 282
Young, D.M. 79,80,81,87, 88,209,210,212,285
Young, G. 61
Yserentant, H. 133,162, 164,165,166
Yun, D.Y.Y. 38,42,52

de Zeeuw, P.M. 133
Zhu, M. 94,112,118
Zlatev, Z. 84

Subject Index

ABS algorithms
   Block formulation   103,104
   General linear systems   93-107
   Linear least squares   107-109
   Nonlinear optimization   94
   Nonlinear systems   94
   Sparse linear systems   111-130
   Scaled class   101-107
   Unscaled class   94-101
ABS algorithms, special cases
   Conjugate direction subclass   104,105
   Hestenes-Stiefel ABS   105,330
   Huang (implicit Gram-Schmidt)   98,99,108,109,126-129
   Implicit LU-LL^T (Gauss-Cholesky)   99-101,105,112-124,128,
      129,331
   Implicit minimum residual conjugate gradient   105
   Implicit QR   105,124-126,128,129
   Lanczos ABS   105
   Modified Huang   99,107,108
   Optimally conditioned   97
   Optimally stable subclass   106
   Orthogonally scaled subclass   105
Algorithms, general classes
   ABS   9-13,93-109,111-130
   Accettably error stable   74
   Conjugate direction   18-21
   Conjugate gradient   22-24,81-83,185-205
   Domain decomposition   185-205
   General sparse systems   277-296
   Linear least squares   57-88
   Multifrontal   278,279
   Multigrid   133-158,161,162
   Multilevel   161-167
   Parallel factorization   217-250
   Single error stable   106
   Stable   74
   Strongly stable   74
Algorithms, special cases
   Aasen   9,13
   Adaptive KJI   240,241
   Additive Schwartz   196
   Biconjugate gradient   211
   Björck-Pereyra   15-17
   Block Cimmino   284,285
   Block Gauss-Seidel   87,195
   Block KJI   239,240
   Block SOR   86
   Block SSOR   86
   CGS   186
   Chebichev iterative   169
   Chebichev semiiterative   81
   Cholesky   8,9,218,230,236
   Craig   107
   D2   282,284
   Damped Jacobi   150
   Divide and Conquer   194

Doolittle 218,230,233
Fletcher-Reeves 321
Gaussian elimination 2-8,28,33,35,172,253,274,278
Gauss-Jordan 29,85,218
Gauss-Seidel 79,80,144,145,150,157
Generalized conjugate gradient 169
Generalized conjugate residual 211
Generalized ND 36,37
Givens 218,242,250
GMRES 186
Gram-Schmidt 99
Greedy scheduling 227,237,238
Greedy SPGS 245-247
Hensel 51
Hestenes-Stiefel 6,82,187,188,202,203,317-322
Householder 71,72
IBFC 175,177
ICVDV 191
ILU 150
Implicit Euler 302
Incomplete LU 169-187
Jacobi 79,86,133,142,150,334
Kaczmarz 150
Lanczos 34,48
LDL^T 230,235,236,285
LDM^T 230,234,235
Levinson 13-15,38
LU 2-8,172,174
Markowitz 35,291
Modified Gram-Schmidt 70-72,74-76
Modified Newton 316,317
ND 112,114
Nyström 156
ORTHOMIN 186
Parallel Gaussian elimination 227-238
Parallel Gauss-Jordan 225-227
Picard 156
PR 270,271
Probing 200
QR 67,68,218
Richardson 79,81,134,142,149,153,157,186
Schur 38
Schwartz 185,194,195
SOR 80,150
SSOR 150
Truncated-Newton 322-327
Arithmetic
Accurate 327,328
Doublet 306,312
Triplet 306,326,332
Asymptotic speed 223
Augmented system for least squares 17
Automatic differentiation 306,326,329,330
Automatic sparse differentiation 332

Benes network 211
Bilinear interpolation 142
Bilinear multiplication 30
Binary segmentation 30

Block cyclic reduction 35
Broyden's stability condition 106

Chebichev acceleration 193
Chebichev polynomials 187
Codes
 BLAS 2 279
 BLAS 3 277-279,281,282,290
 MACSYMA 305
 MA 27 285,288-290,294,295
 MA 28 283,290,292
 MA 47 290
 MA 48 290,291
 PLTMG 162
Complexity of
 Data communication 238-242
 Multigrid method 147
 Parallel algorithms 33,34
 Sequential algorithms 27-52
Condition number estimators 69,70
Convergent splitting 172
Crank-Nicolson scheme 202

Data broadcasting 264-267
Differentiation
 Automatic 306,329,330
 Reverse automatic 310
 Sparse automatic 332
 Symbolic 305
Dirichlet boundary conditions 162,197,202,203,212
Displacement generator 43

Elliptic partial differential equation 211
Equations of
 Elasticity 162
 Euler 287
 Navier-Stokes 204,287,324
 Normal type 58,107,108,285
 Fredholm second kind 156
 Laplace 198,212
 Poisson 133,157,189
 Seminormal type 77
 Stones 317
Error, posteriori estimate 9,10

Fibonacci scheme 244,245
Finite element discretization 162-164
Function
 Discrete Green 170
 Gradient 302,316
 Harwell-Boeing test set 280,283
 Helmholz 307,311
 Hessian 302,313-316
 Jacobian 311
 Lagrangian 301
 Lotke-Volterra 311
 Maany 328,329
 Partially separable 304
 Rosenbrock 303,304

Galerkin product 138,144,152
Gohberg-Semencul formulae 39
Graph
Greedy 237
Precedence 221
Task 221
Gray code 263,264
Grid
Coarse type correction 135,137,138,141
Function 149
Level number 136
Nonuniformly refined 161-167

Hadamard product 179
Hamming distance 263
Hierarchical basis method 162-165
Hierarchy of systems 136,145
Hilbert space 163
Hölder space 157

Image processing 207

LAPACK project 278,280
LINPACK package 109

Machines
ADM array 253,255-257,259,260
Alliant FX/80 287,289
Cray-1 281
Cray-2 293-294
Cray Y-MP 185,202,207,210,213
Distributed memory 253-274
Hypercubes 253,263-274
MIMD 219,239
Shared memory 217-250
SIMD 219,316
Systolic 253-262
Matrices, special types
Banded 112,119-128
Block comparison 171
Block H- 169-175,177,179,180,182,183
Block tridiagonal 185,192,203
Elliptic difference type 170
Generalized diagonally dominant 170,171
Generalized Hilbert 47
Hankel 38,42,48
Hankel like 44
Hessenberg 10
Iteration 138,139,145
Irreducible L- 207,209,210
M- 144,169-172,182
ND- 111,112
Netlib 295
Pointwise H- 169,177
Second normal form 138
Singular H- 178
Strongly nonsingular 94,100,113
Symmetric indefinite 9
Symmetric positive definite 8,9,17,18,111,112,148,180,

186-205,207-213
Toeplitz 13-15,38-48,51
Toeplitz like 45
Vandermonde 15-17,47,48
Vandermonde confluent 17
Z- 170
Minimum degree rule 35
Minimum norm solution 58,108,109
Moore-Penrose pseudoinverse 109
Multigrid methods
Algebraic 158
Compact 49
FAS 155
Frequency decomposition 158
General 35,49-51
NMGM 155
Nonlinear 154,155
Second kind 156,157
Symmetric 152-154
Two-grid 138,142,143,150,151
Multisplitting 208

Neumann series 191
Nine point prolongation 142,144
Nine point restriction 143

Omega network 211
Optimization
Constrained 300
Unconstrained 300-302
Ordinary differential equations 301,302
Orthogonal projectors 61

P-adic lifting 51
Padé approximation 42,52
Parabolic partial differential equations 185
Polynomial acceleration 212,213
PRAM model 32,218
Preconditioners
Algebraic Schur complement 198
Block diagonal 86
Block SSOR 86,88
Bramble-Pasciak-Schatz 185,196,198
Bramble-Pasciak-Xu 162,164-167
Diagonal 84,189
Explicit or G-methods 169,178-182
Hierarchical basis 165,166
Hybrid 182,183
Implicit or C-methods 170-179
Incomplete Cholesky 84,190
INV 191-193
INVV 193
Laüchli 84,85
M-step 213
Polynomial 194
SSOR 84,189,194,213
Postsmoothing iteration 138
Prolongation 136
Property A 87

QR factorization 67-69,108

Rank deficient linear least squares 66-69
Rank revealing QR factorization 68,69
Red-black ordering 189,190
Reduced system 88
Riesz theorem 163

Schur complement 9,17,18,29,199,200,202
Selfadjoint positive definite operator 164,165
Semiiterative smoothing iteration 138
Sensitivity of linear least squares 62-64
Singular value decomposition 59-62,66,67,83
Smith normal form 46
Smoothing iteration 135,138,141,156
Smoothing property 149
Spectral condition number 164
Speed up 222
Splitting 81
Strassen's multiplication 28

Task 220
Task scheduling 223-225
Task system 221
Trilinear aggregating 30
Trivial restriction 137

V-cycle 143,147,154

W-cycle 143,147,154,156
Winograd's multiplication 28
Wolfe's conditions 323

List of Participants

BELGIUM

Dr. Pierre MANNEBACK
Service Informatique
Faculté Polytechnique de Mons
Rue de Houdain, 9
B - 7000 Mons

BRASIL

Prof. Djalma M. FALCÃO
Programa de Engenharia Elétrica
Universidade Federal do Rio de Janeiro
Caixa Postal 68504
21945 Rio de Janeiro, RJ

CANADA

Dr. Shiv Raj VATSJA
Whiteshell Research
Atomic Energy of Canada
Pinawa, MB R0E-1L0

CZECHOSLOVAKIA

Prof. Fridrich SLOBODA
Institute of Technical Cybernetics
Dubavska 9
84237 Bratislava

Dr. Zdenek STRAKOS
General Computing Centre
Czechoslovak Academy of Sciences
Pod vodárenskou vezí 2 - P.O. Box 5
18207 Praha 8 - Liben

DENMARK

Dr. Søren Holdt JENSEN
Electronics Institute
Technical University of Denmark
Bygning 349
DK - 2800 Lyngby

Dr. Zahari ZLATEV
National Environmental Research Institute
Ministry of the Environment
Frederiksborgvej 399
DK - 4000 Roskilde

FRANCE

Dr. Mario ARIOLI
C.E.R.F.A.C.S.
Av. Coriolis, 42
31057 Toulose

Dr. Elena BODON
C.E.R.F.A.C.S.
Av. Coriolis, 42
31057 Toulose

Prof. Claude BREZINSKY
Laboratoire d'Analyse Numérique
et d'Optimisation
U.F.R. I.E.E.A.
Université de Lille 1, Bât M3
59655 Villeneuve d'Ascq Cédex

Dr. Hassane SADOK
Laboratoire d'Analyse Numérique
et d'Optimisation
U.F.R. I.E.E.A.
Université de Lille 1, Bât M3
59655 Villeneuve d'Ascq Cédex

GERMANY

Dr. Thomas ANDRZEJEWSKY
Institut für Mathematik
Universität Augsburg
Universitätstr. 8
D - 8900 Augsburg

Dr. Volker DRYGALLA
Sektion Mathematik
Martin-Luther-Universität
Universitätsplatz 6
Halle

Prof. Claus RICHTER
IH Köthen
Bernburgerstr. 52-57
Köthen

GREECE

Dr. George-Othon GLENTIS
Dept. of Informatics and Telecom.
Nat. Research Center "Democritos"
Ag. Paraskevi - P.O. Box 60228
GR - 15310 Athens

Dr. Stamatis D. VOLIOTIS
Dept. of Computer Engineering
School of Engineering
University of Patras
GR - 26500 Patras

HUNGARY

Prof. Aurel GALANTAI
Department of Mathematics
University of Miskolc
H - 3515 Miskolc

Dr. Csaba HEGEDÜS
Central Research Institute for Physics
Hungarian Academy of Sciences
P.O. Box 49
H - 1525 Budapest

ITALY

Prof. Marida BERTOCCHI
Dipartimento di Matematica
Università di Bergamo
Piazza Rosate, 2
24100 Bergamo

Dr. Silvia BERTOLUZZA
I.A.N. - C.N.R.
Corso Carlo Alberto, 5
27100 Pavia

Dr. Gianluca CAPITANI
C. d. L. Informatica
Università di Bologna
Via Sacchi, 3
47023 Cesena

Dr. Fabio DI BENEDETTO
Dipartimento di Matematica
Università di Genova
Via L.B. Alberti, 4
16132 Genova

Dr. Marco FINARDI
Dipartimento di Matematica
Università di Bergamo
Piazza Rosate, 2
24100 Bergamo

Prof. Maria MORANDI-CECCHI
Dipartimento di Matematica
Università di Padova
Via Belzoni, 7
35131 Padova

Prof. Angelo PICA
Dipartimento di Scienza dell'Informazione
Università di Pisa
Corso Italia, 40
56100 Pisa

Prof. Luigia PUCCIO
Dipartimento di Matematica
Università di Messina
Contr. Papardo - Salita Sperone 31
98166 S. Agata (Messina)

Prof. Michela REDIVO-ZAGLIA
Dipartimento di Elettronica e Informatica
Università di Padova
Via Gradenigo, 6/a
35131 Padova

Prof. Valeria RUGGIERO
Dipartimento di Matematica
Università di Ferrara
Via Machiavelli, 35
44100 Ferrara

Prof. Tamara TOMMASINI
Dipartimento di Matematica e Informatica
Università di Udine
Via Zanon, 6
33100 Udine

Dr. Gerardo TORALDO
Dipartimento di Matematica
Università della Basilicata
Via N. Sauro, 85
85100 Potenza

Dr. Maria Teresa VESPUCCI
Dipartimento di Matematica
Università di Bergamo
Piazza Rosate, 2
24100 Bergamo

Prof. Zun-Quan XIA
Dipartimento di Matematica
Università di Bergamo
Piazza Rosate, 2
24100 Bergamo

NETHERLANDS

Dr. Marcel VAN DIJKE
Faculty of Technical Mathematics
and Informatics
Delft University of Technology
P.O. Box 356
2600 AJ Delft

PORTUGAL

Prof. José CRAVEIRINHA
Departamento de Engenharia Electrotécnica
Faculdade de Ciências e Tecnologia
Universidade de Coimbra
3000 Coimbra

Prof. Joaquim João JUDICE
Departamento de Matemática
Faculdade de Ciências e Tecnologia
Universidade de Coimbra
Apartado 3008
3000 Coimbra

Dr. Isabel VIEIRA
Instituto Superior de Contabilidade e
Administração do Porto
Rua Entreparedes
4000 Porto

SPAIN

Prof. Manuel ALVAREZ
Departamento de Matemáticas Aplicadas
Universidad Politécnica de Madrid
Ciudad Universitaria
28040 Madrid

Prof. Francisco BALLESTEROS OLMO
E.T.S.I. Telecomunicación
Universidad Politécnica de Madrid
Ciudad Universitaria
28040 Madrid

Prof. José L. DE LA FUENTE
E.T.S.I. Industriales
Universidad Politécnica de Madrid
Ciudad Universitaria
28040 Madrid

Prof. José MAS MARÍ
Departamento de Matemática Aplicada
Universidad Politécnica de Valencia
46071 Valencia

Prof. J. M. PEÑA FERRANDEZ
Departamento de Matemática Aplicada
Universidad de Zaragoza
Corona de Aragon 35
50009 Zaragoza

TURKEY

Dr. Haydar AKCA
Department of Mathematics
Faculty of Sciences
Erciyes University
38039 Kayseri

Prof. Mustafa AKGÜL
Department of Industrial Engineering
Bilkent University
P.O. Box 8
06572 Maltepe, Ankara

Dr. Adil DENIZLI
Chemical Engineering Department
Hacettepe University
06532 Beytepe, Ankara

Prof. V. KALENDEROGLU
Department of Mechanical Engineering
Bogaziçi University
80815 Bebek - Istanbul

Prof. Muhammet KÖKSAL
Dept. of Electrical and Electronic Eng.
Faculty of Engineering
University of Gaziantep
27310 Gaziantep

Prof. Osman OGUZ
Department of Industrial Engineering
Bilkent University
P.O. Box 8
06572 Maltepe, Ankara

Dr. Murat OLCAY
Chemical Engineering Department
Hacettepe University
06532 Beytepe, Ankara

Prof. Gonca ONARGAN
Uygulamali Matematik Bölümü
Mühendislik-Mimarlik Fakültesi
Dokuz Eylül University
Bornova - Izmir

Prof. Sennur SOMALI
Elektronic Bölümü
Dokuz Eylül University
Bornova - Izmir

UNITED KINGDOM

Prof. Roger FLETCHER
Department of Mathematical Sciences
University of Dundee
Dundee DD1 4HN - Scotland

Prof. Hugo E. HERNANDEZ FIGUEROA
Department of Physics
Imperial College
London SW7 2BZ

Dr. Marli HERNANDEZ FIGUEROA
Numerical Optimisation Centre
Hatfield Polytechnic
College Lane
Hatfield, Herts AL10 9AB

Dr. Manoranjan JHA
Numerical Optimisation Centre
Hatfield Polytechnic
College Lane
Hatfield, Herts AL10 9AB

Dr. Changjun LI
Department of Computer Studies
Loughborough Univ. of Technology
Loughborough, Leics. LE11 3TU

Dr. David John MILLS
Numerical Optimisation Centre
Hatfield Polytechnic
College Lane
Hatfield, Herts AL10 9AB

Dr. Steve PARKHURST
Numerical Optimisation Centre
Hatfield Polytechnic
College Lane
Hatfield, Herts AL10 9AB

Dr. Stilianos RAFAILIDIS
Chemical Engineering Department
University of Surrey
Guildford Surrey GU2 5XH

Prof. Michael A. WOLFE
Department of Mathematical
and Computational Sciences
University of St. Andrews
St. Andrews
Fife KY16 9SS - Scotland

UNITED STATES

Dr. Karin BENNETT
Department of Mathematics
University of Kentucky
Lexington, Kentucky 40506 - 0027

Dr. Le CHINH
Stevenson Science Center
Computer Center
Vanderbilt University
Box 1577 - Station B
Nashville, Tennessee 37235

Prof. Graeme FAIRWEATHER
Department of Mathematics
University of Kentucky
715 Patterson Office Tower
Lexington, Kentucky 40506 - 0027

Dr. Steven W. HAMMOND
R.I.A.C.S.
NASA Ames Research Center
Mail Stop T045-1
Moffett Field, CA 94035-4000

Prof. Christopher SIKORSKI
Department of Computer Science
University of Utah
3160 Merrill Engineering Building
Salt Lake City, Utah 84112

USSR

Dr. Oleg BURDAKOV
Computing Center
USSR Academy of Sciences
40, Vavilov Str.
117967 Moscow GSP-1

Prof. Valentin VOEVODIN
Department of Numerical Mathematics
USSR Academy of Sciences
Ryleeva 29
119034 Moscow

Prof. Victor IZHUTKIN
Marische Staatliche Universität
Lenin-Platz 1
424001 Joschkar-Ola

YUGOSLAVIA

Prof. Vera VUJCIC
Faculty of Organizational Sciences
Belgrade University
Ul. Jove Ilica br. 154
11000 Belgrade

NATO ASI Series F

Including Special Programmes on Sensory Systems for Robotic Control (ROB) and on Advanced Educational Technology (AET)

Vol. 1: Issues in Acoustic Signal – Image Processing and Recognition. Edited by C. H. Chen. VIII, 333 pages. 1983.

Vol. 2: Image Sequence Processing and Dynamic Scene Analysis. Edited by T. S. Huang. IX, 749 pages. 1983.

Vol. 3: Electronic Systems Effectiveness and Life Cycle Costing. Edited by J. K. Skwirzynski. XVII, 732 pages. 1983.

Vol. 4: Pictorial Data Analysis. Edited by R. M. Haralick. VIII, 468 pages. 1983.

Vol. 5: International Calibration Study of Traffic Conflict Techniques. Edited by E. Asmussen. VII, 229 pages. 1984.

Vol. 6: Information Technology and the Computer Network. Edited by K. G. Beauchamp. VIII, 271 pages. 1984.

Vol. 7: High-Speed Computation. Edited by J. S. Kowalik. IX, 441 pages. 1984.

Vol. 8: Program Transformation and Programming Environments. Report on a Workshop directed by F. L. Bauer and H. Remus. Edited by P. Pepper. XIV, 378 pages. 1984.

Vol. 9: Computer Aided Analysis and Optimization of Mechanical System Dynamics. Edited by E. J. Haug. XXII, 700 pages. 1984.

Vol. 10: Simulation and Model-Based Methodologies: An Integrative View. Edited by T. I. Ören, B. P. Zeigler, M. S. Elzas. XIII, 651 pages. 1984.

Vol. 11: Robotics and Artificial Intelligence. Edited by M. Brady, L. A. Gerhardt, H. F. Davidson. XVII, 693 pages. 1984.

Vol. 12: Combinatorial Algorithms on Words. Edited by A. Apostolico, Z. Galil. VIII, 361 pages. 1985.

Vol. 13: Logics and Models of Concurrent Systems. Edited by K. R. Apt. VIII, 498 pages. 1985.

Vol. 14: Control Flow and Data Flow: Concepts of Distributed Programming. Edited by M. Broy. VIII, 525 pages. 1985.

Vol. 15: Computational Mathematical Programming. Edited by K. Schittkowski. VIII, 451 pages. 1985.

Vol. 16: New Systems and Architectures for Automatic Speech Recognition and Synthesis. Edited by R. De Mori, C.Y. Suen. XIII, 630 pages. 1985.

Vol. 17: Fundamental Algorithms for Computer Graphics. Edited by R. A. Earnshaw. XVI, 1042 pages. 1985.

Vol. 18: Computer Architectures for Spatially Distributed Data. Edited by H. Freeman and G. G. Pieroni. VIII, 391 pages. 1985.

Vol. 19: Pictorial Information Systems in Medicine. Edited by K. H. Höhne. XII, 525 pages. 1986.

Vol. 20: Disordered Systems and Biological Organization. Edited by E. Bienenstock, F. Fogelman Soulié, G. Weisbuch. XXI, 405 pages. 1986.

Vol. 21: Intelligent Decision Support in Process Environments. Edited by E. Hollnagel, G. Mancini, D. D. Woods. XV, 524 pages. 1986.

NATO ASI Series F

Vol. 22: Software System Design Methods. The Challenge of Advanced Computing Technology. Edited by J.K. Skwirzynski. XIII, 747 pages. 1986.

Vol. 23: Designing Computer-Based Learning Materials. Edited by H. Weinstock and A. Bork. IX, 285 pages. 1986.

Vol. 24: Database Machines. Modern Trends and Applications. Edited by A.K. Sood and A.H. Qureshi. VIII, 570 pages. 1986.

Vol. 25: Pyramidal Systems for Computer Vision. Edited by V. Cantoni and S. Levialdi. VIII, 392 pages. 1986. *(ROB)*

Vol. 26: Modelling and Analysis in Arms Control. Edited by R. Avenhaus, R.K. Huber and J.D. Kettelle. VIII, 488 pages. 1986.

Vol. 27: Computer Aided Optimal Design: Structural and Mechanical Systems. Edited by C.A. Mota Soares. XIII, 1029 pages. 1987.

Vol. 28: Distributed Operating Systems. Theory und Practice. Edited by Y. Paker, J.-P. Banatre and M. Bozyiğit. X, 379 pages. 1987.

Vol. 29: Languages for Sensor-Based Control in Robotics. Edited by U. Rembold and K. Hörmann. IX, 625 pages. 1987. *(ROB)*

Vol. 30: Pattern Recognition Theory and Applications. Edited by P.A. Devijver and J. Kittler. XI, 543 pages. 1987.

Vol. 31: Decision Support Systems: Theory and Application. Edited by C.W. Holsapple and A.B. Whinston. X, 500 pages. 1987.

Vol. 32: Information Systems: Failure Analysis. Edited by J.A. Wise and A. Debons. XV, 338 pages. 1987.

Vol. 33: Machine Intelligence and Knowledge Engineering for Robotic Applications. Edited by A.K.C. Wong and A. Pugh. XIV, 486 pages. 1987. *(ROB)*

Vol. 34: Modelling, Robustness and Sensitivity Reduction in Control Systems. Edited by R.F. Curtain. IX, 492 pages. 1987.

Vol. 35: Expert Judgment and Expert Systems. Edited by J.L. Mumpower, L.D. Phillips, O. Renn and V.R.R. Uppuluri. VIII, 361 pages. 1987.

Vol. 36: Logic of Programming and Calculi of Discrete Design. Edited by M. Broy. VII, 415 pages. 1987.

Vol. 37: Dynamics of Infinite Dimensional Systems. Edited by S.-N. Chow and J.K. Hale. IX, 514 pages. 1987.

Vol. 38: Flow Control of Congested Networks. Edited by A.R. Odoni, L. Bianco and G. Szegö. XII, 355 pages. 1987.

Vol. 39: Mathematics and Computer Science in Medical Imaging. Edited by M.A. Viergever and A. Todd-Pokropek. VIII, 546 pages. 1988.

Vol. 40: Theoretical Foundations of Computer Graphics and CAD. Edited by R.A. Earnshaw. XX, 1246 pages. 1988.

Vol. 41: Neural Computers. Edited by R. Eckmiller and Ch. v. d. Malsburg. XIII, 566 pages. 1988.

NATO ASI Series F

Vol. 42: Real-Time Object Measurement and Classification. Edited by A. K. Jain. VIII, 407 pages. 1988. *(ROB)*

Vol. 43: Sensors and Sensory Systems for Advanced Robots. Edited by P. Dario. XI, 597 pages. 1988. *(ROB)*

Vol. 44: Signal Processing and Pattern Recognition in Nondestructive Evaluation of Materials. Edited by C. H. Chen. VIII, 344 pages. 1988. *(ROB)*

Vol. 45: Syntactic and Structural Pattern Recognition. Edited by G. Ferraté, T. Pavlidis, A. Sanfeliu and H. Bunke. XVI, 467 pages. 1988. *(ROB)*

Vol. 46: Recent Advances in Speech Understanding and Dialog Systems. Edited by H. Niemann, M. Lang and G. Sagerer. X, 521 pages. 1988.

Vol. 47: Advanced Computing Concepts and Techniques in Control Engineering. Edited by M. J. Denham and A. J. Laub. XI, 518 pages. 1988.

Vol. 48: Mathematical Models for Decision Support. Edited by G. Mitra. IX, 762 pages. 1988.

Vol. 49: Computer Integrated Manufacturing. Edited by I. B. Turksen. VIII, 568 pages. 1988.

Vol. 50: CAD Based Programming for Sensory Robots. Edited by B. Ravani. IX, 565 pages. 1988. *(ROB)*

Vol. 51: Algorithms and Model Formulations in Mathematical Programming. Edited by S. W. Wallace. IX, 190 pages. 1989.

Vol. 52: Sensor Devices and Systems for Robotics. Edited by A. Casals. IX, 362 pages. 1989. *(ROB)*

Vol. 53: Advanced Information Technologies for Industrial Material Flow Systems. Edited by S. Y. Nof and C. L. Moodie. IX, 710 pages. 1989.

Vol. 54: A Reappraisal of the Efficiency of Financial Markets. Edited by R. M. C. Guimarães, B. G. Kingsman and S. J. Taylor. X, 804 pages. 1989.

Vol. 55: Constructive Methods in Computing Science. Edited by M. Broy. VII, 478 pages. 1989.

Vol. 56: Multiple Criteria Decision Making and Risk Analysis Using Microcomputers. Edited by B. Karpak and S. Zionts. VII, 399 pages. 1989.

Vol. 57: Kinematics and Dynamic Issues in Sensor Based Control. Edited by G. E. Taylor. XI, 456 pages. 1990. *(ROB)*

Vol. 58: Highly Redundant Sensing in Robotic Systems. Edited by J. T. Tou and J. G. Balchen. X, 322 pages. 1990. *(ROB)*

Vol. 59: Superconducting Electronics. Edited by H. Weinstock and M. Nisenoff. X, 441 pages. 1989.

Vol. 60: 3D Imaging in Medicine. Algorithms, Systems, Applications. Edited by K. H. Höhne, H. Fuchs and S. M. Pizer. IX, 460 pages. 1990.

Vol. 61: Knowledge, Data and Computer-Assisted Decisions. Edited by M. Schader and W. Gaul. VIII, 421 pages. 1990.

Vol. 62: Supercomputing. Edited by J. S. Kowalik. X, 425 pages. 1990.

Vol. 63: Traditional and Non-Traditional Robotic Sensors. Edited by T. C. Henderson. VIII, 468 pages. 1990. *(ROB)*

Vol. 64: Sensory Robotics for the Handling of Limp Materials. Edited by P. M. Taylor. IX, 343 pages. 1990. *(ROB)*

Vol. 65: Mapping and Spatial Modelling for Navigation. Edited by L. F. Pau. VIII, 357 pages. 1990. *(ROB)*

NATO ASI Series F

Vol. 66: Sensor-Based Robots: Algorithms and Architectures. Edited by C.S.G. Lee. X, 285 pages. 1991. *(ROB)*

Vol. 67: Designing Hypermedia for Learning. Edited by D.H. Jonassen and H. Mandl. XXV, 457 pages. 1990. *(AET)*

Vol. 68: Neurocomputing. Algorithms, Architectures and Applications. Edited by F. Fogelman Soulié and J. Hérault. XI, 455 pages. 1990.

Vol. 69: Real-Time Integration Methods for Mechanical System Simulation. Edited by E.J. Haug and R.C. Deyo. VIII, 352 pages. 1991.

Vol. 70: Numerical Linear Algebra, Digital Signal Processing and Parallel Algorithms. Edited by G.H. Golub and P. Van Dooren. XIII, 729 pages. 1991.

Vol. 71: Expert Systems and Robotics. Edited by T. Jordanides and B. Torby. XII, 744 pages. 1991.

Vol. 72: High-Capacity Local and Metropolitan Area Networks. Architecture and Performance Issues. Edited by G. Pujolle. X, 536 pages. 1991.

Vol. 73: Automation and Systems Issues in Air Traffic Control. Edited by J.A. Wise, V.D. Hopkin and M.L. Smith. XIX, 589 pages. 1991.

Vol. 74: Picture Archiving and Communication Systems (PACS) in Medicine. Edited by H.K. Huang, O. Ratib, A.R. Bakker and G. Witte. XII, 445 pages. 1991.

Vol. 75: Speech Recognition and Understanding. Recent Advances, Trends and Applications. Edited by P. Laface and Renato De Mori.

Vol. 76: Multisensor Fusion for Computer Vision. Edited by J.K. Aggarwal. *(ROB)*

Vol. 77: Computer Algorithms for Solving Linear Algebraic Equations. The State of the Art. Edited by E. Spedicato. VII, 352 pages. 1991.

MIX
Papier aus verantwortungsvollen Quellen
Paper from responsible sources
FSC® C105338

If you have any concerns about our products,
you can contact us on
ProductSafety@springernature.com

In case Publisher is established outside the EU,
the EU authorized representative is:
Springer Nature Customer Service Center GmbH
Europaplatz 3, 69115 Heidelberg, Germany

Printed by Libri Plureos GmbH
in Hamburg, Germany